KALLITYPE:

THE PROCESSES AND THE HISTORY

KALLITYPE:

THE PROCESSES AND THE HISTORY

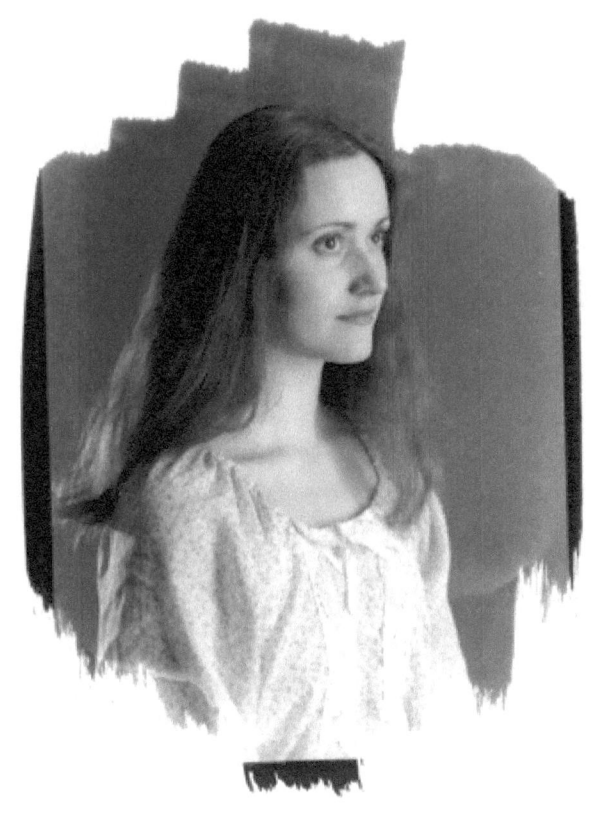

BY

DICK STEVENS

Library of Congress Control Number:		2012920411
ISBN:	Hardcover	978-1-4797-4224-0
	Softcover	978-1-4797-4223-3
	Ebook	978-1-4797-4225-7

To order additional copies of this book, contact:
Xlibris Corporation
1-888-795-4274
www.Xlibris.com
Orders@Xlibris.com
124784

Table of Contents

Acknowledgements..11
Preface..13

Section I.   Kallitype and the War of Process

Chapter I. The Kallitype I From Invention Through Sale
I.   The Very Latest Novelty 19
II.   Nicol: Biographical Details...................................19
III.   The Provisional Specification of Patent 5374..................21
IV.   Nicol's Patented K I Process.................................23
V.   Critical Response to Nicol's Kallitype I Patent 26
VI.   Bothamley's Report on K I 32
VII.   Bedding's Charge of Unjustified Claims 35
VIII. The Claims Controversy and Its Resolution 36
IX.   Announcement of The Sale of Kallitype I Paper 39
X.   Reports of Kallitype Use and "The First False Step." 42
XI.   Conclusion 48
Notes 49

Chapter II. Individual Preparation of Kallitype I Paper
I.   Introduction..51
II.   First Accounts of Individual Preparation of K I Paper 57
III.   Brown's Resolution of the Claims Controversy 66
IV.   Brown's Description of Kallitype I Paper and Process.........68
V.   Kallitype I: Summary and Comment 70
Notes 76

Chapter III. Kallitype II & III: Patent to Manufacture & Sale
I.   The Announcement of Commercial Kallitype II & III 79
II.   Patent No. 7312—Details 81
III.   Designations of Kallitype Processes..........................83

IV.   Nicol's Patented Kallitype II Solutions 84
V.   Kallitype III: A print-out Paper 86
VI.   Responses to Kallitype II in the Market 89
VII.   Kallitype II and "The War of Papers." 106
VIII. Kallitype II: Performance In The Market 108
IX.   Commercial Sale of Kallitype Ends In England............111
X.   Commercial Kallitype II Paper in America 112
XI.   Early Explanations of "The Wane of Favor."................122
XII.   The Lost War of Process: Commentary 124
Notes 130

Section II.   Kallitype and the Amateur

Chapter IV. Kallitype and the Amateur

I.   The Need for Historical Study 137
II.   Notes On This History of the Kallitype 140
III.   The problem of Vintage Kallitypes............................142
IV.   Amateurs true and otherwise 143
V.   The Pictorial Amateur 148
VI.   Social History and the Amateur 150
VII.   Kallitype and Photographic Publications....................152
VIII. Historically Important Writers on Kallitype 155
IX.   The "Progress" of the Kallitype Process....................156
X.   Conclusion 160
Notes 161

Chapter V. The Amateur and the Kallitype

I.   Nicol's And Bennett's Approaches to K II.....................166
II.   W. K. Burton's Refinement of Bennett's Approach........171
III.   Aftermath of the Bennett-Burton Revisions 179
IV.   Summary & Conclusions 189
Notes 191

Chapter VI. Tailoring the Kallitype II

I.   Frederick and Hall Approach..................................193
II.   Henry Hall—More tailoring—The Stock Solution
       Approach...202
III.   After Hall to Early Thomson. 211
IV.   James Thomson's early work on K II 213

V.   Conclusion 216
Notes 218

Chapter VII. Pictorialism
I.   The Historical Context of the Kallitype 1890-1925 220
II.   The Question: Is photography an Art? 224
III. Technique and Art in the early 1900's 229
IV.   Pictorialism and Photographic Technique.......................232
V.   Kallitypists, Artists, and the Pictorial Movement 245
VI.   Platinum Printing and the Kallitype................................248
VII. Commentary on artist Prepared Paper 253
Notes 255

Chapter VIII. Thomson and the Experimenting Artist
I.   The Joys and Sorrows of Photo-Experimentation 258
II.   Thomson and the Kallitype I..260
III. Summary of Chapter VIII 268
Notes 270

Chapter IX. The Brown Print
I.   Introduction to the Brownprint 271
II.   The Chemistry and Processing of the Brown print........273
III. The Beginnings of the Brown Print. 275
IV.   History of the Brown print Processes 1885-1920 276
V.   Conclusion. 296
Notes 298

Section III. The Demise and Rebirth

Chapter X. The Denoument Begins 1904-1914.
I.   Media Responses and Developments 1904-1914 303
II.   Summary and Comment 326
Notes 332

Chapter XI. The Demise and Rebirth
I.   Concluding this History of the Kallitype 335
II.   The Demise? 1915-1925. 335
III. The Progress of the Kallitype Process: 1890-1925 352

IV.   Interpreting the demise of the Kallitype and the
Pictorial Esthetic 354
V.   Changing Times and Changing Taste: 1925-1950. 360
VI. The Rebirth of Kallitype and Allied Processes. 363
VII. Modern Scientific Investigation of the Kallitype............365
VIII. Kallitype in the 21st Century 366
Notes 369

Illustrations 373
Bibliography of Books on Kallitype and Allied Topics....................375
Bibliography of Periodicals 385
Index 391

Dedication

To Becky
Who helped in many ways

Acknowledgements

This project received the support of a number of individuals and institutions.

Phil Davis gave in many ways and over a span of years. He introduced me to the joy and challenge of old photographic processes and taught me a great deal about methods. Peter Bunnell helped with suggestions about libraries and photographic collections.

The University of Notre Dame, College of Arts and Letters was generous with a grant which aided the research for this book.

I thank Rudy Bottei, Department of Chemistry, University of Notre Dame, for his patience in explaining matters of chemistry.

My thanks to librarians in Great Britain, J. Seals, and Dr. B. S. Benedikz, both of Birmingham, for their information on Nicol, the "inventor" of the kallitype.

Elizabeth W. Kraus, Administration Department of the Kodak Research Laboratory and Mary Connolly, Librarian, both helped by providing access to the extensive holdings of the Kodak Research Library.

I thank David Haberstitch and Eugene Ostroff for access to the photographic book collection of the Smithsonian Institution.

Susan Wynegaard and Mary Ann Margolis of the International Museum of Photography at George Eastman House facilitated my work with their collection of books and periodicals.

Elliott C. Finley facilitated access to the ample collection of material on photography in the Library of Congress. The New York Public Library must be mentioned for their helpful collection of historical periodicals on historical photography.

Finally, a number of individuals helped me in my search for vintage kallitypes and publications. Among them were: Jerald C. Maddox, Curator of Photography at the Library of Congress Prints and Photographs Division; Martha Charoudi, Philadelphia Museum of

Art; Clifford Ackley, Assoc. Curator of Prints and Drawings, Boston Museum of Fine Arts; Davis Pratt, Curator of Photography, at the Fogg Museum, Harvard University; and David Travis, Art Institute, Chicago, Illinois.

Without the assistance of these aware and gracious individuals and their cooperative institutions, this book could not have been written.

Preface

In the beginning, I had no inclination to write a history of the kallitype. I only wanted to learn how to make better kallitype prints.

It all started with a one page description of James Thomson's process given me by Phil Davis, a description so complicated and reluctant to work, I began looking for other recipes. I continued searching, xeroxing everything I found. My files grew.

I began to ask questions that went beyond how to make prints. How did the process come to be? Who invented it? When? My first answers came from encyclopedic accounts which didn't go very far. I came across Hall's *PhotoMiniature* monograph on kallitype, a find which gave a brief historical sketch and suggested other published articles. A train of serendipitous discoveries followed, many of them in the University of Notre Dame Library, which I discovered was a fair repository of data on old photographic processes.

By then I was "hooked" on historical study of the kallitype. I discovered and devoured the fine periodical holdings at the nearby Crerar Research Library in Chicago, and wondered what information on kallitype other libraries might contain. A Notre Dame travel grant permitted checking holdings in some of the fine libraries in the East, notably the Boston Public Library, the New York Public Library, the Library of Congress, the Eastman Kodak Company Library, and the Library of the International Museum of Photography, to mention the more helpful. Each revealed important and unexpected material, and my excitement grew. A non-historian, I was amazed at the amount of available material on the history of the kallitype, little of which was mentioned in current publications on old photographic processes.

I took a course in chemistry to understand better what I was reading and what I observed as I applied the new information in the darkroom. I began a systematic study of kallitype procedures. I made hundreds of carefully varied "experiments" on kallitype sensitizers, developers, etc., trying to find the best way of doing the process. I kept the results of all these process investigations in large, ever-growing loose leaf notebooks.

I began to organize my historical files according to the year, starting with 1889, the date of the "invention" of the kallitype. Annual files were opened until 1925, when kallitype activity evidently tapered off. I also began a file on the "prehistory" of kallitype, the period when proto-kallitype activity abounded, and another file on the "post history," for developments that occurred from 1925 to the present. I corresponded with those who might help me in England and Scotland and was delighted with the help I received.

I also searched for vintage kallitype prints, but that venture was not so fortunate. Though I found an extensive list of kallitype "makers," I found only one vintage kallitype print in the photographic collections of major art museums.

As I pursued my study of the process, I found periodicals and books which suggested social, economic, and technological influences that impinged on the process and also indications of the significant role kallitype played in 19th and 20th century photography.

My first response to the burgeoning information was to write a single book recounting the history of the process and the working of the process, like the books on the gum bichromate and the salt print processes recently published. However, by the time I finished writing, the manuscript of the history alone was already book size.

The history had a number of matters to recount: the invention of the several processes: Kallitype I, Kallitype II and Kallitype III and the description of each; the fate of the processes as commercial products marketed by entrepreneurs; and after that, the further development of the processes "in the hands" of amateurs in England and America. I discovered that the kallitype shed light on 19th and 20th century photography and felt that such illumination was worth communicating. As the many bits of information fell together, I began to see the kallitype as a weathervane which responded to and indicated currents in many areas of photography, particularly in technological, social, cultural and esthetic areas. I concluded that what happened to the kallitype and the amateurs who made them revealed changes in attitudes, values, and practice during the quarter of a century before and after 1900 and that such matters merited discussion. Shaped by the data that were found, the book developed into a history of an evolving set of processes in the ongoing stream of cultural, social, and photographic change between 1890 and 1925.

When I began, I thought there was only one kallitype process. As my investigation proceeded, I found many descriptions of kallitype processes in the literature. Each had its own version of sensitizer, developer, fixer, clearing bath, and processing procedure. After a bit of sorting, it became clear that most writers did not invent unique processes, but rather only "varied" a few processes described by earlier writers. I tracked down the writers who first described unique approaches and traced the succeeding descriptions as they evolved from the original creators.

Through such efforts, I arrived at a fairly coherent account of the development of the kallitype from the time it was invented to 1925. The many descriptions of the process have been organized into four distinct approaches, each of which resulted from the creative efforts of specific individuals. The four historical approaches to kallitype are described in sufficient detail so that contemporary photographers could work them. Each will produce charming, but different photographs.

The basic structure of the book follows. Section I recounts the history of the Kallitype I and II as commercial processes. It details the invention, the early announcements, the critical responses by experts, the use and response of amateurs, and the brief success and eventual failure of kallitype paper relative to other printing papers in the "war of process." Section II discusses the extensive kallitype developments that resulted from amateur activity. This section also discusses the social, technological, and cultural influences on the success and failure of the kallitype and the significance of the kallitype in the history of photography. Finally, Section III discusses the decline of the process after 1925 and briefly comments on the rebirth of interest that has occurred since that time.

My hope is that the book will explain what the kallitype is; how it came to be; the forces that controlled its destiny; and the significance of the kallitype in the history of photography. I also hope the book will provide kallitype printers useful information on the several ways the fascinating process can be worked.

A word about quotations. I have not changed spellings of British quotations. Their spelling remains as found.

Dick Stevens
October 15, 2012

SECTION I

Kallitype and the War
of Process

Chapter I

The Kallitype I From Invention Through Sale

I.   The Very Latest Novelty

On the 29th of March, 1889, William Walker James Nicol applied for a patent on "Improvements in and in connection with Photographic Printing."[1] At the London office he left a provisional specification, which he later expanded into a complete specification on 23 December, 1889. The patent was finally accepted on the 15th of February, 1890, nearly a year after the original application. The first patent, for what shall be called here the "Kallitype I" process, was given the patent number 5374.[2] The patent was available to interested parties for a price of 6 d. from the Patent Office. Very shortly afterward, on March 14, 1890, the *British Journal of Photography* published the complete patent # 5374 for all to examine for whatever purposes.[3] Thus was the Kallitype photographic printing process launched into the photographic world of the late nineteenth-century, a world frenetic with amateur activity, technological invention, and commercial exploitation. The *American Annual of Photography* in its Review of the Year 1890 remarked, "the science of photography during the last years has attained the most wonderful perfection and has produced results of great importance in abstract science and in practical work."[4]

II.   Nicol: Biographical Details

The Kallitype patent #5374 contains some information about Nicol, and we may be excused an early digression from the process to relate a few details about the inventor. The patent identifies the patentee as:

"William Walker James Nicol, of Mason College, Birmingham in the County of Warwick, Doctor of Science, and Lecturer on Chemistry." Nicol was at that time 35 years old, and still a bachelor (he was to marry four months after the patent was granted). By all accounts Nicol was a much loved and respected professor at Mason College when he worked out the kallitype process.[5]

While not a great deal is known of Nicol's life, some facts can be given. He was born in Edinburgh in 1855—about the time when the wet plate process began to supersede the daguerreotype. He received his early education in England and his secondary education at Edinburgh Academy.[6] Nicol attended college at the University of Edinburgh, from which he graduated with a Master of Arts degree. Attracted to chemistry, he completed the requirements for a Bachelor of Science degree and finally, the degree of Doctor of Science, receiving several prizes for scholarship along the way. After receiving his doctorate, Nicol did research for a time in Hofmann's laboratory in Berlin. He returned to Edinburgh University where he served as a laboratory demonstrator in chemistry. He was asked by Professor Letts to become a Lecturer in Chemistry at University College, Bristol. Four years later, in 1880, he was appointed Assistant Lecturer in Chemistry at Mason College, Birmingham, under Professor Sir William Tilden, FRS. Nicol held this post during the period 1881-1894.[7]

Nicol's scholarly interests were "centered in physico-chemical problems, especially those connected with the theory of solutions." His research was reported in numerous papers contributed chiefly to the Royal Society of Edinburgh, The London Chemical Society, the Philosophical Magazine, and to various chemical journals in Germany and Britain. His work dealt with such subjects as solution, saturation, super-saturation, coefficients of expansion, molecular volumes, vapor pressures, water of crystallization, and with microscopic observation of the allotropic forms of many salts under varying conditions of temperature and pressure.[8] Nicol was installed as a member of the British Association for the Advancement of Science in 1885 and was the long-time secretary of the Committee on Solutions.

Nicol's obituary notice records "his advanced students [at Birmingham] regarded him as a genius." It reports students often said of him, "If you want to have anything explained, go to Dr. Nicol." Nicol was repeatedly asked to give popular lectures and was also called upon to give courses of lectures in various towns in the South and West of

England, "which points to his success as a teacher." One sentence in the death notice leaves the reader wondering: "His mind was full of ideas which he worked out only far enough to satisfy himself as to their truth." This sentence will come to mind more than once as we pursue the course of his photographic invention, the kallitype. Finally, the obituary informs us that Nicol "was ever full of invention, and his power of quick observation enabled his hands to carry out his thought intuitively, without any outside aid or teaching, so that he made with perfect finish and ease whatever he wished in his fully equipped workshop and laboratory."[9] We learn elsewhere that Nicol was an experienced and knowledgeable photographer whose skills went back to wet plate days.[10] Personally, Nicol was "a man of unfailing gentleness and courtesy, which together with a strong sense of humor, impressed those who know him intimately and made them feel him to be one of the most loveable of men."[11] Nicol died on March 18, 1929 in Edinburgh.

III.   The Provisional Specification of Patent 5374

The provisional specification of Patent # 5374 begins with a general description of the Kallitype I process.

> Suitable paper, such as Rives or Saxe, either previously sized or not with the following salts—Ferric Nitrate, Ferric Tartrate, Ferric Ammonium Citrate or Tartrate, Ferric Sodium Citrate or Tartrate, Ferric Potassium Citrate or Tartrate, or with other salts of Ferric Iron, the solutions of which are not precipitated by Ammonia.[12]

After the paper is coated with one or a combination of these iron compounds which act as a sensitizer, it is dried. The paper is exposed under a negative to daylight until a weakly visible image is produced.

The specification continues:

> The weak iron image is developed by floating the paper on, or immersing it in, a solution consisting of a salt of silver dissolved in a solution of ammonia to which have been added suitable quantities of one, or several, or all, of the following salts: Sodium, Potassium, or Ammonium Citrate, or one of the

oxalates of those metals, or the tartrates of these metals, or the corresponding salts of other organic acids, whose ferric salts are not precipitated by ammonia. The developer may include other salts which alter the tone, or regulate the contrast—the "light and shadow" of the photographic print. [13]

After development, the print is washed in a series of three highly dilute solutions of ammonia to which is added some dilute developer. The ammonia bath acts as a fixing bath and removes the non image-forming silver nitrate from the exposed paper.

Finally, "after the whole of the silver and iron salts has been removed from the paper by the above solutions," the print is washed in water to remove any remaining soluble chemical compounds still in or on the paper fiber. Last, the Kallitype print is dried.

After the process description, Nicol specified what he claimed to have invented in the process he wished to patent. Nicol's claims of invention quickly developed into a controversy. Critics hastened to object in print that the principle of Nicol's invention did not materially depart from the "iron-argentic" paper described by Herschel more than 40 years previously or from the published accounts of iron-silver papers described by Hunt and others during the years before Nicol's 1889 patent. Critics pointed also to the several iron platinum papers patented by Willis in the 1870's and to the definitive description of iron-platinum printing published by Pizzighelli and Hubl in 1883. Herschel's argentotype and Willis's platinotype depended on the same ferric sensitizers that Nicol appeared to claim as his invention. A number of rather heated statements were written which questioned Nicol's claims of invention. In fairness, there were also replies supporting Nicol's claims, some written by leading photo chemists of the day. The matter shall be aired in detail shortly. For now, suffice it to say that Nicol's claims of invention bear close inspection.

Not all of Nicol's claims for invention were disputed. Nicol's innovative fixer, ammonia, is a notable, example. Late nineteenth-century photographers erroneously believed the sulphur compound used to fix silver prints, hypo, was responsible for print yellowing and fading. They also believed, again incorrectly, that sulfur fixing compounds could not be sufficiently removed by washing to prevent eventual print damage.[14,15] Responding to the strong distrust of hypo of the time, Nicol substituted ammonia as the fixer in the kallitype I process, believing it would prevent "image fading and the yellows." In this critical decision he rejected

Herschel's recommendation that sodium hyposulfite is the first choice as an effective solvent of silver salts. Nicol was convinced that the avoidance of hypo "added to the permanency of the print." The effectiveness of this decision will be discussed later. However effective, it is surprising to find Nicol claiming ammonia fixing as his invention, since ammonia was a solvent or fixer of silver salts well known to photo-chemists of the time. Nicol's decision to reject traditional fixing technique had considerable influence on the success of the Kallitype I process, as we shall see.

Nicol also claimed as his invention the production of a satisfactory image color in prints without the heretofore required step of gold toning. This claim was not disputed; it was based on the fact that the kallitype naturally produces a range of print colors that are quite pleasing. By contrast, the most popular print paper in use at the time, albumen paper, required the use of a toner, generally gold, to achieve a pleasing print color, an additional step that was both expensive and time consuming. In the patent Nicol stated the pleasing image color of the kallitype print "greatly simplified the operations involved in photographic printing."

The direct production of an appealing image color and the increase of permanence consequent to the avoidance of hypo are two of the five innovations Nicol claimed in his patent. He claimed three other "inventions":

1. the invention of original developing solutions for the Kallitype process;
2. the invention of specific sensitising, clearing, and washing solutions; and
3. the use of potassium dichromate in the developer to control the contrast of the image.

On the basis of these five innovations, Nicol concluded "I therefore claim as my invention" the kallitype process.

IV.   Nicol's Patented K I Process

The details of working the Kallitype process, as the patent describes them, shall now be reported. Publication of the process as patented may prove interesting to those who want to try the original process. The description will also have utility for those who wish to compare the original with later versions of the process that soon appeared.

Interestingly, the 1889 patent describes not one but two sets of sensitizing and developing solutions. They shall be designated K1a and K1b approaches. The "1" indicates a kallitype process in which the ferric sensitizer is coated on the paper alone. In this approach the silver that makes up the final image, is found in the developing solution, applied after the paper is exposed. The "a" and "b" signify variations in the solutions of the original patent. Anyone consulting the original patent will discover Nicol used a different numbering system. The one adopted here permits a more orderly organization of the various processes he developed over the years.

The sensitizer and developers for the K1a process are compounded as follows:

K1a Sensitizer.
Ferric Sodium Citrate   20% solution
Potassium Oxalate   5% solution

K1a Developer for blue black tones.
Potassium Oxalate   20% solution
Silver Nitrate   1.5% solution
(Add ammonia to the Silver Nitrate till a precipitate forms and and then add more ammonia till the precipitate dissolves.)

K1a Developer for neutral black tones.
Potassium Oxalate   10% solution
Silver Nitrate   1.5% solution (Add
                                       ammonia as above)

K1a Developer for sepia tones from K 1A paper
Borax   7% solution
Silver Nitrate   1.5% solution (Add
                                       ammonia as above)

The K1b sensitizer is compounded as follows.

K1b Sensitizer
Ferric Oxalate   5% solution
Ferric Tartrate   5% solution
Tartaric Acid   1% solution (Or use
                                       Oxalic Acid,)

For Kallitype paper sensitized with K1b sensitizing solution, Nicol proposes one or the other of the following developing solutions.

K1b Developer (1)

Potassium Citrate	15% solution
Sodium Acetate	10% solution
Silver Nitrate	1.5%solution (Add ammonia as above)

K1b Developer (2)

Potassium Citrate	15% solution
Potassium Oxalate	10%
Silver Nitrate	1.5% solution (Add ammonia as above)

Nicol specifies that prints made from sensitizer K1a may be developed in the K1b developing solutions "if the prints after removal from the developing solutions are placed for a short time in a clearing solution." The clearing solution is necessary for the removal of a yellow stain that forms on the paper as a result of the precipitation of "basic iron salts."

K I Clearing Solution

| Potassium Citrate | 20% sol. |
| Ammonia | (Sufficient to render the solution alkaline) |

If desired, Sodium Citrate or Sodium Tartrate or Potassium Tartrate can be substituted in the above clearing formula with the same percentage of chemical to water.

For the control of contrast, Nicol suggested the following solution.

K I Contrast Solution

| Potassium Dichromate | 5 grams |
| Distilled water | 100 cc. |

This solution is added to the developer solution in quantities of from "2 to 10 cc per liter of developer used."

The 1889 patent specifies a final washing solution.

K I Washing Solution
Sodium Citrate 25% solution	75cc.
Ammonia, strong	75 cc.
Tap water	10 liters

The washing solution is compounded of equal volumes of the sodium citrate solution and ammonia. 75 cc of each solution is then added to ordinary tap water to make 10 liters of working solution.

No specification of a final wash in plain water is given in the patent, although such a wash was recommended in the directions for commercial kallitype I paper when it appeared on the market.

V.   Critical Response to Nicol's Kallitype I Patent

The first response to the patent on Kallitype I paper appeared on April 18, 1890, a month after the patent was published. The editors of *Photography* comment on the new process in a short note under the heading of "The Spirit of the Times."

> Dr. W. W. J. Nicol, of Mason College, Birmingham, a chemist known throughout the country for his original research, has devised and applied for a patent for a new method of printing. We gather the paper is sensitized by means of a solution of ammonia citrate of iron, developed by means of nitrate of silver in solution, and afterwards fixed by the application of ammonia. This is so obvious a departure from the present methods—and from the fact that no compound of sulphur is used throughout the process, makes so fair a promise of permanence—that when further particulars are forthcoming, a considerable amount of interest will be aroused. We trust it will be more practical than the new negative process of his fellow-townsman, Dr. Hill Norris, of which we have heard so much and seen so little. [16]

It is notable that the announcement draws attention to the ammonia fixer and the 'promise of permanence.' The reference to Dr. Hill Norris' negative process is an allusion to a complicated method of coating wet collodion sensitized plates with non-drying gelatin as a way of making them into "dry," portable plates, or rather, "wet plates that couldn't dry." The ill-fated Hill Process is detailed in Wilson's *Cyclopaedic Photography*.[17] It is also interesting to note that the first response to the Kallitype was made, not to the process, but only to the patent. Not having seen prints or tried the process, the writer can only "trust it will be . . . practical." Present readers should not be surprised at the implied skepticism. The 1890's were a time when one ill-conceived photographic process followed another out of the patent office, which seemed disinclined, or powerless, to screen out the worthy from the worthless photographic inventions, however much the frequently disappointed public might complain.[18]

The first critical discussion of the Kallitype I process appeared in the May 2 issue of *Photography.* The review followed an interview with Nicol and an inspection of prints he made. The writer began with two observations, one on the probable future of the process and another on the high level of professional research done by the inventor.

> That the new printing process is destined to supplant platinotype and annihilate albumenized paper is scarcely to be expected, but that it will prove a formidable rival to the former, and another nail in the coffin of the latter, we feel safe to predict. It is not a process of variation merely stumbled across by chance, but the result of painstaking and laborious research. The different prints exhibited by the inventor show clearly the various phases through which the process has passed. One by one the points have been thoroughly rounded, and there is no doubt when working details are published they will be such as will entitle the process to be considered at once complete.[19]

The unsigned reviewer's praise of Nicol's research will be worth remembering when the process is commercialized and offered for public use. We shall then learn if the research was painstaking enough and whether the process is "complete."

The review describes the appearance of the first kallitype I prints and provides an indication of the appearance of early kallitypes. The description provides valuable information to the historian, since no vintage prints are available today for visual inspection. The early prints, we are told "present the whole range of appearances of silver and platinotype printing with the single exception of the sepia tone of the latter." The

implication is that the early K1 prints ranged from neutral to warm black.
Apparently the prints shown the reviewer included no brown kallitypes.
The review indicates that variations in image color of the kallitype could
be achieved through simple modifications of the basic process.

> By sensitizing a paper coated with gelatine, a print with
> all the beauty and warmth of a well-toned silver print is the
> result; the layer of gelatine affords protection to the image, the
> silver is deposited at a much slower rate, and this apparently
> affects the colour to a great, or even greater extent than the rate
> of deposit affects the colour of a gold-toned print. By using
> a paper less strongly sized, as well as by varying the nature
> of the latter a whole range of platinotype or bromide effects
> may be obtained; and in answer to the prevailing demand,
> the doctor has sought to perfect this variation. A number of
> prints have been prepared on the same plain paper that is
> used in the platinotype process, and the resulting pictures are
> indistinguishable from high class platinum work; indeed, if
> anything, the fine detail in small prints is better rendered by
> the new method. This may be accounted for by the extremely
> rapid rate at which the development takes place, the silver
> having actually no time to sink into the paper, the whole image
> remaining upon its surfaces; yet with such tenacity does it
> adhere that the most vigorous rubbing with an eraser fails to
> remove the deposit. [20]

"Indistinguishable from platinum"—this is effusive praise indeed,
printed in a journal with the highest photographic standards. It is praise
of a sort rarely given during a period when the platinum print represented
the epitome of photographic esthetics and technology. The reviewer
also discussed the toning of kallitypes. "Toning is entirely unnecessary,
even the warm tones being obtained by simple development." This is
an obvious praise of the natural color of the kallitype over that of the
albumen print, which was considered so unsightly that gold toning was
routinely done. The tones of the Kallitype I, as the patent indicated,
could be controlled from a decided neutral black to a warm black, by the
choice of developer, size, and the paper stock used. When commercial
distribution commenced, the controlled variation of print color became
one of the kallitype's strongest selling points.

The reviewer explained the various apparent colors of the kallitype image in terms of recent research on colloidal deposits. The theory held that a slow rate of deposit during development formed smaller silver particles which made the image appear reddish, while a faster rate of deposit formed larger particles, which appeared bluer.

The reviewer also suggested that the kallitype image appeared more fine grained than the platinotype. Whatever truth this notion had, depended on the version of the platinotype process in use at that time—the "hot bath" process. According to the reviewer, the hot bath process developed the platinum image slowly, and the long development time permitted the platinum particles to "sink into" the paper, lowering contrast and resolution. By comparison, the development of the kallitype image was very rapid, (the image appears fully in about three seconds) and its shorter developing time produced a crisp, well-defined image because the image silver remained on the surface of the paper fiber. The reviewer also praised the K-I process for the firm bond it creates between the image forming silver and the paper surface—one so tight that "a rubber eraser cannot abrade it from the paper fiber." This is a reference to another of the problems of the early platinotype, a tendency for the image "to fall from the paper." Willis solved this problem, at least temporarily, by incorporating some silver in the sensitizer as a means of more firmly bonding the platinum to the paper fiber. It is apparent that the reviewer, whether prepped by Nicol or acting on his own initiative, pointed out the advantages, real or fancied, of the new process over existing ones.

The reviewer drew attention to other details of the kallitype I process, some of them questionable. He stated that the sensitized paper "may be kept a very considerable period." It cannot. Experience demonstrates that the ferric sensitizer is so unstable that it undergoes chemical change in a short period of time, often in a few weeks, without exposure to light. He also stated kallitype paper may be used dry or slightly damp, moisture giving the image a slightly warmer tone. The present writer has found that the very slightest dampness of the paper has a tendency to stain the negative during printing. The review announced surprisingly short exposure times for Kallitype I paper. It reported kallitype requires only one-fourth the exposure of albumenized paper to daylight and is even "more sensitive than albumen in bright sunlight." The paper required as little as one minute to expose a normal negative in bright sunlight. Information such as this called attention to the advantages of kallitype over other print processes in the market.

In this effort the reviewer sometimes goes too far. When he writes "the image is visible on removal from the printing frame, all but the finest detail in the distance being apparent," he stretches the truth. In fact, while it is possible, it is not easy to tell by image inspection when the kallitype has been fully exposed. Depending on how the sensitizer is mixed, kallitype I paper can give very little indication of full exposure. By comparison, albumen paper allowed more certain estimation of correct exposure, an advantage that partially explains its long hold on the Victorian photographic market.

At the time of the invention of the kallitype, fully printing-out photographic papers (POP) were widely preferred by amateurs and professionals, because they were cheap and required little skill to use. With printing-out-paper the operator simply exposed the paper until the image was fully visible—development was not necessary to bring out the image which was fully produced by exposure to sunlight alone. Because albumen paper printed the image fully, it was considered the least troublesome and most waste free approach to photographic printing. The "developing out" papers, (DOP) such as bromide, which were just beginning to enter the market, gave no indication of exposure at all, and therefore were frequently criticized by "operators."

The report that Kallitype I paper indicated when a print had sufficient exposure exaggerates. The reviewer was correct when he said that a visible image is formed, but the image is rather faint. It is true that the incomplete image can be used as an index of exposure, but the indefiniteness of the image is such that estimates of correct exposure based on it are frequently incorrect and costly.

At a time when little was known about the kallitype I, the reviewer may be excused for making a few other questionable statements about the process. He states, for example, that "when inspecting the newly exposed paper on its removal from the (printing) frame, if it be found that overexposure has been made, the print may be placed aside for a time, and the image will gradually and slowly 'go back' until it arrives at a state of correct exposure, and may then be developed with success." The reviewer offered no evidence to support this claim of image regression and none has since been published by researchers or practitioners. In another observation of doubtful value he states that the silver nitrate developing bath can be "starved of silver" as a means of getting "suitable prints from hard negatives." Experience indicates that weakened silver baths do not produce good prints from contrasty negatives, but rather grainy prints.

The reviewer made several important claims about image permanence, a popular subject in the literature of the time, when print deterioration was often rapid and common. He writes without reservation that "kallitype images are as permanent as images in silver can be." He states that Nicol does not hold this as a matter of opinion, but on the basis of "reliable experiments made with the purpose of testing the point." The permanence of the kallitype print became controversial as time went on. At issue was the vulnerability of the silver image, which lay exposed and unprotected from airborne contaminants on the surface of the paper. The writer continues, "if greater stability is desired than silver is capable of, it is quite possible to tone [the kallitype print] with platinum, and . . . afterwards remove the silver" [by bleaching]. This is an effective though costly suggestion, but it may be questioned whether the resulting print can still be called a kallitype print.

The last subject the first review broached is the thorny one of Nicol's debt to others for the basic idea of the kallitype and his right to patent.

> The inventor, Dr. W. W. J. Nicol, has practised photography since the collodion emulsion days, and is well acquainted with the researches of Herschel and also of processes based upon Herschel's in which sensitising by ammonia citrate of iron and development by silver nitrate has been practised, but claims originality on several points, notably his method of fixation. [21]

Without making any judgments, the writer cites Nicol's knowledge of past research on iron silver photographic processes and restates Nicol's "claims to originality" specifying his method of fixation. If his eyebrow is raised, he hardly lets it show. With exemplary British restraint he avoids taking sides on the issue. Later commentators were not so gracious.

They soon questioned whether Nicol should have been given a patent on a process that had been worked out in great part by others. This topic was an especially sensitive one in view of the English experience of Fox Talbot's restrictive licensing of sodium thiosulfate, the most effective fixer for silver photography. Herschel had been the actual discoverer of the fixing properties of "hypo," but Fox Talbot managed to acquire a patent controlling commercial exploitation of hypo in England. Many who knew the facts resented Fox Talbot's patent and right to license hypo, especially since France had earlier "given" the Daguerreotype to the public without restriction.

The objectors were concerned that the patent would enable Nicol to similarly restrict the present and future benefits of a process conceived by Herschel and developed by others. This feeling, if not resolved would have clouded the professional reputation of Nicol and could have damaged the commercial development of the kallitype process. Sooner or later the problem of proprietary rights would have to be resolved.

The reviewer concluded by praising the Kallitype I process for its "strong points of beauty, simplicity, rapidity, and economy." He "looks ahead with considerable interest for further details."[22]

It is interesting to speculate on the identity of the writer of this important but unsigned review. It might have been one of several writers knowledgeable about photo-chemistry such as W. K. Burton, G. E. Brown, or C. H. Bothamley. Each of these writers developed strong interest in the kallitype and wrote popular articles about iron-silver and related processes. Burton and Brown, as we shall see, became outright partisans of kallitype, writing on more than one occasion with considerable enthusiasm about their own researches into the process. Bothamley investigated the kallitype and published several expert reports on it, some of which suggested improvements on Nicol's methods. All of these men were high caliber photo chemists who spent their lives studying photographic processes and writing about them. While one of these men is the likely writer of the review, it must be admitted that there were several other professional photographic scholars in Victorian England, who might have been the author. One conclusion is clear, whoever the writer was, he commended Nicol for a remarkable professional accomplishment, supported his claim of invention, and encouraged the early commercial development of the process.

VI.   Bothamley's Report on K I

A somewhat clearer and more technical report of kallitype I, based on the description in the 5374 patent was written by C. H. Bothamley in the *Journal of the Society of Chemical Industry* of April 30, 1890.[23] Bothamley was a respected photographic chemist, a university lecturer in photochemistry, and a scientific writer who supported himself throughout his lifetime by working the intersection of the three disciplines of photography, chemistry, and professional writing.

Teaching at Yorkshire College, Bothamley wrote articles and reviews on photographic subjects for as many as four journals at the same time, variously packaging the same information and views for the different publics his periodicals addressed on both sides of the Atlantic. As will be seen, his writing on the kallitype is of the highest professional caliber. His report on the K1 process will be given in Toto since it is more clear and direct in a number of respects than the patent itself, which clouds its presentation, perhaps intentionally, with confusing alternatives, arcane chemical description, unclear quantitative measures, and legalistic formalities. Bothamley's report, written by a peer of Nicol in learning and photographic expertise, is offered as an alternative description of the procedure for making Kallitype I. Readers wishing to understand and implement Nicol's description of the kallitype process in the patent will find Bothamley's commentary quite helpful.

PATENT

Improvements in and in Connection with Photographic Printing. W. W. J Nicol, Birmingham. Eng. Pat. 5374, March 29, 1889. 6 d.

Paper, wood, or any other surface is sensitised with a ferric salt, with a salt of an organic acid, which prevents precipitation of the iron by ammonia. It is then exposed to light under a negative, &c., until a faint image is visible, and is developed by treating it with an ammoniac solution of silver nitrate containing a salt of an organic acid. The print is washed with water containing ammonia and an organic salt, and finally with pure water. A solution (1) containing 20 per cent of ferric sodium citrate and 5 per cent of potassium oxalate gives very good results. A solution (2) containing 5 percent of normal ferric oxalate, 5 per cent normal ferric tartrate, and 1 per cent of tartaric or oxalic acid also works well, but the developing solution requires modification in order to prevent precipitation of the iron.

If necessary the sensitising solution may be thickened with gelatin, starch, &c. Paper sensitised with solution (1) may be developed with one of the following solutions: (a) for cold black tones, 20 per cent of potassium oxalate, 1.5 per cent silver nitrate; (b) for neutral black tones, 10 per cent of potassium oxalate, 1.5 per cent of silver nitrate; (c) for sepia tones, 7 per cent borax, 1.5 per cent silver nitrate. In all cases the precipitated silver salt is nearly redissolved by adding ammonia.

If the paper has been sensitised by solution 2) the following developers may be used: a) 15 percent potassium citrate, 10 per cent sodium acetate, 1.5 percent silver nitrate; or b) same as a) except for 10 per cent of potassium oxalate in place of sodium acetate. In both cases the precipitate is redissolved by ammonia. The first series of developers may also be used (with prints sensitized with solution 2), but in this case the print after development must be immersed for a short time in a clearing solution containing 20 percent of an alkaline citrate or tartrate made alkaline with ammonia. The developed prints are first washed in an alkaline citrate or tartrate with an equal volume of strong ammonia solution, and diluting 150 cc of this mixture up to 10 liters with water. They are afterwards washed with water and dried.

Development and other operations are conducted at the ordinary [room] temperature. The process is applicable to fabrics as well as to glass and paper. C.H.B[24]

Bothamley's summary of the patent is faithful to Nicol's original. Ironically, printed directly above Bottamley's report on Nicol's patent, another iron silver process is reported, one invented by an "American amateur photographer." The report reads (in its entirety):

Paper is coated with an emulsion containing ferric oxalate, 7 parts, silver nitrate 7 parts, gelatin 7 parts, tartaric acid, 3 parts, and water, 128 parts. The images are printed out and have a pleasant brown colour, and are fixed in the usual way. C.H.B.[25]

The existence of another iron silver process created by an American amateur a continent away confirms there was no secret about iron-silver photography. Rather, it appears that iron silver printing was an idea whose time had come. By 1890 the basic ideas for this process were well known by chemists and Nicol was only one of several who conceptualized a workable process. Indeed, the literature reports several others, Stebbins and Duchochois in America and Bedding in England had essentially the same process in mind as Nicol. The difference is that Nicol was the first to patent a workable process.

Before leaving Patent # 5374, it is worth mentioning that Nicol applied for a second patent, #4269 on March 19, 1890.[26] This application was subsequently abandoned and declared void and all particulars stricken from the record. No copy of this patent application has been located. In the absence of Nicol's papers, we have no basis for speculating about the subject of the patent, except to wonder if it concerned a breakthrough that did not proove out with further lab work.

VII.   Bedding's Charge of Unjustified Claims

The next significant happening in the story of Kallitype I was the appearance of an article that critically questioned Nicol's moral and scientific right to claim the process as his invention.[27] The article was written by Thomas Bedding, a respected scientist and photographic investigator who later became the editor of the prestigious *British Journal of Photography*. Bedding, as we have seen, had done extensive research on iron photographic processes, or as he called them, "the substitution processes." These processes first formed an image in the salts of one metal, iron, by coating them on paper and exposing them to light. The original metal is then replaced during processing by another metal, for example uranium, platinum, or silver. This is the basis of such processes as the blue print, the uranotype, the platinotype, and the kallitype. Bedding had researched iron-uranium printing for some time and was familiar with early studies in this area made by Herschel, Hunt and others. His article can be reduced to three charges. First, he argued that Nicol had no right to a patent an old idea that had been common chemical knowledge for more than a generation. Second, he argued that Nicol's process was not new but merely an analogue for the platinum printing process, which was already patented by another. Finally, he argued that the chemical innovations made by Nicol were unimportant or unnecessary for a workable iron-printing process.[27]

To understand this controversy, we shall look more carefully at Nicol's claims of invention in the patent and try to establish their basis.

Then we shall examine Bedding's charges in detail and determine if his charges are well founded. Out of such an examination we may come to understand how two such educated and honorable men might disagree, if they were in fact having a genuine disagreement. It may turn out the difference was based on a misunderstanding of patent requirements.

VIII. The Claims Controversy and Its Resolution

As indicated earlier, Nicol ended his patent claiming what he had invented. The statement reads in detail:

> I therefore claim as my invention as follows:
>
> First.—the use of paper or other surfaces coated with ferric salts as described above, to be used for the production of photographic images in silver, along with the class of developing solutions already given and particularly referred to in my second claim.
>
> Secondly—The use of solutions of silver salts dissolved in Ammonia in conjunction with salts of the alkalies or Ammonium with Oxalic, Acetic, Boric, Citric, or Tartaric Acids, for the purposes of producing on paper or other surfaces previously coated with Ferric Salts sensitive to light, images in silver which are of such a colour as not to require toning with gold or other metals.
>
> Thirdly—The combined use of Ammonia and a Potassium Sodium or Ammonium Salt of Citric or Tartaric Acid in a solution used for washing prints obtained on surfaces coated with Ferric Salts and developed with solutions of Silver Salts in Ammonia.
>
> Fourthly—The use of the Solutions
> Sensitising Solutions I and II
> Developing Solutions 1a, 1b, and 1c.
> Clearing Solution
> Washing Solution in the way and manner . . . described and set forth in the foregoing.
>
> Fifthly—The use of an oxydising agent such as Chromic Acid or any of its salts in a developer prepared as described and set forth.[28]

If one reads Nicol's five claims carefully, it is clear that he nowhere claims invention of the iron-silver print process, but rather limits his claims to working out specific chemical solutions that produce a silver print from an iron sensitizer. He claims invention of a specific set of chemical steps, a particular sequence of solutions, not iron photo-sensitizers in general, or substitution printing processes in general. It appears that if another wishes to develop another set of steps that will produce an iron silver print, or an iron uranium print, or an iron platinum print, that is their prerogative.

Now for Bedding's side of the controversy. In an article entitled "Silver Printing by Substitution," published in the **British Journal of Photography,** Bedding wrote that there were "one or two features of the specification" that "seem to call for some explanation." He points out that Nicol's patent as written "closes off all possible commercial exploitation of iron silver printing to any but the holder of the patent 5374." Second, he writes "the toning and fixing difficulties" Nicol takes credit for eliminating "never had any existence in fact; and that the 'means' &c. were not only unknown, but not sought after." Third, he argues that Mr. Willis has claimed for his platinum processes the use of many of the ferric compounds scheduled by Dr. Nicol in his specification. If so, he asks, "can Dr. Nicol subsequently acquire a valid patent for a process which differs from that [of Willis] . . . only in the replacement of the salt of platinum by a salt of silver? Fourth, he argues that "the published experiments of Herschel, Hunt, Draper, Emerson Reynolds, Phipson, Eder, and others . . . compel me to doubt if anybody now living can claim as his "invention" the use of surfaces coated with ferric salts for the production of photographic images. Finally, from the point of view of patent law Bedding argues that, Nicol merely applied "an old or known article to an analogous purpose" which is therefore not subject matter for a valid patent. Applying what he considers to be the patent law to the kallitype, Bedding argues that "ferric oxalate is certainly an 'old' and 'known' article, and that there is an 'analogy of purpose' between Mr. Willis' cold-bath platinum process and Dr. Nicol's silver process—namely the production of positive pictures. Bedding ends on a positive note: "I hope to see Dr. Nicol's kallitype process submitted to an exhaustive trial, as I am convinced from my own attempts that very excellent pictures are to be had by the substitution method.[29]

Bedding's attack drew from Nicol one of only two statements he ever made to the photographic press. His reply to Bedding was published in the ***British Journal of Photography*** on May 17, 1890, in a Letter to the Editor. It said, in full,

> Sir, Allow me to correct an error Mr. Bedding has fallen into in the course of the remarks he made on the kallitype process in your last week's issue. The passage I take exception to is where he quotes from my specification, and in the next paragraph goes on to say," I should take the foregoing to premise that a mixture of ferric chloride &c." to the end of the paragraph.
>
> Mr. Bedding is probably correct, in so far as he states that no such process with ferric chloride has ever been made public, but all the former processes of this kind involved the use of the ferric salts of acids, the silver salts of which are insoluble in water, thus necessitating the use of some fixing agent. For this purpose sodium thiosulphate has been generally recommended, though a dilute solution of ammonia has, I believe, been used. Even in cases, if there are any such, where ferric salts of acids, which do not form insoluble silver salts, have been used, fixing could only be avoided by using distilled water for washing, the expense of which would be fatal to any process.
>
> With regard to toning, the colour of the silver image in all such processes has hitherto left much to be desired. There has been a lack of brilliancy and general greyness that is frequently referred to in accounts of the old processes. I cannot, of course, enter into a discussion of the mysteries of patent law, but I can assure Mr. Bedding that I take his remarks in good part, and shall be very pleased to make his acquaintance and argue the matter if he happens to be in Birmingham and will call upon me here. I am, yours, &c.
>
> W. W. J. Nicol.
>
> (We may here state that we have received (although not in connexion with this correspondence) two specimens of kallitype. The tone is remarkably good.—Ed). [30]

It is evident that Nicol did not answer most of the Bedding's "questions." In fact, he responded only to the most chemically arcane issue. Nicol's reply defends 1) his claim of invention of his ferric sensitizer (ferric oxalate, not ferric chloride) and 2) his invention of a method that produced an agreeably toned image. Nicol, obviously avoids discussing the moral and legal issues involved in patenting a commonly known principle established by the labors of two generations of scientists. Nor does he explain why patent law should give him the right to exclude others from developing further a process in great part developed by the community of scientists. Nor does Nicol discuss his plans with respect to the commercial development of his process or of allied processes. Since Nicol made neither a very strong or detailed defense nor a statement that clarified his and others rights to the exploitation of known and related processes, we should not be surprised when the issue surfaces again. We shall report further developments as they occur.

From Nicol's defense, it may be assumed that he wanted to avoid a confrontation about issues which held little interest for his kind of mind and with which he felt little qualified to deal. He appears to believe that the issue can be settled better in an amicable private conversation than in a public debate conducted on the pages of a journal. One suspects that Bedding's way was not his way. And, it is quite likely that Nicol was also very busy in the laboratory and in the plant, trying to resolve the many chemical and engineering problems that exploitation of his process entailed.

IX.   Announcement of The Sale of Kallitype I Paper

The next development regarding the Kallitype I appears in "The Editor's Work Table," a column of the *Photography Magazine* which routinely announced new photographic products. In the issue dated May 29, 1890 we find the following notice:

KALLITYPE
John Lewis and Co., 100 Gladstone Road, Sparkbrook, Birmingham.

Kallitype is the name of a new printing process which may be said in its general features to be a cross between the bromide and the platinotype, for whilst the image is very often visible before development, like fine platinotype, it has to be developed.

The paper is very sensitive and requires very little printing, a few minutes in good light suffices to make the necessary impression of the image. The prints are developed in a solution of nitrate of silver, citrate of soda, and bichromate of potassium, together with ammonia and nitric acid. The specimens of work done by the kallitype process, which Messrs. Lewis favour us with, show pictures more resembling in their colour and characteristics bromide prints than anything else, having a blue black tone about them. There appears to be no lack of density in the shadows whilst the high lights are perfectly clear, and we should say that with suitable negatives, very good and effective results could be obtained with the new process.[31]

This is the first evidence of the commercial exploitation of Nicol's invention that research has turned up. At the time of writing, the writer has been able to find little about the John Lewis Company or John Lewis the man who headed it. The company, based in Birmingham, where Nicol lived and taught, appears to have been formed to market Kallitype I paper, although its early advertisements list a number of allied photographic products. Early on, it expanded its line of goods to include other print papers, print manipulation accessories, and photographic devices. Shortly after this announcement, the Company changed its name to the Birmingham Photographic Company, perhaps wishing to trade on the city's growing reputation for well-made technical products.

The Company's advertisements appear in the photo-annuals as small one column, one inch cuts. A full page ad is reproduced below in Figure 1 and is the most complete advertisement found. It announces the full line of products the company carried. There is no evidence the company ever became more than a small operation which manufactured, packaged, and distributed the sensitized kallitype paper and a few accessory products. The manufacturing obviously required some expertise and control since a photographic product like kallitype is subject to many influences which can cause it to fail. It is likely that Lewis handled the business end and that Nicol carried the responsibility for product development, manufacturing, and quality control.

Reading the announcement of the formation of the new company and the statement of anticipation about the new commercial paper, one senses the excitement of the new venture and imagines the high hopes that the young chemist and the new businessman had for their product.

Significantly, on the same page which announced the commercial offering of Kallitype I paper, there are announcements of two other photographic papers newly offered for sale. The first, a new matt-surface albumen paper, sold by Mr. Otto Scholzig, is said to have a pleasing purple-gray tone, something between that of a deeply-toned

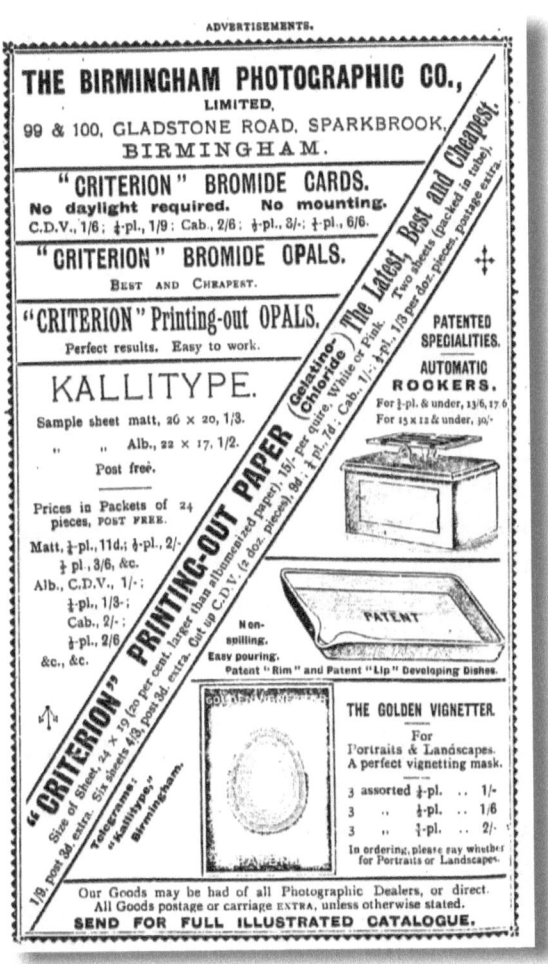

Figure 1. Birmingham Co. Kallitype advertisement

silver print and a platinotype. Also mentioned is a new paper offered by Dr. Jacoby—a direct printing kallitype paper.

We find it prints readily and is extremely easy to handle. It will be . . . doubtless welcomed by those who prefer to print

> the image right out and who cannot get on with the half-visible
> images of the platinotype process. . . . All that has to be done
> with the printed image is to wash the paper in a bath or two
> of dilute acid.[32]

Such notices are indicative of the intensely competitive state of the photographic paper market in 1890. Hardly any issue of the photographic journals appeared without mention of another new paper for sale. Note that the ad mentions several papers offered in competition with the kallitype. The 'half image' spoken of here is the primary iron image formed when the ferric sensitizer turns ferrous under the influence of light. As discussed earlier, the image of both platinotype and kallitype when properly exposed is only partly visible. The rest of the image, in either process, is formed during development.

Although new kinds of papers continued to be offered for sale, albumen paper stubbornly maintained its hold on the market, in spite of continued complaints about its limitations. Ease of estimating exposure, economy, familiarity, and simplicity of handling were the main reasons for its continued popularity. The literature contains repeated complaints about the difficulty of getting correct print exposure when printing platinotype, kallitype, and even silver bromide papers, throughout the last decade of the 19th Century. Such reports indicate the basis of the hold of albumen paper, even though its competitors had greater esthetic appeal.

X.   Reports of Kallitype Use and "The First False Step."

Soon after Kallitype I paper was offered for sale, articles describing camera club demonstrations appeared in photographic publications. Camera clubs in the 80's and 90's were very popular and served an important function—teaching photographers how the many newly introduced products worked. Photographic magazines regularly published reports of demonstrations of a variety of processes at Camera Clubs. These reports are now an invaluable source of historical information about how vintage processes worked when introduced, how users felt about them, and the ways amateurs modified processes they used.

A report of this kind on kallitype I written by J. S. Hodson, F.R.S.I., a member of the Liverpool Amateur Association, appeared in **The Camera** for June 1, 1890. This article is the first account of the use of Kallitype I paper by an amateur. The article examines several processes in current use and compares their advantages. The processes considered were albumen printing, chloride printing, bromide printing, platinum printing, kallitype printing, and "a new method of bichromated albumenized paper offered by J. B. Brown." The context in which the article considers photographic printing is that of amateur competitive exhibitions, in which the rules specify "the photographer may not escape from the drudgery . . . of being your own printer." Hodson declares, "I for one, accept the necessity [of printing] but whether I recognise any semblance of the dignity of labour may be another matter which need not be entered upon." Since the fate of kallitype and other print processes became increasingly tied to amateur attitudes and practices, we shall regularly report on them.

Hodson continues, "there are so many processes from which to select that the decision becomes embarrassing . . . and indeed, a difficulty." Comparing the advantages and disadvantages of the various processes then in use, he finds: "Albumen-silver paper is simple, reliable, and inexpensive compared to aristotype" [gelatino-chloride], but "it is not permanent." Both albumen and aristotype "are now out of fashion" because of "the glazed appearance." They "have given way to the more artistic matt-surface methods of platinotype and bromide which provide no means of knowing whether correct exposure has been given until after it is too late . . ." Platinotype is found "simple in method," but the complicated directions" "presuppose an amount of study that may well deter amateurs from buying it. The matt surface of platinotype is ". . . in its favour, but for many subjects it is . . . deficient in pluck and delicacy of detail" compared to albumen. He concludes that bromide prints are distinguished by their "brilliance and gradation of tone," that they are desirable for enlargement but not for contact printing because of the extremely short exposure.

Hodson reduces the choice of a photographic paper to resolving the claims of two competing photographic esthetics. "Where the artistic effects are produced by broad contrasts and gradations of tone without any considerable minuteness of detail, bromide and platinotype are the methods to be adopted." "When the effects desired . . . depend in great measure upon delicacy and multiplicity of detail—and whatever the new artistic school may say, photography can never dispense altogether with a moderate amount of sharpness of detail—then . . . one or the other of the methods of silver printing must be adopted."

At this point Hodson turns to the "glut of new printing processes on the market" and finds that one or more of these may prove to be useful additions to our printing methods." The Kallitype, he reports, "has been tried in several quarters, but upon its merits opinion seems to be pretty much divided. He recognizes the negative response of some users is to the novelty of the process, saying "it is not always easy to throw overboard habits which have been acquired and are of long standing, to learn something new." Nevertheless, Hodson gave an opinion on the merits of the process, founded upon actual trials, in which a desire has been felt to do justice, as far as possible, to the new process. After reviewing the details of kallitype printing, he presents his conclusions about the processing of kallitype I prints.

> The result of moderately careful attention to the printed instructions have yielded almost from the first a matt print, with a pleasing platinotype tone of color, with all the details of the subject faithfully preserved. The manipulation [processing], it must be admitted, is complex; thus there are no less than six operations [baths] included in the work—printing, developing, washing nos. [sic] 1,2,3, and a final washing in water; but they are all extremely simple, and will offer no difficulties to those who have any acquaintance with other current methods of printing. Another advantage, which counts for something with most of us, lies in the fact that the process, besides being simple, is cheap.[33]

Hodson's views have been reported at length because they indicate the attitudes of the art-oriented, upper-middle class amateur, one of the target markets (the other was the commercial operator) to which the manufacturers of commercial paper, particularly kallitype I paper, hoped to make a strong marketing appeal. Hodson's main judgments, that the Kallitype process resembled the platinotype, that it was easy, and that it was cheap were probably what John Lewis wanted to hear. It is unfortunate that Hodson was so indifferent to the technical details of photographic printing, that we learn little about the specifics of working the early process. It is also regrettable that Hodson omits mention of the important clearing step. Clearing was crucial to the production of a clean kallitype print, one free of stain. Failure to clear away the unused iron, or any iron that precipitated, led to a print covered with an unpleasant yellow stain. The stain was a frequent complaint of amateurs who tried the product.

The June 13, 1890 *British Journal of Photography* reported that Mr. T. G. Hibbert gave a demonstration of Kallitype printing to the Sheffield Photographic Society. The Secretary's report on the presentation is informative:

> . . . having some specimens printed, he proceeded to develop them, somewhat after the manner of the platinotype process, after which the general discussion followed, and it was agreed that as yet it was not superior, if equal to the platinum process, but would no doubt be considerably improved.[34]

What is notable in this report is the observation that the kallitype is not the complete or perfect process it was reported to be. The implication that the kallitype needed to "be considerably improved" is our first clue that all is not well.

Another brief account of a demonstration is reported in the July, 18, 1890 issue of **the *British Journal of Photography***. The article reports that Mr. Farrow of the Manchester Photographic Society showed a collection of kallitype prints and explained the printing and development which gave plenty of half tone, and was considered particularly suitable for portraits. [35]

A report of another first use occurs in *Photography* for September 18, 1890. This time the trial was conducted by an unnamed member of the editorial staff of the journal. The article makes the surprising admission that the "kallitype, or the new printing process, has now been before the public some months, but in a somewhat imperfect condition, as the manufacturers introduced it to the public before they had quite perfected it." The article continues, "Now, however, they have so far succeeded in improving upon their original production that they may consider it a fairly perfect process, and they have submitted to us samples of the paper and materials for the purposes of trial." [37]

What a shocking revelation! The original Kallitype I paper was introduced in an imperfect state! Those familiar with the history of photographic products in the late Nineteenth Century will find this an all too familiar occurrence, particularly likely with manufactured sensitized products, whether film or paper. Sensitized products involved a new technology of which many dimensions were not well understood and therefore were not well controlled. Products that worked well in the inventor's laboratory did not always work under the conditions of commerce or in the darkrooms of amateurs. Even the makers of the much admired platinum paper, Willis and Co., who had a reputation of marketing an exceptionally well engineered product, occasionally sold defective paper without knowing that it fell below their high standard.

To learn what befell the new process, we must make a leap forward in time, to another letter Nicol wrote to the ***British Journal of Photography***, but in July 14, 1891, at least a year after the disaster. We shall quote at this time only the part of the letter that concerns the breakdown, saving for later use the remainder.

To The Editor

Your very full and appreciative articles on Kallitype, the printing process for which I am responsible, lead me to think that you and your readers may feel interested in a short account of the improvements recently introduced, and of some of the difficulties that beset an unfortunate inventor of a printing process.

For years I had believed that it ought to be possible to discover a silver printing process, which in point of simplicity and artistic effect would rival the expensive platinotype, and be able to claim for its results a degree of permanence higher than that possessed by ordinary silver prints. For many months at various times I rang the changes with salts of iron and uranium with but indifferent success; at last, after a long-continued series of experiments in which the effect of each ingredient was systematically studied, I was advanced far enough in 1889 to take out provisional protection. Some months afterwards the process was taken up by a firm of manufacturers, and it was then that my woes as an inventor began in earnest. Results that satisfied me failed to please them, and from week to week the process was altered and changed in detail though not in principle, until, from a comparatively slow printing process requiring only a developer and a single washing bath, it became a rapid one, involving the use of no less than four baths, a fact which, as you justly point out, forms a serious drawback to the process as you know it. At the same time a serious misfortune occurred—a paper was chosen for the process which the makers stated to be pure, but hardly had it been put on the market than it was found that the prints on this paper faded, sometimes in a few days. This was traced to the paper, and another make had to be found; but the mischief had been done. Such very nearly proved fatal to kallitype at the very start. Slowly, however, owing to constant improvements in detail, it has recovered from the effects of this first false step, and now it has completely rallied. [38]

So much for "the first false step." Undoubtedly it was responsible for some of the difficulties kallitype had in entering the market. The letter helps us to understand better what was meant by the remarks, quoted earlier, about the imperfect state of the process and the need for "considerable improvement." It is reassuring to note in Nicol's report that the paper had undergone "constant improvement" and that it had recovered from the "first false step."

In the remainder of the report of his trial, the editor of the "Work Table" gave the improved version of Kallitype I paper a favorable review. He found the "new" Kallitype paper resembled in appearance and working the cold bath platinotype paper. He found it "easy to work and fully satisfactory when used with suitable negatives." He reports that K I paper could be bought from Lewis and Co. or their vendors and that the price, when compared with the price of other papers, "may be considered cheap." The processing chemicals could be bought from any well stocked "chemist" and were also considered inexpensive.

The editor's final evaluation of the new paper was that "Kallitype gives us yet another modification, by which we can vary the method of treatment of our negatives to obtain a desired result." He concludes, "we already hear from one or two of our most esteemed contributors, of very excellent accounts of the process, and we have no doubt it will obtain a considerable share of favour in popular estimation." [39]

It appears that by September 1890, the date of the above review, that the kallitype is on track. An account of a demonstration at the Hackney Photographic Society given by Dr. Roland Smith on November 27 confirms the recovery. The account reads,

> . . . The Chairman called on Dr. Roland Smith for his demonstration of the kallitype paper.
>
> The Doctor proceeded to say that it was very similar in appearance to platinum paper, but was more under control, as the development could be stopped in any part or at any moment easily. It was 50 per cent cheaper, or about the same price as silver paper, was quite a permanent process and prints in about a quarter of the time that silver paper takes. He thought it an advantage in other ways, one of which was that the solution did not want any special way of keeping. It was, however desirable that the fingers be kept out of it whilst developing . . . on account of stains. The demonstration . . . then proceeded . . . and after, questions were asked by members. The process was thought very simple and easy of manipulation.[40]

Dr. Smith called attention to a difficulty occasioned by the use of kallitype I paper—brown stained fingers. This problem was not native to the process. It occurred only when workers put their fingers in the developing solution which contained silver nitrate. Silver Nitrate has the property of making, a dark brown stain on any organic material it contacts, including human flesh. This disadvantage of the kallitype could be avoided by the simple expedient of using tongs instead of fingers to transfer prints into and from the developing tray, the only tray containing silver nitrate.

Shortly after Smith's report, the ***British Journal Of Photography*** printed, under the heading, "Kallitype," that "Messrs. Lewis and Co., Birmingham have sent us some of the later outcomes of the Kallitype process of printing. They are certainly very excellent, showing rich, platinotype-like blacks, good gradation, and pure whites."[40] Apparently the bugs were now safely removed from the product.

XI.   Conclusion

We opened this chapter with a quotation about the many and "wonderful perfections" science contributed to the development of photography in the year 1890. Having detailed the development of the kallitype I, we conclude with the observation that the kallitype I print process surely must be numbered among these "wonderful perfections," and "results of great importance." [41]

Notes: Chapter I. Kallitype I From Invention through Sale

1. English Patent No. 5374; A.D. 1889. The volume of English patents was found in the New York Public Library.
2. Nicol "invented several Kallitype processes which must be distinguished and named. The designation "Kallitype I" or the shorter form "K I" will be used to refer to the earliest version of the process described in patent # 5374, dated 1889. The K I process has variations which shall be referred to by the designations "K Ia" or "K Ib." See kallitype "designations" in index.
3. *British Journal of Photography*, March 14, 1890, 171-2.
4. *American Annual Review of Photography*, 1891, 280.
5. *Mason College Magazine*, Nov. 1890, p. 3.
6. Obituary Notice, *Proceedings of the Royal Society of Edinburgh*, vol. XLIX, no. 49, 1928-29, 369-70.
7. *Mason College Records*. This information was kindly forwarded by Dr. D. S. Benedikz, Sub Librarian, Special Collections, University Library, University of Birmingham.
8. Obituary Notice, Ibid., p. 369.
9. Ibid. p. 370.
10. G. E. Brown, *BJP*, July 3, 1891, p. 420. This article documents Brown's high evaluation of Nicol as a photographic chemist. *Photography*, May 2, 1889, p. 294. "Nicol has practised photography since collodion emulsion days."
11. Obituary Notice, Ibid., p. 370.
12. *English Patent 5374*, 1889, p. 1. All references to Nicol's patents will be to the source described in footnote # 1 above.
13. Ibid. p. 1.
14. "Much has been said about the want of permanence, but I have over two hundred prints mounted with starch that have stood the test of twenty-five and thirty years in a London atmosphere without showing signs of deterioration, except in one or two cases where the hypo has not been effectually removed." Ernest Snyd, *British Journal of Photography*, July 5, 1889, p. 443.
   Some references to this journal will be abbreviated to the short form: *BJP*.
15. "The drawback . . . in these printing methods is the fact that neither [albumen or aristotype] is permanent." J. S. Hodson, *The Camera*, Vol. V, #49, June 1, 1890, p. 38-40.
16. *Photography*, April 18, 1889, p. 266.

17. Wilson's *Cyclopaedic Photography*, Edward L. Wilson, (New York, Edward L. Wilson Co., 1894) p. 259, p. 169.

18. Josef Maria Eder. *History of Photography*, Translated by Edward Epstean, New York, Dover Publications, 1945, p. 440. Eder speaks critically of the "peculiar patent laws in England, which require no proof of originality."

19. *Photography*, May 2, 1889, p. 294.

20. Ibid.

21. Ibid.

22. Ibid.

23. C. H. Bothamley, *Journal of the Society of Chemical Industry*, April 30, 1890, p. 413.

24. Ibid.

25. English Patent # 4269, dated March 19, 1890, records the fact that this patent number, assigned to an application by Nicol, was later abandoned. English Patents, Ibid.

26. Patent 5374, 1889, Ibid.

27. Thomas Bedding, "Silver Printing By Substitution," *BJP*, Feb. 21, 1890, p. 117.

28. Patent 5374, 1889, Ibid.

29. Thomas Bedding, Ibid.

30. W. W. J. Nicol, Letter to the Editor, *BJP*, May 17, 1890, p,117.

31. *Photography Magazine*, May 29, 1890, p. 348.

32. Ibid.

33. J. L. Hodson, "Photographic Printing Processes," *The Camera*, Vol. V, #49, June 1, 1890, pp. 38-40.

34. T. G. Hibbert, "Report of a Kallitype Demonstration," *BJP*, June 13, 1890, p. 382.

35. See also "The disastrous commencement." *Photography,* Aug., 6, 1891, p. 506.

36. Mr. Farrow, *BJP*, July 18, 1890, p 461.

37. *Photography*, September 18th, 1890, p., 600

38. Nicol, "Kallitype Process," *BJP*, To the Editor, May 23, 1890, p.55.

39. *Photography*, Sept 18, 1890, p. 600

40. Report of Dr. Roland Smith's Demonstration of Kallitype, *BJP*, Dec. 5, 1890, p. 786.

41. *BJP*, Nov. 7, 1890, p. 720.

42. G. L. Hurd, *The American Annual of Photography and Photographaic Times Almanac*, "A Review of the Year 1890," published in 1891, p. 280-1.

Chapter II

Individual Preparation of Kallitype I Paper

I.   Introduction

This chapter will complete the history of Kallitype I paper. In the last chapter we watched a young chemist patent what he considered a new iron-silver printing paper and become an entrepreneur starting a business to make and sell kallitype I paper. We described the kallitype photographic paper and its limited penetration of the competitive photographic market of the early 1890's. We noted the tension that arose because of the patent and wondered if it explained some of the initial marketing difficulties.

We also noted some of the players on the photographic stage of that time: the studio operator; the well educated middle class photo-club member, the salon competitor, and the chemically trained photo experimenter. We shall be learning more about these types of photographers and their influence on the story of the kallitype.

In this chapter we shall follow the fortunes of Nicol's K I process as it edged out of the inventor's control and became a paper that amateurs prepared independently. We shall watch as amateur writers, unattached to the Birmingham Photographic Co., publish descriptions of how other amateurs could make kallitype solutions and even prepare kallitype paper in their own workrooms. We shall see the Kallitype I process become involved in a dialectic in which the Company reached for commercial success of their manufactured paper in the marketplace and a group of amateurs tried to secure an independent life for the process. We shall see if the kallitype fared any better as a result of the two forces involved with its destiny. We shall inquire if the amateurs improved the process, made it easier and quicker to work, enhancing its success measured against commercial processes. Finally, we shall revisit the touchy business of Nicol's patent claims and the bad feeling occasioned by them. We shall see if the controversy is resolved in a manner that frees the paper from prejudice.

Our present account of the individual preparation of Kallitype I will resume with the resolution of "the fatal mistake" and continue to July 1891, when the Kallitype I process was superseded by a different commercial process. This chapter will explore three topics. First, it will explain the significance of individual preparation of kallitype paper as distinct from commercial preparation. It will explore the historical literature to discover how and why a number of amateurs coopted a commercial photographic process and what they did with it. Second, this chapter will identify and characterize the early writers who advocated independent preparation of Kallitype I paper and will summarize their approaches and the success their processes attained. Third, it will evaluate what influence the non-entrepreneurs, the experts and the amateurs, had on the progress of Kallitype I paper. The chapter will indicate that their influence was negligible, that Kallitype I paper did not significantly penetrate the photographic paper market of the early 1890's. We will undertake to explain why both commercial and individually prepared Kallitype I paper failed to attract a sizeable following.

This chapter is concerned with the progress of individually prepared Kallitype I paper, so we shall begin by defining what the terms mean. Independently prepared kallitype paper may be defined simply as a photographic paper the user prepares himself from chemical compounds and raw paper stock for his own use from a published or personally researched Kallitype I process. It is distinguished from a paper that is bought ready for use from a manufacturer. That much is clear. What is not so obvious is why or how independently prepared kallitype ever got started in the first place. To explain that, we shall have to discuss "amateur" photographers of the 19th century.

Today, in a digital age, a very small percentage of photographers prepare photographic paper "from scratch" for their own use. A few accept the challenge. Most photographers feel that they have insufficient knowledge and technical resources to accomplish the feat of making their own paper. Most take for granted that the purchased silver printing papers are better than any they could make. They believe that learning to make a photographic paper would be burdensome, time consuming, and profitless. They feel that no one could make a paper better or more reliable than that made by the companies which mass produce photographic papers and sell them at relatively low prices. They reason, how could a nonprofessional individual compete with the scientists, engineers and experienced workers who apply their accumulated expertise in the daily making of commercial printing paper in the great commercial operations. That these presuppositions of the modern photographer are in great part true cannot be argued. Yet it is the opinion of not a few artistic photographers today, who prepare their own paper, that the great commercial producers do not market the most pleasing or beautiful photographic papers possible, but only those papers that can be efficiently engineered, mass produced, stored for long periods, and sold in large quantities for a profit. To say this is not to denigrate the valuable accomplishment and service of established photographic manufacturers. The papers they choose to make and sell are very well made, very reliable, very affordable, and necessarily, in view of the quantities sold to various users, very common. The necessities of the market do not permit the production and sale of special papers suitable to the demands of individual artists at prices they will pay. The progressive withdrawal of fine, artistic photo papers from the market from 1930 onward by the Kodak, Ansco, and Du Pont corporations demonstrates this.

But the conditions that prevail now were not always so. In the last quarter of the nineteenth century, photographic manufacturing was less centralized and less expert. The largest market, then, as now, was for a cheap, effective paper for studio and professional use. The large consumer market, and the auxiliary processing industry we take for granted today, did not exist. However, a sizable number of experimentally minded "amateur" photographers committed to playing an active role in the development of photography, did exist. These photographers were often better educated in the basic sciences that nourished photographic practice than the studio operators and nearly as well educated as professionals employed in manufacture. As amateurs, they found pleasure in applying their technical knowledge to the unresolved problems of photography.[1]

It should be clarified that many of these amateurs were not amateurs in the modern sense of the word, which currently means little more than "consumers who derive pleasure from taking pictures." Many of the amateurs in the last quarter of the 19th century were well-trained professionals in non-photographic fields who deeply loved photography and passionately pursued it as an avocation in which their experimental skills and chemical knowledge could be developed and applied. Such men eagerly attacked photography's unresolved problems, particularly those that concerned the painfully limited films and papers inherited from the previous decades. Photography at the end of the 19th century was ripe for technological extension and many capable amateurs wanted to and did contribute.

Along with the amateur concerned with extending knowledge and technique, there was a sizeable group of amateurs whose primary concerns were artistic. This was a moment in the history of art when many realized that expression of one's personal response to life could be made through the medium of photography. Such concerns led to the growth of amateur photographic societies and, eventually to an elaborately organized system of salon competitions. Artistic amateurs became interested in making prize photographs and believed their chances for receiving awards would improve by extending the print medium. They learned to work a variety of photographic processes many of which involved, if not the complete preparation of one's printing paper, then, at least a personalized manipulation of materiel. Competitive amateurs utilized a variety of individually controllable photographic processes. Eventually these included carbon printing (in color and black and white), platinum printing, gum bichromate printing, bromoil, silver chloride printing, and silver bromide printing, either toned or untoned, to list the most popular. After 1890, the kallitype was added to the list of print process available for study, exploration, or competitive (salon) use. The two types of amateurs, one, experimental and the other, artistic, managed to remain unified with the professionals in the Royal Society of Photography until the 1890's, when a second society, the Linked Ring was formed to emphasize the exclusive goal of artistic interest. After that time artist-photographers tended to separate themselves from the "process mongers" and those driven by commercial motives. The principle focus of the Linked Ring's attention became almost exclusively an interest in the art of photography.

That a large and active class of amateur photographers existed in England, Europe, and America is documented by the number and quality of photographic societies and the number and kind of publications that fed the experimental artistic interests in photography. The "Letters" columns of the major photographic magazines document also the curiosity, erudition and the level of commitment of amateurs. Published articles also document the contribution of amateurs to the growth of expressive photography. This chapter will document their contribution in one area of photography—the extension of Kallitype I. In particular it will document their contribution to making the individual preparation of kallitype I process widely known and used.[2]

The reader of the photographic journals published between 1890 and 1920 soon discovers that experimentally-minded amateurs often felt they were better prepared to understand and extend photography than professional or commercial photographers whom they deprecatingly referred to as "operators." The slighting implication of "hands" not "brains" was probably intended. The amateurs felt that the operators mindlessly applied poorly understood procedures and were only concerned with unchanging routine work and profit. Artistic amateurs frequently considered studio photographers indifferent to the development of the medium, resistant to the improvements of process, and inclined to milk inadequate processes for profit as long as profit would come. Experimentally minded amateurs occasionally bragged that a great deal of photographic progress, for example, the gelatin dry plate, the collodion negative, and gelatin chloride paper, were the result of amateur research.[3] The photographic journals, in effect, supported the contention of amateur superiority by frequently publishing articles written by them that dealt expertly with highly technical photographic subjects. Finally, the photographic societies and clubs of the day, though run by amateurs, were important centers where photographic knowledge and experience were accumulated and disseminated.[4]

A comparable situation exists today in the field of computing. In this field there are many highly informed individuals, who study computer science and technology and apply it wherever their private interest leads. Like the photographic amateurs of the 19th century, there are distinct types of interest, nevertheless the individuals put a great deal of energy, effort, and learning into computing, a field which they believe continues to expand intellectually and practically. As a result, those committed to computing find their endeavors full of intellectual adventure and excitement. Like the photographic amateurs of the 90's, they feel the technology belongs to them as much as the professionals, and they also feel that they share responsibility for developing the field. Like the photographic situation in the 1890's, there are many publications and user-groups in which amateurs and professional individuals share their expertise.

The photographic literature of the late nineteenth century informs us that the artistic amateurs of that time went to great lengths to make creative and individualized prints. To produce unique effects they compounded special processing solutions and often partly prepared their photographic paper. The new competitive salons encouraged these upper middle-class photographers to spare no effort and expense in producing unusual, artistic prints from a variety of exotic photographic processes. The artists as well as the "process mongers" supported the cause of individual preparation of amateur photographic materials.

In the light of these interests and commitments we should not be surprised to find, a year after Nicol's patent had been published, that reports of amateur preparation of Kallitype paper began to appear. The patent, published in *The British Journal of Photography* had made available (although in a somewhat recondite form) the information necessary for the informed amateur to begin experiments. And experiments they had to make, since the patent contained a profusion of chemical possibilities, some of which involved ambiguous statements of the chemicals to be compounded into solutions and others, unclear statements of the quantities. The critical articles published during the first year of the sale of commercial K I paper had not significantly simplified Nicol's confusing description of the process. And the instructions that accompanied the proprietary paper sold by Lewis provided quite limited information, for obvious reasons.

What happened to Kallitype I photographic paper in 1891, then, is not too surprising. Knowledgeable amateur photographers took it upon themselves to work out the process, to make it more understandable and easier to do. They wrote out their findings and submitted them for publication for the benefit of others. The journals published amateur process explanations as a way of disseminating knowledge, which they knew their readers wanted. Their action must have proceeded from the idea that photographic processes in some way belonged to the public, that amateurs had a right to know how to do a photographic process regardless of the fact that the rights to commercial development of the process had been reserved by patent for the inventor. In fairness, it must be said that Nicol and Lewis never publicly objected to the revelations by amateurs on how to do their proprietary process. In all probability they had their eye on the lucrative professional market. Both probably felt that publication of the details of process would lead few amateurs to prepare the paper and that the publicity might increase the sale of manufactured paper.

II.   First Accounts of Individual Preparation of K I Paper

The first account of individual preparation of kallitype I paper, appeared in the **Photographic Times**, an American publication, on February 6, 1891.[5] The article was unsigned. The writer indicated he worked from a kallitype formula presented in the formulary pages of the **British Journal of Photography Almanac** of 1890.[6] He reports his first trial of the process failed "on account of a spurious ferric oxalate," the chemical that made the paper light-sensitive. So he made his own ferric oxalate to assure a supply of fresh chemical at full strength. The article provides one of the first detailed accounts of making ferric oxalate, by an amateur. The details are reported below because the method indicates the amount of trouble amateur photographers were prepared to undertake at that time. The method works and is still recommended for making "home brewed" ferric oxalate today.

> . . . we precipitated ferric chloride with *aqua* ammonia (*fortis*), washed the resulting hydrate of iron thoroughly to remove from it every trace of adhering chloride, dissolved it afterwards with oxalic acid, leaving a portion of hydrate undissolved to be certain of neutrality, reduced the solution to a strength indicating 75 degrees by the ordinary photographic hydrometer, argentometric scale, filtered the solution, and coated it on plain Saxe paper.[7]

A few comments on this approach to making ferric oxalate may be in order. This method, commonly taught in chemistry classes at that time, continues to be a safe, inexpensive, and effective method of making ferric oxalate by amateurs at home, although it is messy and slow. While refinements such as sizing could be added to the process of amateur sensitizing of kallitype printing paper, a simple solution of ferric oxalate, made as reported above, could function as the complete sensitizer. In Kallitype I printing, all that was really necessary to sensitize the paper was to coat it with ferric oxalate. Standardizing the strength of the ferric oxalate solution with an hydrometer and "neutralizing" the solution by maintaining some undissolved hydrate in the bottom of the vessel holding the ferric oxalate are still good methods for amateurs who wish to make and control home-made ferric oxalate. This method and other more modern approaches to the making of ferric oxalate are explained in the book, **Making Kallitypes, A Definitive Guide** by the present author.

The provision of directions on how to make an exotic photographic chemical, one not ordinarily available at the local pharmacy, such as ferric oxalate, was a common subject in the photographic journals of the 1890's. Photographic magazines of that time, as we shall see in greater detail in later chapters, apparently believed that educating the public in the sciences and the processes of photography was part of their mission. They regularly published articles on elementary photochemistry and occasionally offered articles on advanced topics. These offerings included serial publication of dictionaries of chemical terms, discussions of photo-chemical theory and tables of the properties of photographic chemicals. Photographic journals and almanacs routinely published formulas for making photographic plates, processing solutions, as well as toners and bleachers, etc. Obviously the magazines of that time were quite different from those of today which publish little material on basic science and advanced photographic chemistry. Photographers of that time, whether English, French, American, Italian, Belgian, German, etc., whether beset by financial need or inability to buy supplies in remote communities, or driven by pride in doing for themselves, often produced what they needed in their home darkroom, if the content of publications can be regarded as an index of their practice. Magazine articles suggest that numerous amateurs taught themselves what they needed to know to make their work in photography convenient, controllable, and enjoyable. The *Photographic Times* article on kallitype I and an increasing number like it on other processes were written for readers who responded enthusiastically to the chemical process requirements of photography. It is likely that at its inception, the kallitype appealed most strongly to those amateurs who were fascinated by experimental possibilities, one of the compelling attractions of photography since its inception.

The *Times'* article provided other details useful for making individually prepared kallitype I paper. It reported that "brushing the sensitizer onto the paper is preferable to "floating the paper" on the sensitizing solution. It advised that the paper must be "immediately dried after coating, without too much heating." It found "the tone [color] of the print depends on the composition of the developer." For example, "adding 3 to 4% potassium oxalate to any of the published developers produces a sepia or reddish brown color." It included a formula for a clearing solution—20 parts sodium citrate in 100 parts of water made decidedly alkaline by ammonia. It recommended a second clearing bath to rid the paper of any remaining iron which stains the paper yellow if the first clearing bath fails. The second clearing bath is simply a 5%

solution of oxalic acid. Finally, the article detailed how to control "the brilliance and depth of tone" ie, contrast. It recommends adding 1 to 3 grains of potassium bichromate to 8 oz. of any of the developers.

The procedures offered in this first article are, for the most part, practical and sound. Like other articles written by amateurs on individually prepared kallitype, the suggestions remain close to the approach described in Nicol's patent, which was the apparent source of many of them. Nevertheless, this article, and others like it, extended the details of manipulation beyond the inventor's specifications. The amateur descriptions are often simpler and less ambiguous than the inventor's wary ambiguities. They often contain simple explanations of Nicol's sketchy procedures and they often provide additional information on the handling of the process.

The *Times* article documents the experimental spirit that many advanced amateur photographers of Victorian and Edwardian times brought to photography. It suggests how the new process became a challenge for independent research and development.[10] Such articles grew in number and helped to make the Kallitype I process understandable and workable by a large group of amateurs who may never have understood the process or been able to work it without the help of gifted experimental photographers who could write. Though not all of the suggestions made by amateur writers proved helpful or effective, as we shall see, the best made the process clearer and attractive to an audience that continues to this day.

Another report on the individual preparation of Kallitype I paper appeared in the *Photo-American Review* for May 1891. The article summarizes a talk made to the Society of Amateur Photographers of New York by James H. Stebbins Jr., a chemist who later marketed a commercial kallitype paper in New York. Stebbins' article is notable for its presentation of chemical formulas that describe the various steps of the Kallitype I process. Interestingly, Stebbins avoids all mention of Lewis' commercial products. Stebbins' publication appeared a year after the publication of Nicol's patent on Kallitype I and his description of the process bears evident similarities to Nicol's process.

Stebbins directed the photographer wishing to sensitize his own kallitypes to brush on a good paper stock a concentrated solution of ferric oxalate. He wrote that the coated paper, when dry, may be kept for a long while in the dark. He provided developer formulas similar to those recommended by the kallitypist who described the process in the *Photographic Times*. Like that writer, Stebbins discussed the fact that kallitype prints may show a yellow stain after development. He provided a remedy: "should any yellow stains remain upon the paper, they may easily be removed by washing the print in a three per cent solution of oxalic acid."

That discussions of the yellow stain of the kallitype repeatedly appear in articles on both sides of the Atlantic, along with user-developed remedies, raises questions about the adequacy of the commercial information disseminated about the manipulation of the Kallitype I process by Nicol and Lewis. We know from published letters asking for help that Lewis' commercial kallitype paper regularly developed yellow stain. We are led to wonder why the proprietors published so little about the problem and effective remedies to avoid or correct it? One wonders further if the inventor and the company sufficiently understood the cause of the stain and the methods needed to control it. The yellow stain is a problem common to the kallitype which any worker soon encounters. The problem of yellow stain and the difficulties of resolving it will continue to surface in future chapters.

After detailing "the working of the process," Stebbins provides an early description of the chemical reactions of the kallitype process. The first reaction occurs when light hits the sensitized paper. A chemical change occurs in which ferric oxalate becomes ferrous oxalate as a result of the action of light. Stebbins describes that change as follows. I quote his description as found.

1

$$3 \ Fe_2 \ (C_2 \ O_4)_3 \longrightarrow 6 \ Fe \ (C_2 \ O_4) + 6 \ CO_2$$

That is to say, three molecules of ferric oxalate, when exposed to light, form six molecules of ferrous oxalate, plus six molecules of carbonic acid.[11]

The ferrous oxalate formed, when brought into contact with the nitrate of silver of the developer, reacts as follows:

2

$$6 \ Fe(C_2O_4) + 3 \ AgNO_3 \longrightarrow 3 \ Ag + 2 \ Fe_2(C_2O_4)_3 + Fe_2(NO_3)_3$$

That is to say, six molecules of ferrous oxalate, plus three molecules of silver nitrate, form three atoms of metallic silver plus two molecules of ferric oxalate, plus one molecule of ferric nitrate.[12]

Stebbins concludes by directing the reader's attention to another process of making prints "with which I have been experimenting. The prints resemble the kallitype sensitizer, but have the advantage of requiring no silver in the developer. They are developed by a bath of formate, acetate, oxalate, or citrate of soda and the tones vary depending on the developer."[13]

As mentioned earlier, the Kallitype process was not beyond the comprehension of educated professional or amateur chemists of the day. Nor were photographers timid about altering a patented process and calling the resulting process their own.

Peter Duchochois, another chemist-photographer who worked in New York, wrote on the individual preparation of a highly personal version of kallitype. Duchochois, who was well informed on historical photographic processes, was deeply disturbed by Nicol's patent, which he felt was unjustified. In a paper entitled "Oxalitype," Duchochois wrote with such heavy irony that the reader wonders if he intends to commercialize an iron-silver print process of his own or merely aims to ridicule Nicol for exploiting Herschel's ideas on iron printing. Notable in Duchochois' approach is the recommendation of an ammonium sulfocyanate fixer which he claims "possesses none of the undesirable properties of sodium thiosulphate." [14]

In his book on **Photographic Reproduction Processes**, copyrighted in 1891, Duchochois again reported details of another personal version of the Kallitype I process. The subtitle of the book is "Photo Impressions Without the Use of Silver Salts," but the Appendix describes a number of old photographic processes, several of which use silver to form the image. Duchochois details at considerable length how these historical processes were forerunners of the kallitype. He discusses processes developed by Herschel, 1840; Hunt, 1842; Godefroy, 1858; De La Blanchere, 1858; and Borlinetto, 1863. He then describes Nicol's kallitype I process and reproduces Nicol's formulas. In addition he provides Kallitype I solutions and processing methods born of his own research which depart significantly from Nicol's methods. For example, to clear prints of yellow stain, Duchochois recommends citric acid. He also recommends the use of a solution of ammonium sulphocyanate as a fixer "to prevent the paper from being tinged by the reduction of the silver nitrate which is mechanically retained in its fiber."[15] He further recommends that his fixer, a solution of ammonium sulphocyanate, "should be compounded with auric chloride to gold tone the print while it is being fixed.

Kallitype I Gold Toning Fixer

Ammonium sulphocyanate	35 parts
Gold terchloride	0.15 parts
Water	350 parts

He indicates the solution can be used over and over again.[16]

Duchochois book contains other worthwhile photo-chemical information and procedures, for example, how to make ferric oxalate and ferric ammonium oxalate, variant compounds useful for kallitype sensitizers.

In March 1891 C. H. Bothamley, an early reviewer of Kallitype I (see the first chapter) reworked a lecture he gave a year before (April 1890) on "Some Recent Advances in Photography" into a magazine article.[17] In it he published some personal observations on Kallitype I. It is worth noting that the talk introduced three new photographic printing methods, two of them belonging to a new chemistry based "on the effect of light on diazo compounds." The one process was called the "Feertype" and the other, "the Primuline process." The third process, "is a development of . . . an older process which previously had no practical value, the kallitype."

The kallitype, Bothamley wrote,

> "is based on the old observations of Herschel and of Hunt . . . Neither in the hands of Herschel nor Hunt did the method acquire any practical value, and Nicol's improvement consists in the careful working out of the details of the process."[18]

Bothamley here defends Nicol's patent on the ground that Nicol patented a detailed an iron silver process that worked.

Turning to the K-I process, Bothamley informed his readers that "specially prepared paper, coated with ferric salts" is available on the market. He reprints the developer formula that was packaged with the commercial paper for workers who wished to compound their own:

Kallitype I Developer

Sodium citrate	1 oz.
Potassium dichromate	1 grain
Water	10 oz.
Ammonia sol .880 strength	30 minims.

Bothamley also reported the results of his experiments with the developer.

> My experience is that better results are obtained with a solution containing silver nitrate, 1 part, sodium citrate, 10 parts, water up to 100 parts, ammonia, *quant suff.* The silver nitrate is dissolved in five or six parts of water and carefully mixed with ammonia until the precipitate is just redissolved, and the solution is poured into a solution of the sodium citrate in about 80 parts of water and the mixture is diluted to 100 parts. A slight precipitate of silver citrate will form and is filtered off. The developing solution and the solutions used for washing the print should be kept in the dark as much as possible. With the present commercial paper, the use of developing solutions containing potassium oxalate or borax in place of sodium citrate produces no noteworthy difference in the colour of the image. It is stated that the addition of a small quantity of potassium dichromate to the developer gives greater contrasts in the resulting print, but I was not able to observe any such result.[20]

Bothamley's report of modifications of the manufacturer's developer illustrates again the tendency of knowledgeable photographers to experiment and modify received formulas rather than slavishly accept them and their limited results. His remarks on the lack of conformity between the results he was promised by the manufacturer and those he got will not be the last of such complaints. His inability to derive increased contrast from dichromate is puzzling in view of almost universal reports about that chemical's efficacy in altering the contrast of kallitypes. His developer may have had an insufficient quantity of dichromate.

As experts like Stebbins, Duchochois, and Bothamley published their criticisms of the commercial approach and made independent recommendations, one wonders what the effect such revisions had on public confidence in the manufactured product. One presumption is that many photographers held back and waited for the product to reach a final, debugged state.

Bothamley also reported a number of interesting tests he ran on kallitype I prints.

> The image formed consists of metallic silver, or at any rate behaves exactly like metallic silver with various reagents. It is bleached by solutions of mercuric chloride or cupric bromide, but can be redeveloped by treatment with any or ordinary developing solution after the excess of the mercuric or cupric salt has been washed out. The image bleached with mercuric chloride or bromides can be intensified by treatment with a solution of potassium silvercyanide and probably by other methods also.[20]

This information was useful to those seeking methods of "correcting" prints given too much or too little exposure. Since the kallitype does not "print out" all the way, incorrect exposures do not show up until after the print is developed. Bothamley suggests poorly exposed prints can be salvaged by bleaching and intensification methods. Such corrective techniques would provide an obvious economy to the struggling operator or the inexperienced printer, although they were labor intensive. Other writers believed these corrective techniques were not worth employing, cf. Brown below.

Bothamley also tested kallitype I prints for permanence, that is, to determine their resistance to the deteriorating influences of adverse chemicals found in the environment. He reports kallitype I prints are

> not appreciably affected by exposure for 48 hours to the products of the combustion of sulphur in air nor by immersion for the same length of time in a solution of sulphurous acid. A solution of hydrogen sulphide produces a slight change in

the colour of the image, but has no other effect. Ammonium sulphide in dilute solution changes the colour from black to brown without any loss of detail, and with only a slight loss of intensity. The brown colour is not unpleasant and is very suitable for certain subjects. There is no reason to doubt the permanence of the images when exposed to ordinary atmospheric influence.[21]

Coming from a chemist of Bothamley's stature, the above suggestions about manipulation and the assurance of permanence must have been good news to early users of Kallitype I paper. Neither Nicol nor the Birmingham Photographic Company had taken the trouble to publish information on the permanence of the Kallitype I print.

At the same time that articles of substance were written by trustworthy experts like Stebbins, Duchochois, and Bothamley, other material appeared—letters, queries, and answers—written by less knowledgeable amateur photographers. A case in point appeared in the correspondence column of *Photography,* June 4, 1891 when the Birmingham Photographic Co. wrote to correct suggestions from three self-styled chemists who answered an inquiry from an amateur regarding how much citric acid was required to yield 48 grains of crystallized sodium citrate. The amateur had written to request the quantity in order to mix his own kallitype developer. The inquiry drew three answers by amateurs, each of which was published, and each was found to be incorrect.[22] In another instance, this time the fault of a typographic error made by a typesetter, a report of the amount of ammonia used in Kallitype I developer was incorrectly printed as one ounce instead of one dram, a massive error. [23]

With periodicals publishing this kind of incorrect information along with legitimate, it is not surprising that there was confusion about the infant process which may have influenced its struggle for market acceptance. After reading a quantity of such published material, this historian concludes that there were negative as well as positive contributions made by amateurs to the growing dialogue about the kallitype. Workers, then as now, had to be on their guard, for not all published technical information was dependable.

III.   Brown's Resolution of the Claims Controversy

Events soon to occur radically altered the course of kallitype I and it seems appropriate to end the history of the first process with a final appraisal of Nicol's right to patent and an authoritative summary of the working of the process. It happens that such an appraisal appeared in two articles by George E. Brown, one of the most respected photographic chemists and writers of the time. The articles were published in the *British Journal of Photography*, in the July 3 and July 10 issues of 1891. The two articles together contain the most complete consideration of the claims to be made for Herschel and for Nicol as the inventors of the iron-silver process and the kallitype. In the opinion of the present writer, Brown's articles may be taken as a fair adjudication of the issue for all time.

George E. Brown was the ideal person to write the articles. An expert chemist he had written the classic book on iron printing, published in 1888. Brown was for 29 years the editor of the *British Journal of Photography Almanac*, a position in which he gained worldwide respect for his broad professional knowledge of many photographic processes and for his capacity to write clearly about them. It happens that Brown knew Nicol professionally, having collaborated with him on at least one scientific paper they jointly presented to the British Association for the Advancement of Science.[24]

Brown's writing leaves little doubt that he was a photographer of experience and knowledge, a thinker of probity and sagacity, and a man with deep respect for those who pioneered photographic research. He was also a man not likely to be swayed from his best judgment by friendship, professional or personal.

Brown began his first article by noting the "many recent references to the subject of iron printing in our [**BJP**] columns and the frequency with which iron printing forms the theme of papers and discussions at the societies." He expressed the hope that out of such continued interest "great and permanent benefits are thereby likely, in the long run, to accrue to photography in general, and to photographic printing in particular."

Turning his attention to the past, Brown opened the subject of photography's debt to its forbears.

The history of iron printing, it is hardly an exaggeration to say, is the history of photography itself, for the value of iron salts in positive printing was known and understood almost, if not quite, as soon as the Daguerreotype and the Calotype were perfected and practised. It is a fine tribute to the early investigators of the photographic properties of the iron salts that their application to printing purposes should at the present time be so extensive and so successful, and that, notwithstanding all the "discoveries" and "improvements" that have been made during the last half-century both in negative and positive silver processes, the simple principles of iron printing as originally established should remain as valuable, and, indeed, as necessary, as ever. We are, perhaps, as a group rather too prone in these days to forget how deeply we are indebted to dead and gone workers for much that we not only derive benefit from, but also frequently take credit for ourselves, with no show of justice or reason. Photographic patentees are bad enough, . . . under this head, but there are plenty of others.[25]

Approaching the crux of the issue, Brown raises the question whether Herschel or Nicol invented the Kallitype.

Kallitype is possibly the first perfected form of iron printing in which the final image consists of reduced silver. We make the reservation advisedly, because we believe that the late Sir John Herschel, to whom we are indebted for a great deal of our knowledge of the properties of the iron compounds, produced in the course of his experiments positive pictures by a process which we may reasonably regard as the precursor of kallitype. This process consisted of exposing paper coated with ammonio-citrate of iron, and 'developing' with a solution of silver. Unlike most of the author's processes, however, this was never brought to perfection, although it undoubtedly contained the germs of a practicable printing method, as we may now see from what Dr. Nicol has recently achieved.[26]

Brown then discussed in detail the chemistry of Nicol's process. In view of the repeated descriptions of Nicol's process already made, the reader will be spared Brown's account. We shall proceed directly to Brown's final conclusion.

> It will be observed from this that Dr. Nicol has devised a process which, although it depends to a great extent upon some old and familiar reactions for success, nevertheless possesses some highly valuable features of its own, which could only have been reached after a great deal of experiment . . . We believe that, as worked now, the process differs in some material respects from the form in which it was originally introduced, and that the alterations in procedure have been found improvements in practice. The process, both in its original and in its amended state, however, exhibits a most ingenious utilisation of deep chemical knowledge; and although some carping has been indulged in at the expense of the patentee's claims, we should say there is little, if any doubt as to the perfect propriety of them, either morally or legally.[27]

It is clear that Brown, after careful consideration of all the evidence, historical and chemical, concluded that 1) Nicol developed by original experimentation a workable process where none before existed, and 2) that while Nicol profited, as all inventors do, from what was known before, he had a moral and legal right to claim the invention of the particular kallitype process of his patent.

IV.   Brown's Description of Kallitype I Paper and Process

Brown's second article, published on July 10, endeavored to present "a clear outline of the [kallitype I] process" as it existed one and a half years after publication in Nicol's patent. Brown's account of working the kallitype I process can be taken as a relatively accurate description of the process in its later stage, made by a friend of the inventor, (who was undoubtedly well informed by him) and by a photographer chemist who had written the most definitive book on iron printing up to that time. Late in its brief development, in July 1891, the Kallitype I process, involved the following steps, according to Brown, whose description will now be summarized.

1. Coat suitable paper with ferric oxalate

2. Expose the paper in a printing frame till the shadows and highlights are visible.
3. Develop the paper for 10 minutes (by flotation, preferably) in the following solution which is amended from the original developer.

Kallitype I Developer

Silver Nitrate	50 grains
Sodium Citrate	1 oz.
Potassium Bichromate	1-2 grains
Water	10 oz.

4. Over and underexposure are irremediable. Remedies are superfluous. One must learn to expose the paper properly.
5. After developing, a bath clears the iron salts from the paper.

Clearing Bath

Developing solution (above)	.5 oz.
Sodium Citrate	2 oz.
Water	20 oz.

6. After clearing away the iron, "pass the print thru two "fixing" baths of the following solution" to dissolve out of any silver salts that remain in the print.

Fixing Bath

Sodium citrate	1 drachm
Ammonia .880	2 drachm
Water	1 quart

7. Wash in running water for 1/2 hour and dry.[28]

Brown concluded his description of the Kallitype I process with a request that Nicol find a way to reduce the number of processing steps the kallitype required. "The steps are numerous enough to prejudice the process in the eyes of those making a trial of it. We should be pleased to hear that experiment had simplified the washing and fixing operations."[29]

Brown's second article ends with an appreciation of the kallitype's visual appeal and a hopeful assessment of its future success.

> "We have been much charmed with pictures produced by the kallitype process . . . and we have very little doubt that it has a future before it.

But Brown's discussion of Kallitype I soon became academic. Already, on June 4, 1891, almost a month before Brown published his two articles, a guarded announcement had been made by the Birmingham Photographic Co. The announcement read:

> We take this opportunity of saying to the great number of our friends who take a deep interest in the kallitype process that we shall, in a few days, place upon the market a very greatly improved paper, which at the same time that it yields very superior results, is much simpler, perfectly clean working, and will give prints of various tints . . . Due notice will be given of this in our advertisement columns as soon as ready. [30]

V.   Kallitype I: Summary and Comment

The Kallitype I process appeared at a time of intense exploratory activity in both basic and applied products—cameras, lenses, studio devices, negative materials, and printing papers—a time when new products were announced to the amateur and professional with bewildering frequency. The kallitype I process was patented by a young professor of chemistry and marketed by a new firm, the Birmingham Photographic Company. The introduction of the new print material was greeted in the photographic press with eloquent praise for the perfection of its prints and the ingenuity of the process. It drew enthusiastic expectations for its eventual success in the marketplace from professional reviewers of preeminent photographic expertise and reputation. But, in spite of a 50 year period of incubation since Herschel experiments first suggested the possibility; in spite of 40 years of study and experimentation in laboratories and darkrooms within universities and without; and in spite of enthusiastic support by the photographic press, the new paper failed to live up to the expectations held for it.

The paper failed to make inroads into a market securely held at the low end by the often despised albumen-silver paper and dominated at the high end by the justifiably admired, but expensive platinotype. Nor did the new paper compare in sales performance with other new papers, notably gelatin chloride and bromide papers. Sales of the latter papers soared while those of the kallitype I sat. What follows is a discussion of why the Kallitype I, initially so well received, failed to prosper.

On the positive side, Kallitype I had a number of compelling advantages. The new paper had both the black tone and the matt surface then in vogue and was thus responsive to the current esthetic preferences of amateurs hungry for artistic exhibition papers. The kallitype could also produce image colors that varied from brown to sepia by toning. Further variation in tone could be achieved by modifying the developer. The kallitype printed faster than the current market leader, requiring only one-third the exposure of albumen paper.[31] It had distinct economic advantages. It saved time and produced an agreeable image color without the expense of expensive gold-toning. Kallitype required no hypo for fixing. As mentioned, hypo at that time was considered the major cause of the impermanence of silver photographs. In its use of ammonia as the fixer, the kallitype was unique among silver print processes. The freedom of the Kallitype I print from hypo gave it the advantage, real or merely perceived, of being "as permanent as a silver image can be." The cheapness of kallitype paper was a selling point, although in the "toney" world of art competitions and society portraiture, low price may not always have been perceived as an advantage. For Kallitype I there was also a readily available and ample supply of silver, then a relatively cheap metal, compared to the unpredictable and dwindling supply of increasingly costly platinum. On top of these advantages, kallitype was an easy process to work, requiring little exercise of judgment or skill in manipulation, once the paper was exposed. By contrast, press reports inform us that the new bromide-silver paper involved the unfamiliar and difficult to control technology of "developing out" paper in which the exposed image was invisible until development. And in its early stages of manufacture, bromide paper had serious problems with reliability. All of these advantages should have assured kallitype I a favorable, if not spectacular entrance to the market and in time, a continuing and comfortable place in it.

In the secure knowledge that the anticipated success did not occur, we are obligated to discuss the probable reasons for the lack of success of Kallitype I.

One difficulty must have been problems with the early manufacture of the paper. Apologetic announcements occur only a few months after the kallitype was placed on sale, indicating something was wrong with the product. Precisely what was wrong was never announced by Birmingham Photographic Co., although a letter from Nicol, quoted earlier, indicated the trouble originated with a contaminated paper stock chosen by the manufacturer, John Lewis. There are indications that the firm rushed the paper onto the market before it was sufficiently researched.[32] Nicol's letter suggests the chemistry of the process was changed by the manufacturer more than once during the first months of the paper's release to the public.

The difficulties with the design and engineering of the original Kallitype I paper are surprising in view of what we now know about its simplicity. Kallitype I paper was basically a sized, pure rag, acid free, paper that was coated with a simple solution of a water soluble ferric compound. After an initial period, the paper was sensitized primarily with ferric oxalate, according to several reviewers. From the vantage point of the present, a paper coated with a single chemical sensitizer, one which had a long history of successful use in the sister process of platinotype, should have been readily controllable, given Nicol's professional skills to guide the operation. One probable hypothesis of the cause of the "disaster" is that the paper stock chosen by the manufacturer, failed to meet the standards Nicol had maintained as he developed the paper in his laboratory. Failure to discover a wide variation in the pH of the paper or lack of awareness of the presence of a chemical contaminant could cause a proven sensitizer to act unpredictably and produce defective prints. Nicol tells us the problem was quickly solved by finding a new source of paper, but not before the kallitype's reputation, unproved and vulnerable, was stained. The company made every effort to replace the bad paper with good and to rebuild confidence—and for a while business picked up. But, as the old saying goes, customers once burned are twice cautious. It will be remembered that George Eastman suffered a similar misfortune when a batch of dry plate film turned bad early in his manufacturing career.

In spite of wide publicity and strong editorial support by favorably disposed writers, who without exception praised the new process in most of the photo magazines of the day, and despite renewed effort to maintain a high level of quality after "the disaster," commercial Kallitype I paper never became a contender after 1890 and remained little more than "a novelty."[33] The report of the comparative sales of the various printing papers for the year 1890, published in the **Photo Annual,** fails to show kallitype making much of an impact on the market in that year. Rather it suggests the kallitype attracted little interest from operators and amateurs. Apparently interest was high among the relatively small group of photo-experimenters, persons sufficiently educated in chemistry or challenged by it to enjoy experimenting with a new process and to take pleasure from making personal modifications to it. The group of doctors, engineers, chemists, and writers on photography apparently delighted in experimenting with the process and in publishing accounts of their personal variations of it. But the experimenters, apparently, were not a large group. The camera clubs that rushed in with enthusiasm for early demonstrations of the new process show little sustained interest for it. After 1890 the reports of demonstrations of K I taper off.

There was little evidence of much interest in Kallitype I paper, by commercial studio operators, even though their customary paper, albumen, was continuously and universally assailed by the writers of the day for ugliness and impermanence. Part of the operators' disinclination to switch may have been caused by the need for a more contrasty and denser negative for kallitype than for albumen paper, which responds to a longer scaled, less contrasty, generally thinner negative. A studio printer, trying his thin "albumen print negatives" on the new kallitype paper would probably not be impressed. The prints would be, in all likelihood, flat, with grayed whites and heavy blacks—a small argument for change. We are not surprised to find that reviewers repeatedly insisted that Kallitype paper would print well with "a suitable negative." Even granting that contrast of the kallitype print could be controlled by the addition of bichromate to the developer, it would take time for commercial photographers to change their negative making and learn the subtleties of a new approach to contrast control. But how much time would the new process have to prove itself?

The other problem operators had with kallitype paper was its alien manipulation. Little in their experience prepared them to control the exposure of the more rapid Kallitype paper, which gave only a feeble sign when exposure was complete. Operators were used to exposing the old albumen paper until the image was completely visible, and the same was true of the silver chloride papers recently placed on the market. Further, the extensive variation of print-color possible with the kallitype I process was not perceived by the studio printer as an advantage. The studio printer wanted a simple, reliable method that made every print the same. The kallitype was a chameleon print system which produced almost endless color variation, whether desired or not. Its problem was to make two prints that were exactly the same in color and contrast. The kallitype's capacity for variety was not as attractive to the studio printer as it was to the salonist who made but a few prints and had a vested interest in their appearing artistically "different." The studio operators had neither the time nor the inclination to learn the variations the new process made possible and saw little capacity for profit in them. Nor did they appreciate the many baths, six, kallitype processing required, as Brown indicated. It would take time for operators to learn the sensitive but sophisticated system of contrast control of the kallitype. If an expert chemist like Bothamley, familiar with the gamut of new processes, could not get the bichromate to influence contrast, it is certain that less chemically astute and more habit-oriented operators would have trouble getting it to work also. Finally, there is evidence to show that while studio operators may not have been, as a profession, indifferent to matters of permanence, there were many individual operators who were not troubled by prints that quickly faded or changed color. Thus, the vaunted permanence of the kallitype I was a criterion which may not have had much appeal to a group often berated for their professional indifference to poor quality of work.

There was a further difficulty with the kallitype I process. We have encountered repeated mention of the problem of "yellow stain" and have noted efforts by several expert photographers to develop methods of dealing with it. Yellow stain, a precipitate of unwanted and indissoluble iron on the Kallitype I print, proved to be a persistent problem. It was solvable, but thru a level of control that exceeded the knowledge and skills available to many amateurs and operators. It is notable that Birmingham Photographic Co. published so little information about dealing with the yellow stain that amateurs were forced to query publications. Printers who experienced the problem must have felt a certain frustration with a process they could not get to work cleanly, as their frequent letters to the journals show. It was unfortunate for Nicol that the fault lay not in his conception of the process, but rather in inadequate communication about effective processing. When all the steps were properly executed with fresh chemicals and sufficient time, the yellow stain was controllable. But many photographers apparently failed to perform the required manipulation with perfect attention to the necessities of the process every time. One answer to the yellow stain problem would have been additional clearing baths or longer processing times in the six baths usually recommended. But Brown had already strongly urged the reduction of the number of processing steps on the grounds that the present arrangement was already uncompetitive. We shall see that the problem of the yellow stain persists in the history of the subsequent kallitype processes.

The last disadvantage of Kallitype I was "brown fingers." The Kallitype I process developed the image with a solution of silver nitrate. Silver nitrate is a chemical that stains any skin it touches a dark brown. Since it is natural for workers to move prints from tray to tray during processing with their fingers, their finger tips, whenever they failed to use tongs, turned an unsightly brown that would not wash off. G. E. Brown in *BJP* had written, the worker "is advised not to finger the prints . . . but to use forceps."[34] More than one amateur had complained about discoloration attendant on printing kallitype I. In this regard kallitype was no worse off than albumen printing and salt printing, both of which used silver developing baths. But these processes had the advantage of a long tradition of use.

We are therefore not surprised when, one year after the original offering of Kallitype I paper for sale, the Birmingham Photographic Co. removed it from the market and announced the availability of a new, improved, and fundamentally different paper and process, Kallitype II. To the history of this quite different iron-silver print process, we now turn.

Notes Chapter II. Individual Preparation of Kallitype I Paper

1. For just one example of quality amateur thinking see "The Adaptability of Various Printing Processes," **BJP,** July 5, 1889, p. 443.
2. *Photography,* Aug. 6, 1891, p. 506. letters by "HT" and "Sigma Delta." display expert knowledge of kallitype by amateurs.
3. *American Annual of Photography*, 1891, p. 280. ". . . the profession has taken the suggestions and investigations of amateurs . . . ."
4. *Photography,* Ibid.
5. *Photographic Times*, Vol. XXI, no. 490, Feb. 6, 1891, pp. 61-2.
6. **British Journal of Photography Almanac**, 1890, Formulary Section p.100.
7. *Photographic Times*, Ibid.
8. Many examples could be given. For a single article, consider "The Chemistry of Silver and Its Salts and Their Behaviour in Photography" by P. C. Duchochois, *Anthony's Photo Annual*, 1891, p. 583. For small booklets on the chemistry and physics of photography, the *PHOTO-MINIATURE* "magazine" published a number of volumes on Chemical Notions, Chemical Terms, and Photographic Processes, etc. The Issue for September, 1900 on *"Chemical Notions for Photographers"* begins "this number of *The PHOTO-MINIATURE* is an act of faith in the man (or woman) for whom the magazine has been made since its beginning—the man (or woman) who "wants to know . . ."
   "Among photographers I have found no desire so strong as the desire to know "something more" about the wonderful transformations of material which make up the familiar processes of photography." p. 237
9. Several photographic journals ran series on chemical terms and chemical qualities. See *The Camera*, June 1, 1890, page 8 and following for an issue dealing with iron chemicals.
10. There are many references to the pleasures of learning and experimenting in photography. Since this is a relatively important background matter for the kallitype, several examples will be given.

John Tenant, "Platinum Modifications," *PHOTO-MINIATURE,* July, 1902, p. 153 "The wonderland of photographic manipulation grows in interest and fascination as we journey along its highways and byways. The chief charm of photography, in fact, is that there is no finality to its pleasures. No sooner do we turn a corner on the road than a new view opens out before us, revealing new enchantments and possibilities we had not dreamed of on the way."

Tenant, "Blue Print," *PHOTO-MINIATURE*, 1900. "Those who wish to experiment will find Kallitype an extremely interesting and promising field."

Tenant, Ibid. "Experimentation with the salts of iron is very interesting, and surprises await one at every turn."

J. Thomson, "The Blue Print," *PHOTO-MINIATURE,* July, 1904, p. 463. "When I speak of the iron printing processes as interesting from the amateur's point of view, I refer not only to the character of their resulting prints, but chiefly to their interest as processes, offering in their manipulation innumerable opportunities for gettng a practical knowledge of many things which will be found profitable in general photographic work."

11. James H. Stebbins, Jr., *The Photo American Review,* May, 1891, p. 40. This is a report of his talk to the Society of Amateur Photographers of New York.
12. Ibid.
13. Ibid.
14. P. C. Duchochois, "Oxalitype," *International Annual of Photography* and *Anthony's Photo Bulletin,* Vol. IV, 1891, p. 127-9.
15. P. C. Duchochois, P. C., *Photographic Reproduction Processes,*" A Practical Treatise of the Photo-Impressions Without Silver Salts" (New York, The Scovill and Adams Co) 1891.
16. P. C. Duchochois, Ibid., p. 110-115.
17. C. H. Bothamley, **Journal of the Society of Chemical Industry**, June 30, 1891, p. 523-4.
18. Ibid. p. 524.
19. Ibid.
20. Ibid. Note that G. E. Brown also denies the practicality of intensifying and bleaching in *B.J.P* July 10, 1891.

> "In remedies for both errors [underexposure and overexposure] for kallitype, we confess we have no confidence . . . On the whole, we are disposed to regard serious under or over exposure in iron printing as practically irremediable." p. 434.

21. Bothamley, Ibid.

22. Letters, *Photography*, June 4, 1891, p. 365

23. Ibid. p. 364

24. ***British Association for the Advancement of Science***, 1894 Annual Report. The title of the Jointly published article was "On the Action of Potassium Permanganate on Sodium Thiosulphate and sulphate."

25. G. E. Brown, ibid, July 3, 1891, p. 420

26. Ibid.

27. Ibid. p. 420

28. G. E. Brown, *BJP*, July 10, 1891, p. 434.

29. G. E. Brown, Ibid., p. 435.

30. Announcement, *Photography,* June 4, 1891, p. 365

31. G. E. Brown, *BJP*, July 10, 1891, p. 434.

32. W. W. J. Nicoll, *BJP*, July 24, 1891, p. 479. Also see the *Photography Annual* of 1891, p. 494. "It was not perfect when first put on the market."

33. "This paper would probably be better known than it is at present had it not had a disastrous commencement." *Photography*, Aug. 6, 1891, p. 506.

34. G. E. Brown, *BJP,* July 10, 1891, p. 434.

Chapter III

Kallitype II & III: Patent to Manufacture & Sale

I.   The Announcement of Commercial Kallitype II & III

On July 24, 1891, in a letter to the editor of **the *British Journal of Photography*,** Nicol wrote

> Your very full and appreciative articles on Kallitype . . . lead me to think that you and your readers may feel interested in a short account of the improvements recently introduced

He continued.

> Early in this year, the desire on the part of many users of the paper for a process which is capable of yielding warmer tones, and which does not involve the use of a silver bath, with consequent staining of the fingers, as in No. 1 Kallitype, led me to make further experiments, with the result known as No. 2 Kallitype.
>
> In this modification of the original process, the silver is present in the paper. The developer consists simply of Rochelle salt, with or without the addition of borax or other salts. A ten per cent solution of Rochelle salt gives yellow prints, which become darker and darker as more borax is added, until equal parts of the two salts are present, when the tone is a cold black. I have much pleasure in sending you specimen prints illustrating three of the tones thus obtained. The washing is much simplified. As the developer is very cheap, it is used as the first washer.

The prints are left in it for fifteen to twenty minutes, and then washed in dilute ammonium, one ounce to the gallon of water. One bath of this will do, but two are better. A slight rinse in plain water follows, and the prints are ready for drying.[1]

But Kallitype "2" is not all the news.

I may add that I have not yet attained my ideal paper . . . which is one that prints out in the frame and requires only washing in dilute ammonia. I have nearly reached it . . . and hope soon to perfect it.[2]

This would be the consummate iron-silver paper, Kallitype "3," a "print-out" paper, which is exposed till the image appeared as desired and which required only minor processing in water and ammonia.

Nicol ended his letter with some remarks about "how much I am indebted to former experimenters."

In conclusion, no one realizes more fully than I do, how much I am indebted to former experimenters. Herschel, Hunt, and, later, Willis, laid the foundation of all iron processes. I have done little more than use reactions known to all chemists, but it is the very use of what is old in a new way that is the essence of invention.[3]

The last statement by Nicol must be viewed as a public admission of his debt to past researchers who laid the foundation on which his work depended. It is clear that Nicol here openly and generously stated his obligation to previous researchers—and may have, perhaps, overstated it. Belittling his contribution by saying he did no more than use "reactions known to all chemists" may err on the side of excess humility. The letter may make too abject a response to Brown's editorial on the "debt" all inventors owe their forbears. Nicol had worked out detailed iron silver printing processes for both K I and KII, which were ingenious and innovative in their particulars and worthy of patent protection, even if the basic chemical reactions of iron-silver printing were known. As a rhetorical stance the contrite remarks of the letter were well designed to placate critics at the moment when his new process was announced. The remarks salved the feelings of those who were piqued at his earlier seemingly exaggerated claims. The letter had the effect of ending reports of ill feeling and no more complaints were forthcoming. The new statement cleared the air so the new process, Kallitype II, did not have to struggle against a tide of ill will as had Kallitype I.

The editors of **BJP** report that Nicol sent to the journal, along with his letter, an admirable and most instructive collection of prints in which "the influence of borax is shown by three sets of examples." They describe the print color as varying from brown to "rich warm black," and from "dark purple" to "brownish red, depending on the amount of borax added. The editors congratulated Nicol "on the high state of perfection to which he has brought his process in so short a time."[4]

K II. Patent No. 7312—Details

Nicol had applied for a patent on Kallitype 2 on April 28, 1891, almost three months before the above letter was written. The application was given the number 7312. The provisional specification states the patent is for improvements in a photographic print process in which the sensitizer solution (heretofore only a ferric salt dissolved in water) will now contain, in addition, the silver salt which had formerly been contained in the developer. It is worth noting in passing that the simple name "Kallitype" never appeared in the #7312 patent. The paper described in the patent was called "Kallitype 2" in press and commercial writing to distinguish it from the earlier Kallitype I paper. The new K 2 paper employed for development a solution compounded of Rochelle salt and borax to produce image colors that varied from red to black depending on the quantity of borax added. A second developer was offered which produced an invarying black-toned image.

There was some urgency about getting the improved process patented. One month after the application, in May 1891, James H Stebbins., in an article reviewing Kallitype I, called "attention to another process of making prints with which I have for some time been experimenting."

> The prints closely resemble those obtained by the Kallitype process, but have the advantage of requiring no silver bath to develop the image. The prints, after exposure, are simply developed by being immersed in a bath of formate of soda, acetate, oxalate or citrate of soda, and the tones obtained range from pure black and white to sepia brown according to the developer which is used.[5]

Earlier, we suggested that Nicol had no monopoly on the ideas leading to a workable kallitype process. It appears Stebbins was getting too close for comfort. Stebbins did market a kallitype II process in America, as we shall see.

In the 7312 patent, Nicol announced a second new Kallitype paper, also left unnamed, presumably a Kallitype III paper. It was a print paper on which would be coated a) the iron sensitizer; b) the silver salt; and c) the developer—all on the same paper stock. Nicol wrote in the specification: "In this manner the image may be made to print out and thus development may be avoided." After exposure to light under a negative, Kallitype III paper would require no further processing than a slightly ammoniated water wash to clear away residual chemicals.[6]

Nicol's letter of July 14 states he began working on K2 and K3 "early in this year," that is, January or February, 1891.[7] He must have made good progress, since he applied for a patent on the new process on April 1, 1891. The complete specification was "left" at the patent office on 27 January, 1892, more than six months later. It is notable that the announcement of the "new printing process soon to be offered for sale" by the Birmingham Photographic Co. was published June 4, 1891, a full six months before the patent application was complete.[8] Nicol must have been a busy man with lectures to write, demonstrations to prepare, and innumerable experiments to perform in order to work out the details of two new photo chemical processes, which had already been promised for sale "in a short time." In his letter to the *BJP* dated July 14, 1891, Nicol wrote the new albumen kallitype paper "will be on the market in a few days."[9] The problems of Kallitype 2, at least, must have been solved by that date, in view of the fact that he forwarded excellent prints in various colors to the journal. The final patent for K2 and K3 was granted on the 27th of February, 1892 and the full patent was published in the April 1, 1892 issue of the *British Journal of Photography*. Copies of the patent were available for 6d. from the patent office. [10]

The complete specification of patent 7312 for what shall hereafter be designated Kallitype II and III differs in manner and content from patent 5374. The patent for Kallitype I, Patent 7312, was remarkably more business like, made fewer and more modest claims, and was more direct about chemicals and quantities. Nicol's claims about what he invented were much more restrained. Nicol claimed "the developing and washing solutions," "particularly" described. There was nothing here to disturb the partisans of Herschel and Hunt. Nicol claimed credit only for the solutions he worked out. We may note in passing that the patent was applied for before the appearance of Brown's article on fairness to historical sources.[11] Whether Brown's editorial on the moral obligations of patent writers led to Nicol's apologetic remarks we will never know. We do know that Nicol had six months to consider Brown's moral criticism before making the final submission of the new patent.

III.   Designations of Kallitype Processes

Before discussing the newly invented papers we must establish a system of designation so that references to different papers and processes will be clear. In the last few pages various designations have been applied to various processes. In the 7312 patent, Nicol simply called the three processes I, II, and III. While these designations help to distinguish the three processes discussed in the patent, they do not help to distinguish them from the processes dealt with in the older 5374 patent. It appears a system of designations is needed to keep track of the several processes Nicol invented.

In the preceding chapter we discussed a variety of processes called "Kallitype I" processes. All had simple iron sensitizers coated on the paper before exposure and all developed the image after exposure by immersing the print in a silver nitrate bath. K I sensitizers could be compounded of various iron salts and Kallitype I developers were compounded of various alkaline salts mixed with the silver nitrate—all such are here classified as K I sensitizers and K I developers and the prints are called Kallitype I prints. We have already noted that various chemicals were added to the silver developing bath to change image color and contrast. Prints made from these variations in development will be called "Kallitype I a" or "Kallitype I b" prints or given short names such as "K I a" or "KI b."

In this chapter the second patented print process called "Kallitype II" shall be discussed. Here and in the remainder of this book the name "Kallitype II" will be used to refer to any kallitype process in which the solution coated on the paper is a combination of the sensitizer, a ferric salt, and the image forming material, a silver salt. Kallitype II, then, is a class of print processes in which the maker coats on the paper a solution containing a combination of iron and silver salts. Kallitype II prints are developed in a solution of Borax and Rochelle Salt or in a solution of Sodium Acetate. Finally, Kallitype III is a print process in which all the chemicals required to form the print image are coated on the paper before it is exposed: the iron salt, the silver salt, and the solvent necessary for the formation of the image. After the K III paper is exposed to light, only water and fixer are necessary to "process" the image. The different approaches of the three kallitype processes are diagrammed below:

IV.   Nicol's Patented Kallitype II Solutions

Schematic Comparison of Kallitype I, II, and III Print Processes

	Sensitizer	Developer	Clearing	Fixer
KI	Ferric Oxalate	Silver Nitrate	Pot. Citrate	Ammonia
KII	Ferric Oxalate	Rochelle Salt	Pot. Citrate	Ammonia
	Silver Nitrate	Borax		
KIII	Ferric Oxalate	Ammonia		
	Silver Nitrate			
	Potas. Oxalate			

The diagram oversimplifies the situation slightly, but it is an accurate indication of the different chemical approaches used in the three processes.

In practising my invention according to the second method, (KIIa) I use by preference solutions of the following composition

Kallitype II a Sensitizer

Water one hundred	(100) cc
Ferric Oxalate fifteen	(15) grammes
Silver Nitrate three	(3) grammes

Kallitype II a Developer

Water One hundred	(100) cc.
Rochelle Salt Ten	(10) grammes
Borax Seven	(7) grammes

To this is added one tenth (0.1) to four tenths (0.4) of a cubic centimeter of a five (5) per cent solution of Potassium chromate.

Potassium dichromate, the last named chemical is added to control the contrast of the print. The amount added depends on the contrast of the negative. The less contrasty the negative, the more dichromate had to be added to produce a print with a full scale of print tones.

Nicol specified that K II prints are immersed in a developer solution for fifteen to thirty minutes. The long developing time was necessary to remove residual iron, which if not removed, left an unsightly yellow deposit on the surface of the print. The developing time was considerably longer than was customary with other processes of the time, and would become the source of complaints.

After development the prints are "washed" in a fixing solution.

Kallitype II Fixer

Water	one liter
Ammonia (sp. g. .880)	three cc.

In the patent, Nicol specified the prints are washed in two changes of the fixer, solution, but he does not give the duration of each wash.

The Kallitype II b solutions are:

KII b Sensitising solution

Water One hundred	(100) cc.
Ferric citrate ten	(10) grammes
Oxalic acid three	(3) grammes
Silver Nitrate three	(3) grammes

After exposure to light the K II b print is immersed in the following developing solution for an unspecified time. K II b developing solution is used to achieve black image prints.

KII b Developing solution

Water	One liter
Ammonia (.880)	six (6) cc.
Sodium Citrate twenty	(20) grammes [12]

After development, the prints are washed in the dilute ammonia fixer described for the K2a prints above.

There is some question about whether a final wash in water is required after the soak in the ammonia bath. Nicol doesn't mention one for the K II b process, but he does for the KII a. The presumption here is that a final wash in water is required but was omitted in the specification. Whatever the case, it is clear that the K II a and K II b solutions and operations are significantly different. While the reason for the different developing solutions was not given in the patent, Nicol elsewhere explained they produce variations in the color of the K II print.

V.   Kallitype III: A print-out Paper

Patent #7312 introduced a "print-out" version of the kallitype, called in this text, "Kallitype III" or, "K III." Nicol had written as early as 1890 that a print-out version of the kallitype was the ultimate target of his researches. The belief was widespread at the time that whoever produced an attractive print-out paper would quickly capture a good share of both the professional and amateur market. A print-out paper would enable the printmaker to expose the sensitized paper under a negative until a finished-looking print was achieved from the action of light alone. The printer would process the exposed print in a single solution, but the processing would not change the print's appearance produced by exposure. The intention was to employ as few processing solutions as possible so that the print-out process would not only provide a foolproof indication of proper exposure, but also a brief processing time. A single water bath would enable the chemicals present on the coated paper to interact and make a developed and fixed print. A final wash in water would remove any residual chemicals from the paper.

Print-out versions of photographic papers were the dream of photographers during the 80's and 90's. The **British Journal of Photography** for April, 1, 1892, assessing amateur preferences, had written "The larger number of amateurs prefer systems . . . which produce the image without developing."[13] Pizzighelli and Jacoby, had already managed to create a print out version of platinotype, which worked to some extent, although some said not as well as they would like.[14] Willis, the successful English engineer who made platinum papers that worked flawlessly, attempted to create a print-out platinotype paper but he remained dissatisfied with his best results.[15] Gelatin chloride papers, increasingly available at the time of the kallitype, printed out their image fully, but required a developer and hypo fixer step to rid the print of unused silver salts and a final water wash to get rid of the fixer.

Nicol wanted an unambiguous sign to the printer indicating when the print was properly exposed. Such a paper would be foolproof, quick, easy, and inexpensive to use.[16]

We have already seen that Kallitype I had been criticized for its uncertain indication of sufficient exposure. It is clear from several statements Nicol made that his K III paper was intended to remedy this defect.

Nicol's patent #7312 does not openly call K III a "print-out paper," but it does specify "after exposure to light, the prints should . . . attain . . . their full image." It should also be noted that Nicol made no claim in the patent that he *succeeded* in producing a working Kallitype III paper. Print-out Kallitype paper was never put on the market by the Birmingham Photographic Co. or successfully marketed by any manufacturer, although a few tried. Interestingly, N. C. Hawks, an American businessman we will meet shortly, used Nicol's K III formula to prepare a commercial paper, but he sold it as a Kallitype II paper with a developer. [17]

Extensive research has turned up no photographer who claimed success at making Kallitypes from Nicol's K III formula or by KIII formulas of his own invention. A number of amateurs tried, and wrote that they were close to succeeding. Will J. Brooke, N. Gray Bartlett, Professor Emile Boivan, and O. Prescott Bennett all tried. But when last heard from, all of these intrepid experimenters reported they were still working to get KIII under control. For example, Brooke wrote that Nicol's kallitype III formulas,

> gave very fair results, but not quite equal to developed ones. Besides the exposure needed was two or three times than that for the others.[18]

Editors who saw the K III prints these workers sent, agreed that further work was necessary. For example, the editors of **The Photo Beacon** wrote, in response to K III prints sent by O. P. Bennett:

> The prints are exceedingly interesting. Number I is a print-out Kalitype [sic] print as full and as perfect as in ordinary print-out paper [silver albumen]. The others show the result of fixing. The tones are all sepia, some of them very warm but all pleasing. Where an effort has been made to get a black tone the prints are rather muddy, but Dr. Bennett hopes to overcome this at an early date—Ed.[19]

When Nicol included K III in the patent, he had not yet perfected a workable process. Had he succeeded, the history of Kallitype would almost certainly have been different. As several writers of the time pointed out, a working K III print-out-paper along with the impressive KII developing-out paper would have assured the kallitype a place in the market. The fact that Birmingham Photographic Co. never announced such a paper for sale, and the fact that no one in the more than 100 years since the publication of the KIII idea has claimed a completely effective print-out process suggests that the possibilities of realizing a practical K III process are slim.

In any case, for those who might wish to experiment, Nicol's specified solutions for K III paper are:

Kallitype III Sensitizing Solution

Water One hundred	(100) cc.
Ferric Oxalate	(fifteen (15) grammes
Potassium Oxalate	three (3) grammes
Silver nitrate three	(3) grammes

After exposure to light, the prints, "which should have attained their full depth, are washed in the following."

Kallitype III Washing Solution

Water	100 cc.
Sodium citrate	3 grammes
Citric acid	0.5 grammes [20]

The prints are then washed in a dilute ammonia solution, after which they are given a final wash in water and dried.

Two platinum print-out papers reached the market about this time, but neither succeeded.[21] Writers reported problems arose when the ferric oxalate, the platinum salt, and the developing chemicals were coated in close proximity on the paper. One writer noted,

> A further variety of the platinum process, the print-out, is devised to allow the image to be seen of full depth in the printing-frame . . . But the process has never been satisfactory, owing to the fact that the degree to which printing is visible

depends on the state of the moisture of the paper. The damper the paper—and some damp is necessary—the more the image prints out. In other words, what is printing to a correct depth one day is not so the next, and therefore the process is bound to be irregular.[22]

Sometimes the chemicals on print out paper reacted prematurely, without exposure to light, because of moisture getting to the paper. The presence of the slightest amount of moisture—even that in the air at the moment of inserting the unexposed paper in the printing frame—could be enough to cause the chemicals to react before the print was exposed to light. Those parts of a sensitized paper that were more moist would produce darker image areas than those there were less moist. Such uncontrolled reactions produced spotty, uneven images. The prevention of moisture variations before, during, and after the exposure, was practically impossible outside of a laboratory. Ironically, while the KIII process fascinated investigators, the KII process was the one that worked reliably and well. It achieved the widest acceptance by photographers in the past and is the one most kallitype printers use today.

VI.   Responses to Kallitype II in the Market

Kallitype II, or "2" as it was occasionally referred to in the early days of its release, was offered for sale by the Birmingham Photographic Company sometime between June, 4, 1891 and July 10, 1891. The earlier date is suggested by an announcement in the June 4, 1891 issue of *Photography*.[23] "We shall in a few days place upon the market a very greatly improved paper." The later date comes from another announcement in the issue of *Photography* dated July 30, 1891.

In an interesting letter which Dr. W. W. J. Nicol writes to the columns of a contemporary [actually the *BJP*] upon the subject of his patent "Kallitype" paper, he makes a decidedly interesting disclosure that informs us of the ideal he has in view. The ideal paper which he is still striving after, but has not quite reached yet, is one that prints out in the frame, and requires only washing in dilute ammonia. Dr. Nicol says he has nearly reached it, and hopes soon to perfect it. Should he succeed, as it is by no means unlikely that he will, and the process prove capable of producing as good results as the "Kallitype No. 2" paper which he has just introduced, the extreme simplicity of the method will ensure for it, amongst amateurs at any rate, an enormous sale.[24]

By the end of July, 1891, Kallitype II paper was certainly on the market. But photographers had already been teased by releases like those above suggesting that "kallitype 3" print-out paper was soon to come.[25] A moment's reflection suggests that publicizing a forthcoming K III paper was not the most prudent way to enhance sales of the K II product. One must remember that in all probability unsold K I paper still sat on the shelves of both dealer and user. Publications describing commercial KI paper as well as formulas for individually prepared K I were still coming out in photo journals at the time when Kallitype II and III papers were being discussed. It is likely that users were in for a period of confusion. One wonders how many decided to wait till the KII and KIII products settled down.

An early response to the new KII paper came in July 1891 from Bothamley, who spoke again on new printing processes, this time before the Photographic Convention at Bath. The Convention was the major photographic art, science, and marketing event of the time and was attended by professionals and amateurs alike from all over the country.[26] Bothamley revised his earlier report, made no mention of Kallitype I, and confined his remarks to the K II process. He presented new information acquired while trying the process and displayed prints he made for public examination.

Bothamley reported the Kallitype II process had a number of advantages over the original process. He found the new process wasted less silver, avoided staining the fingers of the operator, and required no control of the amount of silver in the developing bath, a perennial problem with kallitype I as well as the popular albumen print. He observed that the revisions of the Kallitype process followed in reverse the order of the revisions of the development of the platinotype process. In the latter, the platinum salt was placed first in the sensitizer, but in the latest modification it is found in the developer. Bothamley did not yet know that another modification would occur in April, 1892, in which, as in K II, the platinum would be placed again in the sensitizer.[27]

Of considerable interest is Bothamley's revelation of the contents of the Kallitype II sensitizer. He writes

> The specification of the patent of kallitype No. 2 is not yet published, but Dr. Nicol has kindly informed me that the paper is coated with a solution containing ferric oxalate, ferric nitrate, silver oxalate, silver nitrate, and nitric acid.[28]

We are provided here an insight into the difference between the chemicals an inventor specifies in a patent and those he actually coats on the commercial paper. Some of the chemicals tried by Bothamley were not mentioned in the specification. They and their quantities remained proprietary secrets of the Birmingham Photographic Co. which made and sold the paper.

An explanation of the "extra chemicals" used in the commercial sensitizer but not specified in the patent may be desirable. The extra chemicals perform a service to the sensitizer or they wouldn't be used. The theory of the process suggests that sensitizers require only ferric oxalate and silver nitrate. Now a supply of these chemicals would vary in purity and other characteristics, as would the paper they were coated on. Ferric nitrate may have been added to the ferric oxalate solution, as a buffer, to make sure there was an excess of ferric ion. Nitric acid was probably added to adjust the acid level, or ph. of the sensitizing solution to a standard level and also to inhibit the tendency for the silver to precipitate.[29] The silver oxalate may also have been added as a buffer to assure a plentiful supply of oxalate ion.

Bothamley discussed certain details of Kallitype II manipulation, notably how to achieve various print colors by varying the amount of borax in the Rochelle salt developer and how to control the contrast of the print by varying the amount of dichromate added. He reported dichromate which hadn't worked for him when he tried it with K I paper, did work with the K II process. He showed examples of the use of dichromate in the print and stated they "demonstrate very clearly" the effect of the oxidizer in the production of clearer whites while maintaining strong darks.

It is notable that Bothamley recommended the use of an additional bath to clear the prints of yellow iron stain. None had been stipulated in the patent.

> My own experience indicates that, even after prolonged immersion in the developer, there is danger of iron salts remaining on the paper, with of course, loss of purity in the whites. I also recommend that, after removal from the developer the prints should be immersed in a 10 per cent solution of Rochelle Salt before being put in the ammonia.[30]

Bothamley's suggestion would not be the last recommendation for solving the problem of yellow stain.

Bothamley ended his talk with a discussion of the permanence of Kallitype II prints. He found Kallitype II to be as permanent as K I, which he had tested under the most grueling laboratory conditions he could devise to simulate severe atmospheric contamination. Bothamley admitted he made "no prolonged experiments on the behaviour of mounted prints Future testing of the the mounting paste and the cardboard used for mounting remained to be done. The products then in popular use for mounting were known to be enemies of print stability. Bothamley concluded "there is no reason to doubt that under ordinary conditions kallitype prints are permanent." His final remarks on permanence include the following curious statement:

> It is, of course, not to be expected that kallitype can as yet equal platinotype, but it is by no means improbable that in time the difference will be chiefly a possible difference in permanency.[31]

Bothamley's implication is that platinum prints are probably more "permanent" than "permanent" silver prints. The former will always be more permanent, in theory, because the platinum metal which makes up the print image is chemically inert while silver is not. This fact had been labored for some 20 years, in the marketing of platinum paper by Willis. What Bothamley's statement probably means is that Kallitypes are as permanent for practical purposes as those made by any silver print process.

An early amateur response to K II paper, perhaps the first in the American photographic press, appeared about this time in *Anthony's Annual*. It was signed with the photographic pen name, "Talbot Archer."

> Dr. Nicol has improved his Kallitype printing process, and the "Kallitype No. 2" gives results much resembling platinotype, at half the price and with less trouble. This process clearly has a future before it. The paper is exposed in contact with a negative for five minutes; then is placed in a weak ammonia bath for ten minutes; after which it is finally rinsed in water and dried. The resulting image is in silver, and any tone, from red to black can be obtained.[32]

The processing described by this writer is incomplete and would not work on K II paper. It appears Archer wrongly mistook K3 water "development" for the complex developer of the K2 process. Nevertheless, he reveals his positive attitude toward the new paper. He suggests the new process is sure to win friends "among those who prefer ordinary silver printing" [printing on matt surface paper] to silver printing on albumenized paper, "with its meretricious glaze, accompanied by the need for troublesome gold toning and a bath of hypo, which adds the seeds of yellow fever to the print."[33]

The concerns of amateurs using the new paper were also indicated in the questions and answer columns of the photo journals. One month after the new Kallitype paper had been released, 'Dick Swiveller' wrote to request information about it. His query drew three answers which were printed in **Photography** on August 6, 1891. The first read:

> New Kallitype—"Dick Swiveller" is not very clear in his question. Does he want to know the features and advantages of No. 2 as compared with the No. 1 process, or does he want general information? H.T.

H.T. answered his question with a discussion of the advantages of the KII process over K I—greater permanency, increased control over color by development, less deterioration by damp, less danger of staining the paper by iron deposits during development, and greater command over density and contrast. He then provided particulars of working the process and details how to achieve the black, purple, and sepia tones. He warned that "the developing solution serves the double purpose of developing the image and of rendering the iron used in the sensitizer soluble in the ammonia bath. If the developing solution is used for too many prints, they will show "yellow stains when finished." He advised the operator "to keep the prints moving about freely during the whole time they are in the (developer) solution."[34] It is clear the yellow stain continued to be a problem with K II as it had been with K I.

"Sigma Delta" also replies to Swiveller's query about Kallitype II paper.

> The paper would probably be better known than it is at present, had it not had a disastrous commencement, and the brand now on the market is infinitely superior to No. I. The want which it was intended to supply is a similar paper to

the platinotype at a cheaper cost. It is printed in much the same way as the platinotype and a result much resembling that paper can be obtained.[35]

Delta also lists advantages of kallitype II paper—"cheapness, variety of tone, and greater permanency for a silver process." He ends with a discussion of Bothamley's experiments on the resistance of K II prints to chemical attack.

When the new year, 1892, arrived, published reports on KII paper come from a wider and more diverse group of approving correspondents. But critical responses occur. In the "Queries" column of the January 29 issue of the *Amateur Photographer* the following appears.

> I have followed the instructions given for Kallitype II but cannot get rid of the yellow tinge. Will any reader help me? Signed Borax.[36]

A reply appears in the February 5 issue of the same magazine. It reads,

> I have used the above paper successfully and obtained the three tones, black, purple and sepia. The yellow tone you complain of is caused by removal [of the print] from the developing bath too soon. By the by, have you read Mr. Herbert Thompson's lecture on the Kallitype #2 at Holborn, reported in the *Amateur Photographer* for January 15, 1892. Cyanin [37]

Unfortunately no copy of the *Amateur Photographer* for January 15, 1892 has been found, so that report cannot be included here. The February *Amateur Photographer* provides an editorial report "of our inspection of Thompson's specimens of the Kallitype II process." It says,

> The matt surface prints could not be told from the finest example of platinum printing and the albumen surfaced print is very pleasing. I recommend K #2 as the only process that will enable the amateur to attain this frequently desired result. For Professionals this ought to lead to increased business.[38]

While these remarks may sound so laudatory they raise doubts about their source, they are informative of amateur concerns about kallitype II six months after its introduction.

It is interesting to find in *Anthony's Annual,* an American publication, another report of the Holborn kallitype demonstration. It reads

> At a recent meeting of the Holborn Camera Club, a very interesting paper was read by Mr. Herbert Thompson, on Kalitype [sic] No. 2, in which the speaker said that the paper strongly resembles platinotype in many ways and should be printed until the detail in the high lights was faintly visible. [39]

The remainder of the short article gives standard information about exposure, development, contrast-control, fixing and washing.

More of Bothamley's remarks on Kallitype II appeared in a summary report contained in a new British publication, ***Photography Annual*** for 1892. The entry begins with an overall favorable evaluation of Kallitype II and continues with a discussion of the details of processing.

> While potassium dichromate is of very great value in keeping the whites clear, too large a proportion destroys the details and half tone. The sepia developer gives somewhat uncertain results, but frequently the colour of the image is good. Keeping the paper dry seems to be a point of considerable importance.
>
> The black image in both processes consists of metallic silver, and can be bleached and redeveloped, or intensified. It is not appreciably affected by exposure for 48 hours to the products of combustion of sulphur in air, nor by immersion in a solution of sulphurous acid for the same period. A dilute solution of sulphuretted hydrogen slightly alters the colour of the image, but produces no further change; a strong solution changes the colour of the image to brown. A dilute solution of ammonium sulphide changes the colour of the image to a not unpleasant brown, without any loss of detail and with very slight reduction of intensity. The sepia image is unaffected by the products of combustion of sulphur, but sulphuretted hydrogen or ammonium sulphide changes the colour to black and afterwards to brown. There is no reason to doubt the permanence of kallitype prints under all ordinary conditions.[40]

It is evident that this report covers some of the same ground as Bothamley's previous report on K II. But the additional details on testing for permanence justify the repetition.

Three snippets appeared in the *BJP* in early 1892, which suggest amateur attitudes about printing. The first, from the January 8 issue, reports, "it is very general with the London Houses, when platinum prints are ordered, to supply bromides instead when the light is bad for printing."[41]

The next issue reports that "Willis is having many difficulties" making a printing—out paper in platinum. He finds it "almost impossible to get a perfect process of this kind." [42]

The third is from the January 22 issue.

> The general opinion among practical photographers is that albumenized paper is now meeting the most formidable rival it has ever had, in gelatino-chloride printing out paper. On several occasions when fresh print papers have been introduced, the knoll [sic] of albumen has been said to have been rung. Yet it survives and is as yet the most popular process of the day, and the one most extensively used in commerce. There is doubtless a big future for the new paper, but albumen will die slowly, notwithstanding all that is predicted. [43]

On April 1, 1892 *The British Journal of Photography* ran a full column editorial on "Advances in Kallitype Printing." The editor reports that at the same time that Willis revised the Platinotype process to make a paper sensitized with ferric oxalate and platinum, he has been informed of some "improvements recently effected in kallitype process, which, from the point of view of simplicity of working, we are disposed to consider as advances." He reports the processes through which the kallitype has passed, Kallitype I, "the two modifications of the process now published" and finally K III, "which prints out in the frame." The editors find the first process, K I "has been abandoned" and "that the possibilities of . . . {KIII} are not much more than in their initial states of realisation."

The editor concluded with an evaluation of the Kallitype II process.

> Assuming the last mentioned KIII process to be perfected for commercial purposes, kallitype will offer the advantages of supplying both a developing and a printing-out process. Of the comparative merits or demerits of the two methods we do not seek occasion to speak at present, but it may be permissible

to submit that probably the larger number of modern amateurs prefer those systems of printing which produce the image without the need of development—a fact to be considered in noting the vitality of the much-condemned and attacked albumen process.

Kallitype is the only silver printing process extant which does not entail the use of hypo as a fixing agent—which taken in conjunction with the variety of ways in which the image may be produced, disposes us to believe that it has secured a permanent place among modern printing processes. Of its artistic capabilities we have spoken before in a favorable strain, while as for the probable "permanence" of the results we think there is little to be feared. [44]

The editor's hopes for the new papers are very buoyant here. It is notable that the success of the kallitype is said to depend on the successful development of the K III process. Unfortunately, this was the one process Nicol had not worked out sufficiently well for the Birmingham Photographic Co. to market. While such a paper was theoretically possible, at this time, there was no assurance that a practical printing-out-paper could ever be made. We have already seen Willis' lament about the difficulties of making a workable print out platinum paper. So long as the manufacturer could not control all the conditions of moisture that might influence the paper, during and after production, just so long would it be impossible to produce a successful printing out iron silver or platinum paper. Looking back to 1892, one regrets seeing the future of the kallitype made so dependent on a paper that had little chance of being perfected—even by the resourceful Nicol.

A French version of Kallitype was described in 1892 at a meeting of the *Societe Francaise de Photographie* under the heading of "An Imitation Platinum Paper," The ***BJP*** reported that M. A. Pavard read a paper for Professor Boivin, describing his many experiments with the salts of iron.

Boivin has prepared a paper with those salts as a base, which keeps well, and is cheaper than either platinum or silver paper. It is printed out until the details of the shadows are seen, and when removed from the frame, the image is either steamed or breathed upon. This moisture develops the picture, which, after washing, is fixed in the following solution

Fixing and Toning Solution

Water	1000 parts
Hypo	80 "
Gold Chloride 1 in 1000	20 to 40 cc.

This bath gives purple, sepia, or warm black tones, according the length of the immersion of the print.[45]

This notice reported only a toning bath and withheld any information about the sensitizer or developer . . . Since these were not published, it appears reasonable to conclude Boivin intended to market his K III process. It is also interesting that the statement does not designate the described process as a kallitype process. Many years later, L. P. Clerc, the great French photographic writer on photographic technique, mentions Boivin's process, but, unfortunately provides few practical details.

> Boivin's paper is sensitized with a mixture of ferric oxalate,
> a silver salt, and a large excess of alkali oxalates. The image
> is developed in plain water by immersion or by steaming.[46]

No more than this is known. We do not know if Boivin's paper was ever sold in France. Another overseas item about Kallitype paper appeared in the July 1892 edition of the German publication *Photografish Archiv*.

> *Die Kallitypie is ein von dem Englander James Nicol
> ausgearbeitetes Eisencopiverfahren, bei welchem das
> Papier mit einer Mischung von oxalsaurem Eisenoxyd mit
> Silbernitrat praparirt und nach der Belichtung in einem Bade
> von weinsaurem Kali-Natron and Borax entwickelt wird.
> Fixirt wird in verdunnter Ammoniaklosung. Eine Genaue
> Beschreibung enthalt das Phot Archiv, 1892, Nr.I.*

A translation of these and the following remarks is given in the notes. In the Fragen and Antworten section of the German **Amateur-Photograph**, in the same publication, we find the question,

> Kallitypie-Papier.—Was is "Kallitype paper" and wie is
> das selbe van Amateuren am besten Herzustellen? [47]

Unfortunately, the answer was given in the following issue, but that issue has not been found.

Kallitype II was discussed at some length in W. Jerome Harrison's, *The Chemistry of Photography*, published in 1892. Harrison lived in the Birmingham area and was quite likely a professional acquaintance of Nicol. An honored and innovative science educator, Harrison was the author of three important early books on photography. The first, written in 1881, was entitled *The Literature of Photography*. It was a bibliography of all English books on the subject of photography. The second book, *The History of Photography*, was published in 1888, the year before Nicol finished his research on the Kallitype, so the book contains no reference to the inventor or his paper. Harrison's third book, *The Chemistry of Photography*, was published in 1892. This book has a chapter entitled "Printing with the Salts of Iron", which gives a rather extensive discussion of the chemistry of Kallitype I, an abbreviated treatment of the Kallitype II process, and no information on Kallitype III.[48] There is some evidence Nicol and Harrison occasionally worked together. Harrison, as president of the Birmingham Photography Society, arranged a demonstration of KII by Nicol for the membership. The demonstration occurred on March 4, 1892. The *BJP* reports:

> From the Birmingham Photo Society we have received some specimens of Kallitype Printing No. 2, which illustrate in a favorable degree the varied capabilities of this beautiful process. The albumen like gloss on some of the pictures certainly conduces to the provision of the finer details, while the matt surface on others is as near an approach to the characteristic beauty of platinum as could be obtained. We welcome Kalitype as a distinct and agreeable advance in silver printing.[49]

In view of the proximity of the two men and their common interests, Harrison's information on the kallitype may be more than ordinarily credible.

Harrison explained the derivation of the name, "Kallitype."

> The Kallitype process takes its name from the same two Greek words, signifying "beautiful picture," from which Fox Talbot derived the name of his "calotype" negative process . . .

Harrison was not impressed by Nicol's second choice of the name.

"As the two words are very similar, they are liable to be confounded, and it seems a pity that some more distinctive name was not chosen." [50]

One is inclined to agree with Harrison, especially when the other sound-alike name, "Collotype" is remembered. People have been confusing Calotype, Kallitype, and Collotype for nearly one hundred years and will probably continue to do so. The three names designate three quite distinct printing processes which have very little in common beyond the sound of their names.

Harrison provided an early authoritative discussion of the chemical reactions of the K I and K II processes. The basic principle of both processes depends on the "action of light to reduce ferric oxalate to ferrous oxalate." He also provided descriptions of the chemical change that occurs in the kallitype printing process.

$$Fe (C_2 O_4)_3 \longrightarrow 2 Fe (C_2 O_4) + 2 C O_2$$

Ferric Oxalate produces Ferrous Oxalate and Carbonic Acid Gas [sic].

Carbonic acid gas is now known as Carbon Dioxide.

Harrison published a formula for compounding a kallitype I paper developer.

Harrison's K I Developer

Nitrate of Silver	50 grain
Bichromate of Potash	1 grain
Water	10 oz.
Strong Ammonia	.5 drachm

Harrison reported the chemical reaction that occurs during K I development. He believed that the "chemical action of this developer can hardly be represented by equations; but it is plain that the ferrous oxide contained in the ferrous oxalate reduces the silver oxide in the silver salt to the state of metallic silver." His approximate equation follows.

$$2\ FeO + 2AgO \longrightarrow 2\ Ag + Fe2O3$$

Ferrous Oxide and Silver Oxide produce Silver and Ferric Oxide.[51]

Equations for the chemical reactions of the kallitype I developer that are more current will be found elsewhere in this book.

Harrison also published formulas for compounding the solutions used in kallitype I washing solutions. After developing the KI print, he wrote,

> It now only remains to wash everything out of the paper, except the black metallic silver which forms the picture. This is effected by soaking the print for ten minutes in each of the following.

> Washing Solution No 1.

> | Kallitype developer | 1/2 oz. |
> |---|---|
> | Citrate of soda neut. | 2 oz |
> | Water | 20. oz. |

> Washing Solution Nos. 2 and 3.

> | Citrate of soda (neut.) | 1 drachm |
> |---|---|
> | Ammonia .880 | 2 drachm |
> | Water | 1 quart. |

Harrison remarks that

> the object of the citrate of soda in this and in the washing solutions is to prevent the precipitation of the iron by the ammonia used for dissolving the silver salts, that is, to remove yellow stain. [52]

> No comments on the chemical activity of these solutions are given, other than "they [must] always smell distinctly of ammonia."

> Harrison concluded, "as no hypo is employed for fixing, the prints should be more permanent than ordinary silver prints."[53]

Harrison had much less to say on Kallitype II than on K I. He reported "the new paper is coated with two iron salts—ferric oxalate and ferric nitrate and also with the corresponding two silver salts—silver oxalate and silver nitrate." As in the K I process, when K II paper is exposed, the light reduces the ferric oxalate to the ferrous state. Harrison provided no chemical equations or description of the solutions used for the K II process. Perhaps there wasn't time, since Harrison's book went to press about the time Nicol applied for his patent. Harrison does make one instructive comment on the chemistry of the K 2 developer.

> The ferrous oxalate combines with the Rochelle salt, and reduces the silver to the metallic state; the Rochelle salt also combines with the iron to form ferric tartrate Fe2 ($C_4H_4O_6$) 3. The first reaction precipitates silver; the second prevents the precipitation of iron.[54]

Harrison's statement that the Rochelle Salt acts as a solvent which prevents the precipitation of iron that causes yellow stain is a helpful explanation of the chemical activity of the developer.

It is regrettable that Harrison provided no more information on the chemistry of the Kallitype than this, considering his expertise in chemistry and the difficulties with yellow stain that were so prevalent.

In 1892 *The Photo Annual* published a full page advertisement for the Birmingham Photographic Co. The ad provides valuable information about commercial Kallitype II paper sizes, surfaces, and prices.[55] The ad is re-produced in fig. #1., above. In the ad we learn that the Birmingham Photographic Co. marketed bromide papers, gelatine chloride papers, and kallitype, but no albumen silver paper. The Kallitype paper was sold in two surfaces—matt and albumen. Matt surface paper was sold in large sheets, 26 by 20", for 1/3 and albumen surface was sold in 22 by 17" sheets for 1/2. Smaller sizes cost less and the paper came precut for Cartes de Visite and Cabinet size mounts. Free samples were available.

The ad indicates the company also marketed photographic equipment—trays, vignetters, and even an "automatic print rocker" machine. The latter piece of equipment helped to assure thorough chemical reaction in the developer tray and helped to eliminate the yellow stain caused by insufficient development. The ad also informs us that the telegram code for Birmingham Photographic Co. was "Kallitype."

The directions supplied with kallitype paper sold by the Birmingham Photographic Co. are reproduced below in their entirety.

Kallitype No. 2 Birmingham Photo Co.

This process is slowly but surely making its way into public favor and the latest introduction, "Kallitype: No. 2," is aiding the fact considerably. In the first place, it does away altogether with the use of nitrate of silver with its finger-staining propensities, a feature that at once commends it to amateurs. Moreover, it is now more certain in its working, and the results are better. It very closely resembles a good platinotype print. Two kinds of paper are sold, viz., matt surface and albumen. The process has certainly a future before it.

INSTRUCTIONS

PRINTING: Kallitype is not a printing-out process; the image before development is only a faint one. Care must be taken that the paper is quite dry when put out to print, or it will not be so easy to determine the proper exposure, and the resulting print will not be so good a colour.

EXPOSURE: Five to ten minutes in a good light, or from two to three minutes in sunlight, is an average exposure. When the detail in the densest parts of the negative is faintly indicated.

DEVELOPMENT: The prints are immersed, one at a time, in a solution of

No. 1 For Black Tones.

Rochelle Salt	1 oz.
Borax	3/4 oz.
Water	10 oz.
12 to 15 minims of potassium bichromate solution 20 grains to the oz.	

<u>No. 2 For Purple Tones.</u>

Rochelle Salt	2 oz.
Borax	1/4 oz.
Water	10 oz.

10 to 12 minims of potassium bichromate solution 20 grains to the oz.

Be careful to remove any air bubbles that may be formed on immersing the prints in developer; a touch with the finger will easily do this. A considerable number of prints may be developed at the same time, and development will be complete in about thirty minutes. If greater contrast is desired, the addition of one drop of a 10 grain solution of potassium bichromate may be made to each 10 ounces of developer, and more if necessary. Too much bichromate will destroy the half tone. Development must be conducted in a subdued light, and if the developing solution is kept carefully from the light, it will serve to develop a great number of prints. Five or six dozen 1/2 plate prints may be developed with 10 ounces of solution, then it will be well to throw developer away and make fresh. The developing solution serves the double purpose of developing the image and also that of rendering the iron soluble in the ammonia fixing bath, so if it is used too long the prints will be liable to show yellow stains when finished. The developer is very cheap, and it is recommended to use plenty of it, and to keep the prints moving about freely during the whole time they are in the solution. If the prints are removed from the developer too soon, the yellow colour will not entirely disappear in the ammonia baths.

FIXING: The prints are next fixed by immersion for 10 minutes in the fixing bath

<u>Fixing Bath</u>

Strong Ammonia	4 drams
Water	1 quart

NOTE—With albumen prints, use only 8 to 10 drops of bichromate of potassium solution in 10 oz. of developer, and fix in ammonia bath half the strength of that used for the matt paper. To ensure perfect fixation, it will be best to pass them through a second ammonia fixing bath, made in the same way as above—allowing them to remain ten minutes, moving frequently. After washing in several changes of running water for about a quarter of an hour, the prints may be blotted off between clean cloths or blotting paper and dried in the air.[56]

A few comments on these directions are in order. First, no sensitizer formula is given for the obvious reason that the company wishes to sell prepared paper. Secondly, the formulas for developer and fixer are nearly identical to those given by Nicol in patent # 7312. One difference is that the patent expresses quantities in metric terms while the directions express them in terms of familiar "English" measure. Finally, the directions do not include a clearing bath but rather suggest the print be held in a fresh "developing bath for a long enough period to remove the iron that causes yellow stain.

Finally, a table of processing times for each step will make clear the length of the total sequence.

KII processing Times

Step Time

Developing	30 minutes
Fixing (1)	10
Fixing (2)	10
Washing in water	15
Total processing time	65

Hardly a speedy process. It is easy to see how the developing time might be more honored in the breach than in the keeping and the same is true of the second recommended fixing bath. Cutting these times surely must have been responsible for some of the complaints about faulty prints. We have already noted that these processing times were found excessively long by Brown—whose data supported the temptation to use shorter times.

VII.   Kallitype II and "The War of Papers."

This is as good a place as any to suggest the "war of papers" that was going on when K II was placed on the market. In the 1890's there were so many commercial photographic papers for sale that publications made efforts to help readers keep track of them. Annual publications listed papers by manufacturer and type of sensitizer. As a further service to photographers the annuals reprinted directions for processing and formulas for mixing the processing solutions used with the different kinds of papers.

The paper listings of the ***Photography Annual*** provide an insight into the crowded state of the photographic paper market in England in 1892.

PRINT-OUT PAPERS

Sensitised and Albumenised

Brand	Name
Albion	Ridgway
Cartwright	Rivot
Edwards	Scholzig
Fallowfield	Scott
Ford	Self-toned
Hogarth	Sun
Iris	Thomas
Newton	Thorns

Other Papers

Aristotype
Obernetter
Ferro-Prussiate[57]

The following list of papers and the location of their manufacture were described in the *Photography Annual* for 1891.

Agathos	Vevers
Argentic Platinum	Marion
Beta	Hardcastle

Blanchard's Platinum
Celerotype   Scholzig
Direct Platinum   Hardcastle
Jacoby's platinum   Scholzig
Fallowfield's Silver Emul.
Jacoby's Platinum   Scholzig
Jacoby Silver   Scholzig

Sensitized Albumen

Blackfriars
Elliott
Tear
U.S. Co.
Verel
Whelpton Wilten

Sensitized Silk
Marion

Sepia type
Sharp
Thula Platinum
Rudowsky
Transparency
Vevers [58]

These lists of the kinds of print material indicate the quantity and the variety of papers for sale in 1892.

Both lists of papers and manufacturers indicate the market was divided among a profusion of manufacturers, brand names, and types of printing materials. The sources of supply had not yet consolidated and consisted of many companies, most of which were operations of small scale. It should be noted that the above lists do not exhaust all print materials in use. Important types of photographic print papers, such as carbon and gum, are not mentioned, and within the classes, variant papers of each type are not itemized. It is also notable that Kallitype papers are not listed, although their processing was detailed in the formulary section of the publication.

VIII. Kallitype II: Performance In The Market

The editors and reviewers had been very enthusiastic about the revised estimate of KII's success, as indeed, they had been about the success of earlier kallitype I paper. We ask, how did kallitype II do in the war of papers? Let's examine the available evidence.

The *American Annual*, which published a "Record of Photographic Progress" may be of some help in answering our question. The 'Record' provided a valuable year by year appraisal of the relative success different papers experienced in the market along with reports on the relative gain or loss of public acceptance during the previous year. It sums up the sales of print papers for the year 1891, published in 1892, as follows:

> Though the various aristotype papers have increased in popularity throughout the country during the last year, albumen paper is still largely used. There is a certain charm about a good albumen print which is rarely obtained on any other paper. The greater simplicity of working the aristotype paper method commends it to a large number of photographers. Aristotype paper has been much improved during the year, and therefore justifies its increasing popularity. There would be more certainty about its keeping qualities, however, were all emulsion prints toned and fixed in separate solutions. Being in most cases subjected to the action of combined baths, their permanency is doubted by many on account of the conversion of the silver deposit into sulphide of silver.
>
> Bromide printing, except for enlarging, does not seem so popular as some time ago. Enlargements on bromide paper, however, have been very successfully toned to warmer colors by means of uranium nitrate, and large editions are printed on bromide paper by automatic machines.
>
> Platinum process continues to be in favor with many professional photographers of the higher class. Since the cold process was introduced, it has gained in popularity among amateurs.
>
> Kallitypy has made rapid progress since the silver nitrate formerly acting as a developer has been incorporated with the sensitive solution.

Excellent matt-surface papers are being turned out by the various makers of aristotype papers. This paper rivals platinotype in beauty of effect.[59]

The aristotype, reported as coming on strong, was a print-out, gelatin silver-chloride emulsion paper. This paper is still sold today as an inexpensive proofing paper, one often used by beginning photographers to play with. Exposure is slow and produces a reddish image. The print must be fixed in hypo and toned if a pleasing color and permanent image are desired. The response to the Kallitype seems enthusiastic—"rapid progress." One wonders if the report is entirely objective. Could it be boosting the beleaguered Kallitype a bit?

In the report for the following year, 1893, which summarizes photographic sales activity for 1892, the success of the different papers is quite different. The year 1892 appears to have been a year of decision for more than one process. The report begins,

Photo-Chemical Printing has undergone enormous changes during the year. The old orthodox and reliable printing on albumen paper is to a great extent superseded by the chloride of silver emulsion, aristotype, with either gelatin, collodion, or cellodine, as the basis. The popularity these papers enjoy, however, is not in proportion to the results obtained. The uniformity of tones is not at all what could be desired, neither is the permanency of the prints made on these papers an established fact . . . There are still, however, many professionals who prefer the evenness and regularity of the albumen print.

Platinum Printing, is in ascendancy again, but a successful rival to this very much admired process is and ever will be printing on plain or matt-surfaced silver paper.

Kallitypy, although paper of this kind is now manufactured in this country in excellent quality, it has not made signal progress, and the process is practiced only by a few amateurs."[60]

This is hard evidence that, already in 1892, commercial kallitype was experiencing difficulty. Up to this point, published reports had all been optimistic, speaking of a future that was all but assured. This report suggests photographers had chosen processes other than kallitype to replace albumen, whose market share was finally beginning to decline. The evidence indicates the big winner of the year was gelatin silver chloride—aristotype. Behind gelatin silver chloride, the bromide enlarging papers were poised to move ahead. As early as 1893, it was already clear that the successor to albumen was not to be kallitype.

Now the question became one of how the Birmingham Photographic Company would react to the difficulties the paper was experiencing. Two pieces of evidence suggest the response was to pull back, not fight. The first had to do with the *Photo Annual's* handling of the directions it had been printing for Kallitype II.

Strangely, in 1893, the *Photography Annual* did not reprint the Company directions for Kallitype II, but rather printed process data taken directly from the K II patent specification.[58] No explanation was given. The *Annual* also began publication of the KII sensitizer, which it had heretofore avoided printing, in view of the proprietary nature of that information. The decision to publish the sensitizer formula along with the developer and fixer could not help commercial interests, but rather facilitated the individual preparation of the paper. The obvious question arises, why would a company permit the publication of proprietary information, heretofore kept private, which could only threaten the sale of the paper it was in business to sell? One wonders if this is the first sign of the Birmingham Photo Co.'s fading interest in Kallitype paper.

There is some evidence for this speculation. The Company's large advertisement featuring kallitype does not appear in the 1893 *Annual*. Nor did advertisements for kallitype appear in other English photographic publications after 1892. The absence of advertisements, even small ones, suggests that Birmingham Photo may have decided sometime during 1892 or early in 1893 that the Kallitype process wasn't doing well enough to justify further support.

One other happening may have influenced the Company's changing attitude. Part of the future of the kallitype, as manufactured and marketed by the Birmingham Photographic Co. depended on the successful development of a Kallitype III process. But recent indications increasingly suggested that the optimistic hopes for the success of print-out iron silver processes were not likely to be realized. By the end of 1892, it was clear that, if the problems of K III were solvable in a laboratory, it was not likely they were solvable in a way that depended on sales to amateurs. Amateur and professional printing practices involved too many uncontrolled variables. Further, the possibility of a K III paper depended on Nicol's continuing commitment to finding a solution. But about this time Nicol left Mason College and Birmingham, the location of the Company, and moved to the University at Edinburgh. And, from that time until the end of his life Nicol made no further discoverable public statements on the kallitype. It appears that when Nicol moved, sometime after 1893, he cut all ties with the Company and the process.

Whatever the explanation, the Birmingham Photographic Company ceased the manufacture and sale of kallitype paper in late 1893 or early 1894. No document has been found that sets the precise date when the Birmingham Photographic Company ended manufacture and sale of kallitype paper. The company continued to sell other photographic products after its termination of kallitype activity.

IX.   Commercial Sale of Kallitype Ends In England

The only known documentation that sheds light on the discontinuance of the manufacture and sale of kallitype by the Birmingham Photographic Company is a letter published in *Photography* dated January 2, 1896 signed by "Evan." The letter is written in response to an inquiry seeking information about the maker of Kallitype II paper and requests the address where the paper can be purchased. The response reads:

> The original and only maker of this paper, under Nicol's patent, was the firm trading as the "Birmingham Photographic Co.," of Criterion Works, Great Charles St., Birmingham. It is not, I believe, now manufactured by them, or by any other firm. Some of the large dealers may still have some . . . I saw some mentioned in a second hand list of Marion's last winter.[61]

The letter suggests the normal sale of kallitype II paper ceased before the end of the winter of 1895, when some paper was seen being remaindered. This does not deny the possibility that the manufacture could have ceased a year or two before. No notice by the Company of the termination appeared in the journals. Neither did Nicol or Lewis make a public statement to their photographic clientele about their reasons for discontinuing business. We have only a letter, written by an amateur to a magazine. However, in the letter that informs us of the termination of commercial kallitype, we have an indication of the future. It is notable that the letter indicates the possibility of further life for the process. Evan's response concludes

> If the querist is desirous of employing this most beautiful process, I may recommend him to prepare some paper himself. This is by no means difficult. [62]

The letter ends with a formula for mixing the K II sensitizer.

X.   Commercial Kallitype II Paper in America

Earlier it was mentioned that Kallitype paper was made and sold in the United States. The first notice of American-made Kallitype II paper occurred in an article in *The Photographic Times,* published on July 21, 1893. It announced a kallitype II paper, "improved by Mr. James H. Stebbins." The article was written by W. G. Oppenheim. The *Times* editor found the writing so full of errors that he provided corrections in footnotes. Nevertheless, the article contains some useful information on the American paper. Oppenheim provides descriptions of Stebbins' developer and fixer, which, when examined, prove to be the same as those published by the Birmingham Photo Company for their paper. This leaves little doubt about their origin. Oppenheim does not describe Stebbins' sensitizer.

In an article obviously trying to promote the American product, Oppenheim directed most of his attention to describing the advantages of the kallitype over the platinum print. It is clear he believed platinum printers were a likely group for conversion to the iron silver process. Oppenheim noted the kallitype's ability to secure a variety of tones, the cheapness of the paper and developer, and a few other "advantages" of doubtful value. For example, he comments that "overexposed kallitype prints may be reduced by immersion in a dilute bath of hydrochloric acid" and that development of the print can be "stopped by plunging the partially developed print into water." (Both of these techniques are impractical.) Oppenheim claimed one advantage that can be believed: cost. He reports the price of Stebbins' kallitype paper is 36 cents a sheet (presumably 26" by 20") compared to the cost for a similar sheet of platinotype, which is 95 cents.[63] Oppenheim writes "the object of the present article is to call attention to the kallitype paper, now improved by Mr. James H. Stebbins, Jr., of this city" (New York). The word "improved" suggests the paper being sold is K II. The article did not make clear whether Stebbins, a practicing chemist, sold his own improved paper or whether he was a licensee of the Birmingham Photo Co and sold their paper.

Peter Duchochois, another chemist-photographer from New York was also involved in kallitypy. With an eye for selling some kallitype paper himself, he questioned whether Nicol's patent applied in the United States.[64] J. C. Strauss wrote the "patent was granted to him [Nicol] in the British Empire," implying the patent did not apply in America.[65]

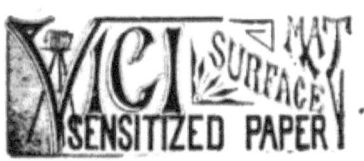

Sample Package 4 x 5 sent upon receipt of 25 Cents.

. PORTRAITS OR VIEWS,

AMATEUR OR PROFESSIONAL.

It is also now being largely used for line drawing as per the following :

CHATTANOGA, TENN., June 2, 1893.

MR. C. E. HOPKINS :

Heretofore pen drawings for Photo Process reproduction have usually been made on salted paper sensitized with a solution of nitrate of silver. The present advanced state of the art of Photo Engraving demands that the methods employed should be economical, speedy and sure. A trial of Vici Paper, a ready sensitized paper, manufactured by C. E. Hopkins, Brooklyn, N. Y , for the above purpose, was so satisfactory that I now use it exclusively in the following manner :

An exposure of from one to five minutes by electric lamp or sunlight, according to the density of the negative, until the shadows are faintly visible. Develop according to the directions which accompany the paper, wash and paste the print to a piece of glass. When dry, cover main lines with ink as when using a plain silver paper, using a waterproof ink. When the drawing is completed, bleach with a solution of bichloride of mercury; wash under the tap and dry. If a slight yellowness remains after bleaching, it can be removed in a bath of weak nitric acid. The drawings can be easily removed from the glass by soaking them in water.

R. E. M. SUVERKROP,

Artist and Engraver for the " Times."

C. E. HOPKINS,

Figure 2. Vici Paper ad.

A second kallitype II paper manufactured in America, "VICI" was also sold in New York between 1893 and 1896 as a paper primarily useful for "graphic arts, portraiture, and landscape photography." The C. E. Hopkins Company advertised the paper for "drawing from photographs"—a process that involved

1. making a photographic print, for example, a landscape, then
2. drawing the image on the photographic print with insoluble ink, and

3. next, bleaching away the photographic image so that a drawing
    on plain matt paper remained.
4. finally, transferring the drawing to a printing plate.

Vici paper and process were used in the graphic arts industry to
quickly generate a drawn image for line reproduction. The value of
kallitype paper for this process was that there was no gelatin or albumen
medium on the surface of the paper to make difficulties for the ink pen.
The developer sold by the Hopkins Co. appears identical to that sold by
the Birmingham Photo Co. The Hopkins' ads mention that a platinum
toner was available.[66] The toner presumably "dressed up" kallitype
prints and made them visually indistinguishable from platinotypes. By
making simulated platinum prints, the studio operator could save about
two thirds the cost of platinum paper.

Another Kallitype II paper sold in America was named "Celerite
Paper." This ambiguous name was a play on 1) the speed of development
of Kallitype II paper, which is instantaneous, and 2) on the anticipated
sales performance, (sell right). Celerite paper was manufactured and sold
in Chicago. *The Photo Beacon* for 1895 reported that the first owners of
the company, who began operations in July, 1894, were dilatory in their
operations and as a result could not pay their advertising bills, much less
meet the demands of their clients. The editor complained

> . . . We were considerably annoyed by the laxity of the
> management in attending to mail orders. Our subscribers
> complained very strongly to us, knowing that we should put our
> own end of the matter right by stopping advertisement—which
> we at once did. Now that a responsible firm has undertaken the
> manufacture of the paper, there should be no more trouble. The
> price is placed at about thirty per cent lower than formerly. [67]

In July 1894 Celerite paper was taken over by James H. Smith and
Co., which was located at 261 Wabash Ave, Chicago. The company sold
a variety of photographic chemicals and accessories in addition to the
paper. (Incidentally, this company continues in business today under the
name of Smith Victor Co., a well known name in studio lighting, etc.)
The announcement of Smith's K II paper follows.

> We noticed the fact that (Smith and Co.) were beginning
> to manufacture this paper as long ago as July 1894, but as

they wished to place on the market a perfect article, they have refrained from putting it on sale. The sample prints they send us show the wisdom of this decision. In color they run from dark, velvety blacks to a very warm sepia, with all the gradations of color between. The whites are pure and the shadows luminous. We have no hesitation in recommending this paper as especially suited for landscape and big portrait work.[67]

No details about processing appeared with this announcement.

Another commercial kallitype II paper entered the American market in 1895. The paper, called Mirrotype" KII paper," was described in detail in the February issue of *The Photo Beacon* of that year. The name presumably suggested the accuracy with which the paper reproduced detail. According to the writer, the unique feature of the paper was a "new" approach to contrast control. "This paper was so prepared that the use of bichromate of potassium was wholly unnecessary when printing from negatives of average density." The writer considered the change "a valuable improvement" over older varieties of kallitype, as it eliminated "the greatest cause of uncertainty in the printing process." When printing "very thin negatives, bichromate must still be employed in a small amount," however.

Mirrotype Paper was prepared and sold in two varieties, black and sepia. The prints were developed in a solution of Rochelle salt and borax, the proportion of borax determining the color of the print. The fixer employed continued to be Nicol's weak solution of ammonia in water.

Mirrotype paper achieved proper contrast by adding a certain amount of dichromate to the paper sensitizer during manufacture, thus freeing the photographer from that responsibility. The manufacturer added enough dichromate so the paper would make a correct print from an average negative. Mirrotype management apparently believed the addition of dichromate to the developer to achieve proper contrast was a step too tricky for the amateur to control. However, the photographer still had to add some dichromate to the developer whenever his negatives were short of proper contrast.

The announcement and description of Mirrotype was contained in a report of a demonstration that is notable on two counts: the audience receiving the demonstration and the man giving it. The audience was the Chicago Society of Amateur Photographers,—*not* the Chicago Association of Professional Photographers. The officers of the amateur society are recorded and examination of their identities indicates what was true of many members of that time. The club membership was made of upper middle-class people, many of them professionals—doctors, lawyers, and businessmen. [69] Examination of the social level of the members of photo clubs at that time reveals that amateur photography club members were generally successful people. This was the group that kallitype II entrepreneurs hoped to win over.

The demonstrator—*Photo Beacon* began a new column headed "the Demonstrator" in the 1895 Issue, which reported new appointments, problems, and funny stories of the new profession of photographic sales representatives[70]—was Professor N. Gray Bartlett. We are not told where Bartlett taught, but it is obvious from the detailed transcript of his talk provided by the secretary of the club that he was exceptionally well versed in the history of photographic process and photochemistry in particular, especially the processes dealing with iron-silver and iron-platinum. Bartlett's discussion of the history of kallitype and the chemistry of the kallitype was accompanied by test tube demonstrations of all the major chemical reactions of the process. His discussion of the chemistry of the kallitype remains outstanding to this day for clear presentation of the basic chemical facts of the process. They are worthy to be presented here.

Bartlett's demonstration proceeded along the lines of a chemistry lecture. With test tubes containing prepared chemicals, Bartlett demonstrated the reduction of ferric compounds to ferrous by the action of light, drawing attention to the slight change in color that occurs when the sensitizer is exposed to light. He demonstrated the need for the presence of a solvent of ferrous oxalate. Without the solvent, the ferrous oxalate produced no reduction of silver salts to silver metal. Bartlett showed nothing happened—i.e, no silver metal was produced or deposited—when a solution of silver nitrate was added to a test tube containing the ferrous oxalate produced by the action of light. When Rochelle salt, a solvent of ferrous oxalate was added to the test tube containing ferrous oxalate and silver nitrate, an immediate precipitate of black silver metal formed in the test tube. This precipitation of silver metal is what occurs on a Kallitype II print, when the exposed print is placed in the developer.

Bartlett also demonstrated the action of the contrast control agent, potassium dichromate, by adding some of it to the developer in a test tube. The dichromate acted as a restrainer and partially inhibited the formation of the black precipitate of silver.

After he completed the chemical demonstrations, Bartlett gave a hands on demonstration of the manipulation of Kallitype I, II, and III printmaking with detailed explanations of each step. He exhibited exposed prints that were not yet developed so members could form an idea of the extent to which print exposure should be carried. Subsequently, he developed the prints. Finally, he submitted for the examination of the audience a number of prints of various colors and contrasts on papers of various weights and textured surfaces.

In his general remarks on K II, Bartlett said "the kallitype process had not received the recognition it deserved, judging from the few current references made to it in the English and American photographic journals and annuals." He believed that against all its advantages of cost, long shelf life, permanence, freedom from gloss and yellowing, and its variety of colors, "one disadvantage has been brought forward by some experimenters, namely, that there is difficulty in determining the exact extent to which the image should be printed out, and that success or failure is dependent thereon." In opposition to this view, he argued,

> It should not be forgotten ... that in albumen and aristotype papers the printing is carried to an indefinite point, determined by individual experience, but is, in all cases, several shades darker than it is desired to have the finished prints, while the toning which must follow is notoriously uncertain in its results.[71]

The report of Bartlett's demonstration of Kallitype II was exceptionally well written (by the secretary of the Chicago Society of Amateur Photographers, Chicago,) and was highly informative. Interestingly, it was entitled, "A Neglected Printing Process." The title aptly describes Mirrotype Paper. In the periodical literature of the time the paper's name never appeared again—neither in letters to the editors, nor in commercial announcements, advertisements or articles. Apparently more than a superb demonstration was necessary to boost kallitype in an indifferent market.

Yet another Kallitype II paper was offered for sale in America. This paper was manufactured in the far western town of San Francisco, California and was marketed at least until 1903 by Nelson C. Hawks. The paper was called "Polychrome," after the variety of image colors it could produce by varied development. Ads for Polychrome appeared with some regularity in *PHOTO-MINIATURE,* an American magazine popular with amateurs interested in experimentation.[72]

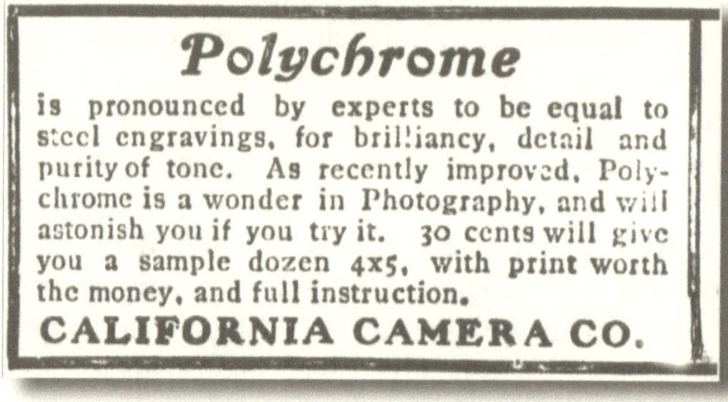

Figure 3. Polychrome ad.

An article in **Camera Craft**, by Sigismund Bluman told the story of Hawks' development of Polychrome. Bluman, as we shall see, was a skilled writer and a committed kallitypist who wrote frequently on the process.

We learn from Bluman that Hawks retired from a type foundry business in 1894. He became interested in photography, first as an amateur, and later, as an entrepreneur of a small photographic supply business. His interest in the kallitype came about quite accidentally thru a chance meeting with a Frenchman, Armand Rubin, a retired Parisian dyer. Rubin was an enthusiastic amateur photographer with a professional knowledge of photographic chemicals and was experienced in the use of iron salts, which were used in the French cloth-dying industry in which he worked. Bluman reports

> Armand Rubin had an excellent basic knowledge of Kalitype [sic] and practical experience as well. "This fine gentleman," says Mr. Hawks, "had many new and not a few original formulae. On our first and many subsequent meetings we talked them over. Only after several months did I succeed in getting him to come to my home. That day he wrote out for me all his formulae and left late in the most cheerful frame of mind. No one ever saw him again; and, beside myself, no one seems to have made any inquiries. [73]

Hawks continued his experiments with the data Rubin gave him. He had not been trained as a chemist with the result his "tryouts were often wanderings far from the beaten tracks." He "put in any old salt that was handy and got all sorts of results, mostly mediocre." After wasting a great deal of time, "he got down to trying for the fewest ingredients and the simplest methods, eventually arriving at [his own] way of doing the thing."

Hawks' formulas for the solutions used in making Polychrome paper follow.

Hawks' K II Polychrome Sensitizer

Water	4 oz.	118 cc.
Ferric oxalate	400 grains	26.5 grams
Potas Oxalate	100 grains	6.5 grams
Nitrate of Silver	100 grains	6.5 grams

The amount of oxalate of potassium and silver nitrate may be reduced to ninety grains each, but like amounts must be used.[74]

Hawks' sensitizer bears comparison with Nicol's formula for K III paper. In Nicol's patented formula, the proportion of ferric oxalate to potassium oxalate to silver nitrate is 5:1:1. Hawks' ratios are 4:1:1. Hawks apparently stumbled upon a K II formula almost the same as Nicol's K III. Hawks sold his paper as a K II paper that required development. His developing solution follows.

Hawks' K II Polychrome Developing Solution

Water, hot	8 oz.
Borax, powder	1 oz.
Sodium Tartrate	1.25 oz.

The tartrate is added when the borax solution cools.
Rochelle salt [Sodium Potassium Tartrate] may be substituted.[75]

Bluman reports this developer "gives a rich velvety black." He adds that the use of less borax produces a warmer print. Hawks fixed his prints in water to which he added strong ammonia. He ended processing with a plain water wash.

Hawks was 75 years old when he related his reminiscences to Bluman. Hawks spoke with warm nostalgia of the days when he traveled the country "demonstrating" Polychrome paper. But, he added, "for all the adventure and experience, there was little profit in it."

Hawks' "Polychrome" paper was said to have a longer commercial life than most kallitype II papers, if not the longest life of all—about five years. Perhaps Hawks' venture lasted that long because he was content with a return of "experience and adventure" instead of financial profit. Interestingly, Hawks' 4:1:1 formula became a classic in the literature on individually prepared kallitype II, as we shall see in the next chapter.

The photographic literature reports the sale of a few other kallitype II papers. M. H. Wilde, writing in *Photography* in 1899, noted that "Verotype paper" has been offered for sale, but offered no details.[76] John Tennant, in the February 1903 issue of *PHOTO-MINIATURE* provided a short list. He reported that "Sensitol, Soline, and Platinograph were all modifications of Kallitype."[77] He does not report where these papers were sold, America or England. The list of kallitype processes seems to grow without end, and success appears to depend on how lucky one is at finding published reports.

We shall end this account with a description of a commercial KII paper made in Belgium in 1899. The *British Journal of Photography* reported a Mr. Van Loo, introduced a new photographic paper at a meeting of the Association Belge called "Simili Platine." This paper, according to the *BJP* was "a modification of Nicol's old Kallitype" and was one of several recently offered substitutes for platinum paper. Platinum substitutes proliferated in the first quarter of the twentieth century as platinum became increasingly expensive. Entrepreneurs sold less expensive kallitype paper and recommended the prints be toned with platinum salts, which, if properly done, provided a paper almost indistinguishable from real platinum paper. Van Loo's sensitizer formula is:

Simili Platine Kallitype Sensitizer

Ferric Oxalate	15 grams
Oxalic acid	3 grams
Silver Nitrate	3 grams
Water	100 cc.

After the sensitizer dries, the paper is exposed "like platinum, till the image is distinctly printed out." It is then developed by immersion in

Simili Platine Developer

Borax	60 gram
Sodium tartrate	60 gram
Water	1000 ccs

A few drops of a 5% solution of potassium dichromate are added to the developer for contrast control. After development for five or six minutes, the print is well washed in running water. Then it is toned in

Platinum Toner for Simili Platine Paper
Potassium Chloroplatinite	1 gram
Citric Acid	10 gram
Salt	10 gram
Water	1000 ccs.

The print is left in the toner until the desired tone is obtained.

Finally, the toned print is fixed for five minutes in a 2 percent solution of ammonia (.880) to rid the paper of unused silver salts.[78]

The quantities in Van Loo's formulas, with the exception of the toner, bear a striking resemblance to Nicol's patented formulas. The platinum toner is one commonly found in formularies published at that time.

This completes our survey of commercial kallitype activity in America and elsewhere. Our study supports an observation made about kallitype II paper by "Wingrod" in the January 2nd, 1895 issue of **Photography**. Wingrod observed that Kallitype II paper "is much more used in America than in this country [England]." [79]

Our search has revealed a number of Americans who tried to commercialize the K II process. Few managed to create enduring, profitable enterprises. In some cases the business ventures were only sketchy conceptions of what might be done, and the business hardly got off the ground. In others the business did not prosper, but had lasted for up to five years. From reading between the lines, one is inclined to conclude that no entrepreneur became rich selling kallitype paper. One cannot help but wonder at the charm of a photographic paper that was so successful in attracting investors, one after another, but which delivered so little in the way of profit.

At this point the account of commercial exploitation of Kallitype II paper ends—unless suppliers who currently market kits and old process materiel shall be included. For the record, the Photographers Formulary Co. of Condon Montana supplies kallitype print materials in kits to the amateur market as this is being written in 2012.[80] For other sources google Alternative Photographic Processes.

XI.   Early Explanations of "The Wane of Favor."

Readers properly expect from the historian some kind of accounting, or explanation of what happened. In the case of the commercial kallitype I and II three questions arise. The first can be phrased as follows: why did the kallitype fail as a commercial product when it had so many attractive photographic qualities and when so many important writers predicted "a future for it." The second question can be stated: was the failure a result of inherent flaws in the process (was the paper "flawed" or "no good"?) or was the failure caused by something else, for example, poor management or insufficient investment by the business owners? We ask a final question, did the photographers who watched the progress of the kallitype offer any explanations of its passing?

We can answer the last question in the affirmative, the photographers did, and we can use their words to provide some answers to the first three questions. We begin with Mr. W. H. Smith, an "expert of long standing in the iron-silver process." Smith provides the first explanation why the kallitype failed in the paper market. In remarks on a talk to the Croydon Camera Club in May, 1916, Smith traced the failure of the kallitype to its variability.

> Commercially, kallitype has never achieved success, probably in large measure due to the fact that print uniformity is most difficult to secure; indeed, the demonstration and prints on the walls aptly illustrates the gulf which may exist between a charming home-prepared printing process and the possibilities of putting it on a commercial basis . . . Compared with Platinotype and Satista Papers, [the latter was a kallitype process that produced a platinotype look-alike print] a far wider range of color is available with kallitype, but it is difficult, if not impossible to obtain uniform prints, or a pure neutral black. If a batch of, say 50 prints is developed in one lot of developer, the first and final print will show a marked difference, as the constitution of the developer alters with each print passed through it . . . Small differences are of little moment to the amateur, but otherwise to the professional. [81]

In a note appended to the article, The editor of *BJP* agreed with Smith's attribution of commercial failure to print variability.

John Martin Hammond, an American expert in old photographic processes, gave another explanation of the kallitype's failure. Hammond wrote an excellent series of essays for **Photo Beacon** on how to work various photographic processes which included the kallitype. He also published informative articles on how to work kallitype in several other publications including the **BJP**. Hammond believed the failure of kallitype had to do with

> "the wane of favor which all sun-printing papers have suffered in recent years." [82]

Hammond's explanation points to the rise of a new method of making prints by the use of enlargers with fast bromide papers. He suggests that a new printing technology superseded the older method of contact printing. The kallitype, like its competitor silver-albumen paper, could not be used with enlargers because it was too slow. Contact printing depended on sunlight, which everyone knows is a variable and undependable light source. Contact printing papers required exposures to sunlight of 5 minutes or more. The new enlarger printing papers required only seconds of controlled artificial light. The kallitype paper was rendered too slow by the new printing technology.

J. C. Strauss in 1895 found a different cause: the "process became, through all the improvements, complicated and confusing, so that few operators, even though careful, could use it with success."[83] Strauss was no mere complaining amateur. He had been both a photographic entrepreneur and a photo-educator in Germany before coming to America. His explanation suggests the many changes of kallitype process in a short period of time created more confusion and difficulties for the workers than they cared to put up with. Finally, James Thomson, the most published writer on kallitype in the 20th Century offered the following explanation of the kallitype's failure in the market.

> . . . Kallitype has heretofore never been much of a commercial success, and what effort has been made has been usually under a name other than its own. In any event, as a process it has never been taken up by any of the big interests, which of course today is necessary to make any mode of printing a success. Under various fanciful names, I have recognized the process and I have an idea that present effort in that line

is being made in a small way, but the product is not exploited
under the legitimate name. Some years ago, a paper was on
the market having a name suggesting a platinum base. It was
really kallitype and was given lasting qualities by gold plating
the image in the combined bath. When one counted the cost of
the paper and added the expense for borax, Rochelle salt, and
finally, gold chloride, it dawned upon the consciousness that it
would have been better to buy platinotype in the first place.[84]

To summarize, early interpretations of the failure of commercial
kallitype paper attributed the failure in the market to the near impossibility
of achieving uniform results, more efficiency and economy of processing
by competing papers, excessive change in the product and consequent
confusion about the process, and the lack of development and promotion
by a major photo marketing manufacturer.

XII.   The Lost War of Process: Commentary

We have reached the point when the the kallitype ceases to be a
commercial process—when both kallitype I and II have clearly lost "the
war of process."

Looking back on the three chapters just completed, we feel compelled
to comment on the fate of the process from the viewpoint of the present.
We believe that there are two levels on which the failure of the process
may be considered, which for want of better terms we shall call the
process view and the managerial view. The process view considers
physical and chemical considerations and values such as efficiency,
reliability, and cost. In these matters, Nicol and Lewis failed to play
their parts as well as they might and as a result the kallitype missed an
opportunity for a short term victory.

The limitations Nicol imposed on the kallitype, if they can be called
that, have to do with the limits he imposed on the process by 1. the
lateness of the time when he invented it, the rather inept way he reported
his invention, and his decision at a crucial moment to abandon the
process. As reported earlier, Nicol's two commercial kallitype processes,
as processes in competition with other processes, were deficient in
simplicity, reliability, speed, and control. Nicol developed a charming
process that had so much flexibility that studio photographers found it
was impossible to control. The three processes Nicol developed were

chemically innovative, powerful, and charming, considering the short time from conception to marketing. How technically proficient they were can be measured by the universal respect he earned from the well reputed photo-chemists who wrote approving critiques on the processes. But Nicol's processes were unable to compete for long with the vastly more simple, reliable, and controllable processes that evolved beyond direct silver paper. One specific mistake Nicol made was to release the paper for marketing before he had full understanding and control of it. Another was to release three processes in little more than three years. Finally, Nicol may have overcommitted himself to the K III print out process, a process that on the drawing board appeared to solve all the problems of a print-out approach to kallitypy, but which had a history of inconsistency and uncontrollability in printing rooms. Perhaps the final and greatest limit Nicol imposed on the kallitype came when he abandoned the process before its problems were fully solved.

Lewis, the businessman who ran the Birmingham Photographic Co. which marketed the kallitype, also contributed to the failure of the kallitype. He apparently scheduled the too-early marketing date for the kallitype I, which led to the early disaster of defective material being sold. Lewis also can be faulted for his premature announcement of the future delivery of K III paper, the wonder material, before KII had a chance to prove itself. The early announcement of a superior product, K III, soon to be released, in all likelihood undercut the healthy development of K II, which was, and is, a fine if difficult to control, printing process. I believe Lewis, as the marketing director, was responsible for much of the confusion that existed in the public mind about how the process worked. As the marketing manager, he should have paced product change and assured an acceptable level of public information. Reading the literature one senses too much confusion about the different processes and their many solutions. Clearly more advertising and public information were necessary to achieve acceptable levels of customer information, success, and good will.

The above difficulties attributable to Nicol and Lewis help to explain why the performance of kallitype II in the market was mediocre. But there were other factors. The unfamiliar technique of the kallitype process—the problem of the uncertain amount of exposure, the unfamiliar chemistry, the prolonged processing sequence, print variability, the persistent yellow stain, and the brown fingers—all of these process problems created resistance which along with the very poor marketing decisions limited the progress of the process. These difficulties proved too much for a successful outcome in the competition with the more progressive, faster printing and quicker processing bromide silver papers.

To summarize, from the vantage point of history, the iron silver process was born too late and was developed too slowly. Had extensive research and development of iron silver methods started in 1840 when a prototype process was first announced, and had the development of the process continued throughout mid-century, the kallitype may have had a chance to establish itself before silver-albumen and its successors achieved a stranglehold on the market. The kallitype may then have played a greater part during the middle years of the century as a competitor of silver-albumen, when 'sun printing' was the universally accepted method of printing. But the invention and marketing of kallitype in 1890 left it too little time to develop before photography changed radically to the more efficient methods of taking and printing pictures that would dominate twentieth century practice. Given more favorable product development and management, and an earlier arrival in the market, Kallitype might have played a more successful part on the commercial stage. But it would have been a short part in any case, one destined to end when photography entered its modern, highly technological phase. There was a moment in time when kalliltype papers might have penetrated the market. But that moment was in 1840 or 50, not in 1890, when contact printing methods had time to grow, before being replaced by the more efficient method of printing with enlargers. Even had Nicol and Lewis grasped the opportunity in 1890, and responded to it with consummate technical and commercial skill, the progress of kallitype would have been closed off quickly by products with superior reliability and convenience which were to evolve into today's direct-silver processes.

By1890, albumen paper had been the market leader for a quarter century. Its limitations had encouraged the development of two modern papers—gelatin silver chloride and gelatin silver bromide. These two papers had great capacity for improvement—and through inspired and prolonged research and development, they eventually commanded the black and white paper market, with silver chloride and silver bromide enlarging papers emerging as the clear leaders. Both papers in time became fast and simple to use and incredibly reliable. They were made so by great research and engineering, great investment and inspired marketing. But the direct silver process from which they evolved had the right potential for such development. And the men who guided the development were ingenious, resourceful, funded, and persistent.

Perhaps the judgment of history should be that the lovely kallitype process, with its infinite capacity for delicacy and variation, may not have lost the war of papers, but that other methods won it. The other methods had some particularly strong advantages working for them and those advantages increased with time. In 1890, for those who wanted a print-out paper, the silver chloride papers carried the day with their simpler, quicker, cheaper, more reliable and familiar processing sequence. For those who wanted elegance, there was the permanent grace and extended scale of tones of the platinum print. The new, fast bromide-silver papers, while, not as charming as kallitype, met a new demand for reliable, fast, enlarged prints made from small negatives. The contest was over by the turn of the century.

I reproduce some of the advertising logos of papers that played their short moment on the stage of photographic paper making. There were many photographic paper products put out for sale. These are but a few of those offered to eager amateurs. They have a certain appeal in their design and symbolism.

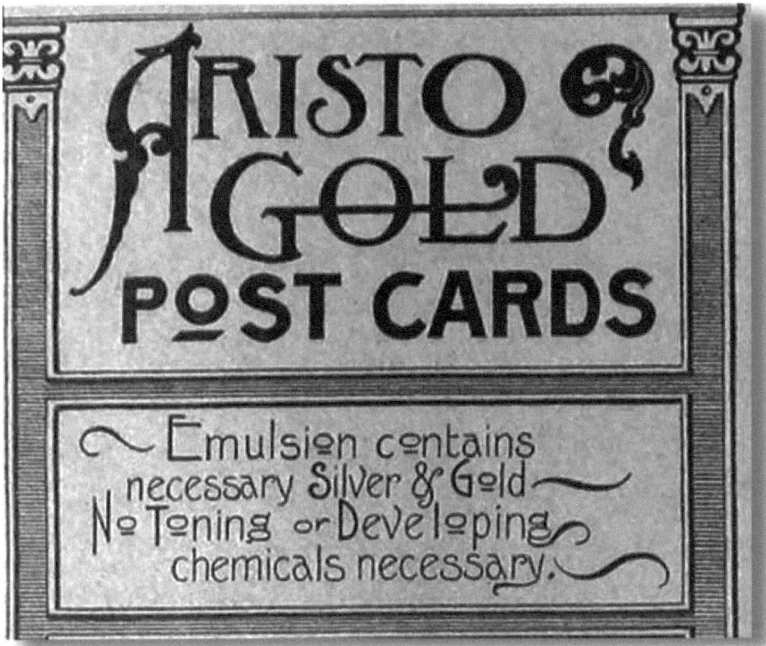

Figure 4. Aristo Gold Post card ad. Mailing photographic post cards was something of an amateur fad during these times. Note the implication of ease of use.

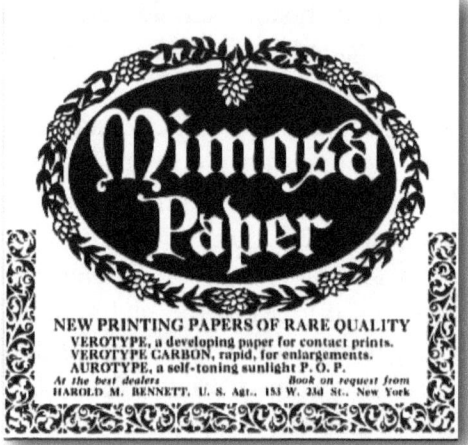

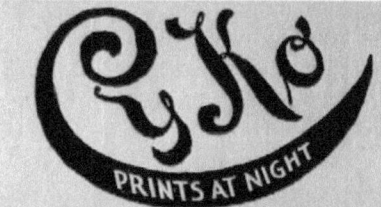

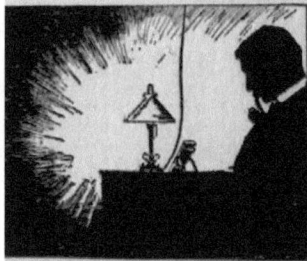

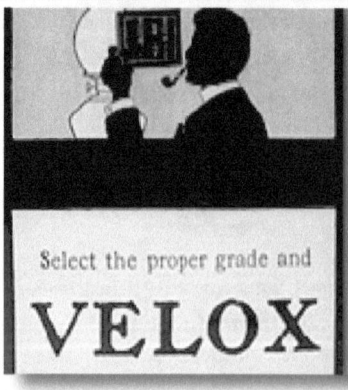

Figure 5. Mimosa, Cyko, Velox, Eastman paper ads.

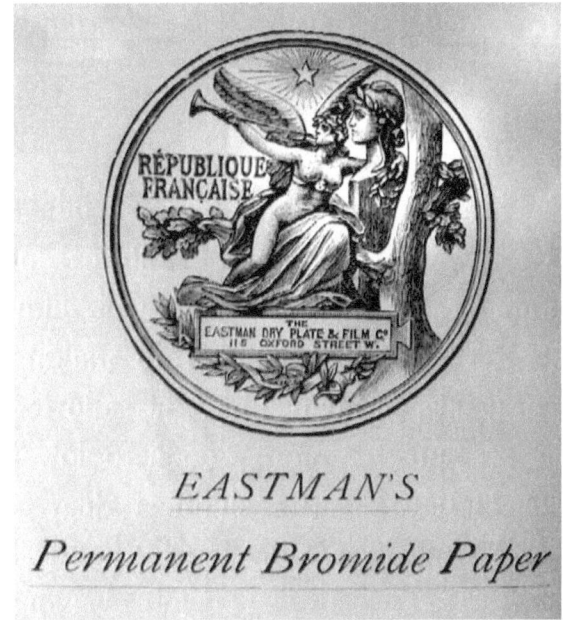

By the mid-point of the twentieth century, the period of miniature cameras, mass marketing and consumer convenience had arrived. Inexpensive handheld cameras and convenient roll films grew astronomically popular. The development of "automatic" cameras and miniature films were soon followed by machine processing. Then the old printing-out approach—an approach that required large expensive negatives made with bulky cameras was superseded by the simpler, more convenient technology which inexorably changed photographic interests and methods.

But did all this change signal the end of the kallitype?

Maybe not! As we shall see in the next chapters, the kallitype developed a second life. Cut adrift from commercial management, the process found a new life among amateurs. The desire for profit and success that drove the commercial process was replaced by new creative energy borne of amateur love of inquiry, experiment, and shared knowledge. The next chapter will continue the narrative of the kallitype under a new standard bearer. The reader is invited to discover if the process fares any better under the leadership of amateurs.

Notes: Chapter III. Kallitype II From Patent through Sale

1. W. W. J. Nicol, "The Kallitype"—To the Editor, *British Journal of Photography* July 24, 1891, p. 479. The letter was dated June 14, 1891.
2. Ibid.
3. Ibid.
4. Ibid.
5. James H. Stebbins Jr., "The Kallitype Process," **Photo American Review,** May 1891, p. 40.
6. *British Patent # 7312*, London, Printed for her Majesty's Stationery Office by Darling and Son Ltd, 1892. Reproduced in *British Patents*, p.2., New York Public Library.
7. Editor, **BJP**, March 4, 1892, p. 156.
8. Birmingham Photographic Co., Ltd., "Kallitype, A Correction," **Photography**, June 4, 1891, p. 365.
9. *BJP*, July 24, 1891, n.p.
10. G. E. Brown, "Advances in Kallitype Printing," *BJP,* April 1, 1892
11. "Kallitype," *BJP*, July 3, pp. 419-420.
12. *British Patent # 7312,* p. 2.
13. G. E. Brown, *BJP,* April 1, 1892, p. 210
14. Josef Maria Eder, *History of Photography*, Dover Publications Inc., New York, N.Y., 1972, tranlator, Edward Epstean, p. 545.

    "Capt. Pizzighelli . . . made the first experiments with the 'direct platinum printing out process' without developing and . . . he sent the first successful 'direct platinum prints' to the author in 1887."

15. William Willis, **BJP**, Jan 15, 1892, p. 38. speaking about platinum printing-out paper: "many difficulties make it almost impossible to get a perfect print of this kind."
16. *Photography*, July 30, 1891. p. 481.
17. Sigismund Bluman, "A Patriarch in Kallitype," *Camera Craft*, 1916, pp. 263-70.
18. Will. J. Brooke, "Kallitype Printing," *Photography,* Dec. 5, 1895, p. 779.
19. O. P. Bennett, M. D., "Kalitype," [sic], *Photo Beacon*, July 1895, p. 226.
20. **Patent #7312**, p. 2.

21. Editor, *Photographic Annual*, 1892, p. 430
    "Direct Platinum" and "Jacoby's Platinum."
22. John Tennant, *PHOTO-MINIATURE*, vol. X, no. 115; May 1911, p. 314.
23. The earliest mention occurs in *Photography*, June 4, 1891, p. 365. "We shall in a few days place upon the market a greatly improved paper."
24. *Photography*; July 30, 1891, p. 481.
25. *Anthony's Annual*, 1891, p. 585.
26. C. H. Bothamley, "Recent Developments in Printing Processes, *Scientific American Supplement*, #815, publ., Aug. 15, 1891, p.1302.
27. **British Patent # 2011**, June 5, 1873; #2800, July,12 1878; #1117, March 15,1880. All Platinum Printing patents.
28. Bothamley, *Scientific American Supplement* #815, ibid.
29. Julius Schnauss, "On Kallitype," **American Annual Photography**, Vol. 9, 1895, pp. 186-189.
30. Bothamley, Ibid.
31. Ibid.
32. Talbot Archer, *Anthony's American Annual*, 1891, p. 583. Archer has obviously confused the K II and K III processes. The K II (not K III) was developed in Rochelle Salt etc.
33. Ibid.
34. "Answers" New Kallitype, *Photography,* Aug 6, 1891, p. 506.
35. Ibid.
36. *The Amateur Photographer* January 29, 1892, p. 88.
37. Ibid. *Amateur Photographer* February, 5, 1892, p. 107.
38. Ibid.
39. *Anthony's Annual*, 1892, p. 100.
40. *Photography Annual*, 1892, p. 103.
41. *BJP,* January 8, 1892, p. 20.
42. Ibid., January 15, 1892, p. 38
43. Ibid., January 22, 1892, p. 52.
44. Ibid., April 1, 1892, p. 210.
45. Ibid., June 10, 1892, p. 374.
46. L. P. Clerc, **Photography, *Theory and Practice***, Pittman and Sons, Bath, 1930, p. 387.
47. *Photographic Archiv*, Nr. 8, July 1892. These remarks can be translated:

The Kallitype is an iron-printing method worked out by the Englishman James Nicol in which the paper is prepared with a mixture of oxalic acid, iron oxalate, and silver nitrate. After exposure, the print is developed in a bath of potassium salt and borax. It is fixed in a weak ammonia solution. An exact dilution is found in **Photo Archive**, 1892, # I. Kallitype paper—What is Kallitype paper and how is it best prepared by the amateur?

My thanks to Margot Otter of South Bend, Indiana for this translation.

48. W. Jerome Harrison, *The Chemistry of Photography*, Scovill Adams Co., N.Y., N.Y., 1892, p;. 266-71.
49. "Our Editorial Table," *BJP,* March 4, 1892, p. 156.
50. W. J. Harrison, Ibid., p. 269.
51. Ibid. p. 269. This equation may not be entirely acceptable according to current descriptive practices.
52. Ibid. p. 270.
53. Ibid., p. 270.
54. Ibid., p. 271.
55. *Photo Annual*, London, 1892, p. cliv.
56. Ibid., p. 429-430.
57. Ibid., p. 430.
58. Ibid., 1891, p. 412.
59. *American Annual of Photography and Photographic Times Almanac*, Vol. 7, 1893, p. 264.
60. Ibid., Vol. 8, 1894.
61. "Replies," *Photography*, Jan. 2, 1895, p.16.
62. Ibid.
63. Wm. Geo Oppenheim, "Kallitype Paper Vs. Platinotype Paper," *The Photographic Times*, 1893, p. 387-8.
64. P. C. Duchochois, "Kallitype No. 2," *Photographic Times*, May 6, 1892, p. 240.
65. J. C. Schnauss, On Kallitype, **American Annual of Photography**. Vol. 9, 1895, p. 186ff.
66. Advertisement, *American Annual of Photography and Pho graphic Times Almanac*, 1896, p. 139.
67. Editor, "Printing Room," *Photo Beacon*, July 1894, p. 252.
68. Ibid., March 1895, p. 104.

69. *The American Annual of Photography and Photographic Times Almanac* listed American Photographic Societies and their officers. The incidence of titles such as "Dr." and "Prof." among those serving is high.

70. "The Demonstrator," *Photo Beacon*, March, 1895, p. 104.

71. N. Gray Bartlett, "A Neglected Printing Process, **Photo Beacon**, Feb. 1895, pp., 60-62.

72. Polychrome Advertisements appear in *PHOTO-MINIATURE*, January 1900, May 1900, Sept 1900, and July 1902. Tennant and Ward Co., New York.

73. S. Bluman, "A Patriarch in Kallitype*," Camera Craft.*, July 1916, p. 264.

74. Ibid., p. 266.

75. Ibid., p. 269.

76. M. H. Wilde, *Photography*, 1899.

77. John Tennant, *PHOTO-MINIATURE*, Feb. 1903.

78. "Simili Platine **Paper,**" *BJP*, 1899, p. 102

79. "Replies," *Photography*, January 2, 1895, p. 16.

80. The Photographers Formulary Co. address 7079 Mount Highway 83, Condon Montana, 59806. The company offers many old process kits and stocks photo chemicals used in older processes. For other sources, google **Alternate Photography.com**

81. W. H. Smith, "The Kallitype Process," *BJP*, May 26,1916, p. 16.

82. John Martin Hammond, *BJP*, May, 26, 1916.

83. Dr. Julius Schnauss, On Kallitype, *American Annual of Photography and Photographic Times Almanac*, vol. 9, p. 186.

84. James Thomson, "Possible Substitutes For the Platinum Print," *American Photography*, November 1918, p. 648.

SECTION II

Kallitype and the Amateur

Chapter IV

Kallitype and the Amateur

I.   The Need for Historical Study

> Amateur photographers are now counted by the thousands,
> and in the different cities are organized into flourishing and
> growing societies.
>
> *Photographic Times* 1885.

The history of the kallitype is the history of a photographic process which for a brief period, from 1889 to 1895, was controlled by an inventor and a manufacturing company. In 1895 both controlling parties, Nicol the inventor and Lewis, the entrepreneur abandoned the process. The company terminated the manufacture and commercial sale of the kallitype in England in that year, and the inventor discontinued further public, and presumably private activity with the process at that time. After 1895 no further publication on the kallitype from either source surfaces in the literature. As reported in the last chapter, a number of businesses tried to manufacture and sell kallitypes but none enjoyed genuine successful ventures.

One might think that this would be the end of kallitype's history. But surprisingly an impetus for continued life is found from another source. From 1895 to the present, a period of over 100 years, a sequence of amateurs experimented with the process and published their findings. What is surprising about the amateur involvement is the fact that kallitype, while a relatively simple chemical process, is not easy to understand or control.

Kallitype is not the only photographic process to receive such attention. Amateurs played an important role in the development of a variety of photographic processes including platinum, gum, cyanotype, carbro, bromoil, the salt print, the brown print and the kallitype. The period during which amateurs explored these individually prepared processes began in the 1880's and continued at a high level until 1925, when the quantity of activity tapered down reaching a low after 1940 when a new taste in "pure" photography, influenced by Stieglitz, Weston and Adams, asserted itself. The concern with individual process reasserted itself after the 1960, probably as a result of programs in photography at universities which emphasized artistic creativity with all aspects of the photographic medium. Sustained amateur involvement with kallitype occurred in several Western countries, notably, France, Germany and Italy. Amateur photographers engaged in exploring various photographic media included both men and women. Photographers of all stripes came to investigate print media from a wide variety of educational and professional backgrounds. Most had little, if any, connection with professional or commercial photography.

While excellent studies of the history of some photographic processes, for example gum bichromate and the salt print, have been published recently, none has given due attention to the contributions of amateurs to the development of photographic process in general.[1] It is time to ask how such an extensive tradition of amateur involvement with individually controlled photographic processes began, why it continued, what it accomplished and finally, what it signifies.

Historians of photography have exhibited varying degrees of indifference to the study of photographic printing processes and have given small attention to amateur involvement with them. Eder, whose **History of Photography** is an incomparable source of basic process information, is an important exception.[2] Newhall's history of the daguerreotype is a model history of a process invented by an amateur. Its excellence will not soon be matched.[3] Outside of these two shining examples, most histories of process have been written by authors more concerned with teaching readers how to achieve control of a process than they were with creating a detailed historical account of the happenings. In the latter quarter of the Twentieth Century, the history of photographic printing processes suddenly and unaccountably changed. Without obvious cause, excellent works on the history and methods appeared, one after another: the salt print by Reilly, the gum print by Scopick, and the platinum print, by Haffey and Shillea.[4] Admittedly they contained more process than history, but they began at least to fill a prolonged void.

The indifference of historians is notably apparent in their lack of study of the kallitype, which rarely is mentioned at all in photographic histories, much less discussed. No historical study of the kallitype as a process has been made, and amateur artistic involvement with the kallitype, as with other processes, has been largely ignored by historians of photography. Until recently, historians have rarely discussed in any detail the important category of iron-sensitized print processes, which include platinotype, blue print, brown print and kallitype. Inevitably they mention Herschel's discovery of the cyanotype and then proceed to the development of the daguerrotype, silver print, etc. Ignoring processes in general, photo historians have also shown little interest in the amateurs who investigated and promoted the variety of individually prepared print processes before and after the turn of the 20th century. For example, the identities of the amateurs who initiated the "takeover" of the kallitype or the sequence of workers who continued to write about it are ignored. In particular, we are given no sense of the time span during which amateurs were active with kallitype or any idea of the number of investigations into the process they generated. While some histories mention the kallitype processes and almost every encyclopedia of photography offers at least one formula, the amateur investigators who created the formulas receive no credit for their contribution.[5] Little, if anything has been written about the social, economic and esthetic factors that influenced the amateurs whose labors sustained the kallitype and other processes. Most histories of photography, including those dedicated specifically to the late Victorian period when process development was most active, fail to mention the kallitype as a process at all, even in their glossaries, much less discuss its place or significance or contribution to the history of photography.[6]

The curators of collections of photographs in Art Museums have done no better. Museum collections do not contain examples of vintage kallitype prints for scholars to study. Even collections whose purpose it is to collect and preserve specimens of photographic process, materiel, and literature, such as the International Museum of Photography or the collection at the University of New Mexico do not possess or are prepared to show visiting scholars even one example.[7] It is regrettable but true that at this time we do not know a) if any vintage kallitype prints exist or b) the location where they might be found.

The above observations underline the historical needs relative to the kallitype. The order in the list is not meant to suggest the priority of the items. And it is obvious that the list is applicable to other processes.

1) We need to discover how many kallitype processes were originated by historical workers, to identity of each process, and to clarify the differences among the processes.

2) We need to discover the authors of the several kallitype processes and to locate the papers they wrote about their investigations and the prints they made as a result of their explorations.

3) We need to evaluate the contributions of the various workers who pursued the development of the kallitype.

4) We need to locate vintage kallitype prints and initiate procedures that will preserve them. We must also catalog their locations so they can be studied.

These needs have importance to contemporary printmakers as well as historians. Both can profit from knowledge of what happened. Artists working the process today can derive real benefit from the information past workers accumulated and might also benefit from study of prints they made. Contemporary historians of photography could benefit from increased awareness of the exciting part the kallitype played in the history of photographic art. This writer believes that the history of the kallitype is a key to better understanding of certain developments in the history of photography, specifically the growth of amateurism and the esthetics of pictorial printmaking. Whether for improved printmaking or for increased historical understanding, a better knowledge of the kallitype is long overdue.

II.   Notes On This History of the Kallitype

The kallitype, as it exists today is the product of more than 100 years of rather wild growth sustained by a number of unaffiliated amateurs who loved the process, wanted to understand it, studied it, experimented with it, worked with it, and wrote about it. Amateur contribution to the kallitype, as we shall see is sporadic and erratic, but, remarkably continuous. In the following chapters, we shall trace the amateur development of the process with the hope that some order and meaning can be found in the wandering progression of events. This history shall endeavor to establish the identity of the several kallitype processes, trace the sequence of individuals who formulated in writing each process, detail the working of each process development, and finally, shall evaluate which amateurs made the most significant contributions to the evolving process. We shall conclude with a discussion of the historical significance of kallitype activity.

Amateurs coopted the kallitype more than a century ago. Since then, they have made studies, published formulas, and recommended methods of working that continue to this day. Taken as a whole, the writers display varying levels of photographic, chemical, and esthetic competence. Though a few writers recorded their reminiscences of the history of the kallitype, their recollections were generally unanalytic and uncritical, as we shall see. Unlike what happened in the development of the silver bromide print, no company with extensive resources became involved with the kallitype. Had such an institution taken over the process, it might have collected, maintained, and organized the scattered lore. Since none did, with the possible exception of the Kodak Research Library, the task of collecting, compiling, and creating an order out of raw source material awaits being done. With no guide to point the way, the historian of kallitype has a daunting task

The following chapters on the history of the kallitype will survey in more or less chronological order all the writing on the process that was discoverable during the 90 years it was most active. The writing will aim at a relatively complete report of all discussions of kallitype in journals, annuals, books, etc., hoping by such inclusive coverage to inform the reader of the number of kallitype workers who worked the process and the variety of process descriptions which arose. The succeeding chapters shall spotlight the individuals who created the processes of continuing importance. They will inquire whether the theory and technique of the process evolved progressively during the century of amateur involvement, or whether individual researchers merely wandered in the same desert, never finding the promised land of a really controllable and reliable process. One underlying purpose of these chapters will be to discover if a perfect understanding and control of the kallitype was ever achieved. The remaining chapters of this book will address these concerns and provide as much detailed information as the available research material permits.

But before continuing, we must discuss several matters.

III.   The problem of Vintage Kallitypes

At the outset we must admit that we shall not be able to discuss one of the key concerns listed above. Much as we would like to do so, we have little to report on the actual kallitype prints known historical amateurs made. We cannot report any data or evaluations based on current study of vintage kallitype prints. The absence of this invaluable material results not from a lack of inclination or effort. A diligent search of major photographic print collections that began in Chicago, proceeded through major museums of the Northeastern and Southwestern United States, and which ended at the Museum of Science and Technology in Washington turned up but one authentic vintage kallitype. If kallitype prints exist, they were not found in the major institutions visited.[8]

It could be that these institutions had kallitypes but did not know that they had them, ie, that some prints in their collection were in fact kallitypes but were not identified as such. This is quite possible, since kallitypes are easily mistaken for other types of silver prints on plain paper. It is possible that prints labelled platinotypes or palladiotypes may have been kallitypes, which can be made to closely resemble them. It is likely that a good number of kallitypes, produced by the thousands who made them, lie undiscovered in attics or basements of the houses where their makers or their progeny lived.[9] The closets and files of local historical societies are another likely location where unidentified kallitypes may lie undiscovered.

From already made but unrewarded searches, two conclusions are likely: 1) since there are virtually no known vintage kallitypes prints to study at this time, historical scholarship of the process must proceed, with serious limitations, without them; and 2) scholars must make haste to find what kallitypes remain before unfavorable conditions of storage destroy them. Authentic prints are needed as primary evidence for discussions about the technical and esthetic qualities of prints made by the processes developed by historical kallitype workers. Vintage prints are also necessary to enable technical assessment of the kinds of changes that have occurred to kallitypes during the passage of time, especially to evaluate the often mentioned problem of impermanence.

There can be no doubt that historical amateurs made beautiful kallitype prints. There exist a number of published reproductions of kallitypes,[10] which can be offered as examples, in spite of the fact that an ink printed reproduction provides only an indication of the qualities of the original print. There also exist a number of creditable statements made by editors which praise the quality of kallitype prints that workers submitted to them. There exists also the indirect evidence of remarks made in published letters praising the prints of specific workers and the announcement of awards for prints acknowledged to be kallitypes. There can be no doubt that beautiful kallitypes were made—it seems futile to question that fact. So the mandate is clear: we must take whatever steps are possible to discover what prints exist and let them speak before they are irretrievably lost. Vintage kallitypes would have much to say of value to photographic historians and curators.

IV.   Amateurs true and otherwise

We should like to introduce several topics that relate to the role amateurs played in the development of photographic print processes. We shall discuss first Seiberling's study of amateur involvement with photography during mid-Victorian times, a period when photography was in its earliest stages. We shall go on to describe the changing roles a different category of amateurs played when photography flourished in the last quarter of the nineteenth and the first quarter of the twentieth century. These glimpses of the changing role amateurs played during what is commonly called the pictorial period of photography will provide illuminating background for our understanding of the kallitype.

The kallitypists who contributed most to the development of the process came to photography from many occupations and from a wide range of educational backgrounds. They worked as doctors, engineers, designers, printers, chemists, editors, and so on. Their occupation generally had little to do with their initial fascination for kallitype or what they eventually did with it. Clearly the willingness to learn something about photographic process and the enjoyment of experimenting with a variety of chemical compounds were a factor in their commitment. The fascination with kallitype and with the creative challenges of photography are in general grounded in part in the undefineable magic of chemistry and the unfathomable mystery of image-making.

Even though it began as a commercial product, the kallitype very early took on the identity of being an "amateur" process. That is, amateurs developed a special relation to it, not unlike the relationship many amateur hackers displayed toward the micro-computer in the 1980's. But the amateurs of the 1890's were not the first to exhibit this coopting behaviour in the history of photography. What developed in the

1890's was only an extension of the special 'relationship' that amateurs had established with photography from its very beginning.

The earliest amateur photographers acted as if they had a right to the process, as if photography, in a special way, belonged to them. That perceived right led them to a sense of responsibility and a special involvement with photography's development. Though amateur involvement in the 150 years of photographic history has not always displayed the same kind of commitment or level of intensity, amateurs have always been an important force in the development of photographic materials, equipment, and esthetics. Therefore discussion of the role amateurs played in the development of photography before the kallitype appeared on the scene is desirable. This will allow observation of early amateur attitudes and practices. Subsequently, we will be prepared to evaluate the changes in attitudes that occurred in later times, with reference to the the kallitype and other "alternate processes."

The flowering of amateurism in photography is generally reported to follow the invention and sale of the dry plate. Taft traces the flowering of amateurism to about 1880.[11] Freund puts the date a bit later, in 1888, when the Eastman roll film cameras and machine made prints became available.[12] The import of both inventions made it infinitely less troublesome to move about and take photographs. Henry J. Newton, an amateur who was active at this time recounted the ordeal of earlier photographic activity.

> To succeed then meant hard work and study. You were required to know how to make almost everything connected with the production of a photographic print. You must know how to make collodion; how to coat a plate and how to sensitize and develop it; how to construct the silver bath in which the plate was sensitized; how to make the developer; how to clean the plate; how to prepare the nitrate of silver bath for sensitizing the albumen paper; how to fume, print, tone and fix the prints; how to make paste, and how to mount the prints. The amateur of those times was further required to make himself familiar with the chemistry involved in all this work, first, in order that what he did he might do intelligently and successfully; and, second, to be qualified to determine with a degree of certainty what was the matter when his chemicals gave unsatisfactory results. The negative bath was one of his most treacherous friends; he could not predict, with

any degree of certainty, what would happen to the next plate
by the result on the one immediately preceding it. [13]

After the dry plate became available, photographers no longer had
to carry to the site where they photographed, a portable dark tent well
stocked with solutions, in which wet plate negatives would be sensitized
and developed. Later they could develop exposed dry plates in their
darkrooms at their pleasure. After the roll film camera became available,
the amateur had even greater convenience. Kodak's slogan, "You push
the button, we do the rest," was born at this time. A new link was forged
between late 19th Century amateurs and convenience. The kallitype was
invented at this time, when convenient cameras and negative materials
were already in place and easier processing was available. When we
remember that both inventions, the dry plate and roll film, were developed
from the research of amateurs, it becomes clear that the kallitype enters
the scene at a critical moment in the history of photography, a moment
when amateurs are pulled in two somewhat divergent directions: one
toward understanding, control, and even invention, and the other, toward
convenience and freedom from responsibility for materials.

What were the amateurs like before convenient materials and
cameras came on the scene? Seiberling made an extensive study of this
subject in her doctoral dissertation entitled, ***Amateur Photography and
the Mid-Victorian Imagination***. Writing about amateur photographers
in mid-Victorian England a generation before the kallitype was invented,
she observes that the "early" amateur saw photography as an art-science
in which the interests, education, and leisure of a gentleman could be
put to productive use. She found that the early amateur was a person
from the higher classes, either independently wealthy or nearly so, with
sufficient leisure, education, and resources to pursue photography as
a scientific, cultural and social phenomenon. She offers Fox Talbot as
the exemplar of the amateur during the early period of photography's
development. She detailed the qualities early amateurs shared:

The early amateurs had been a relatively homogeneous group. The
experimental nature of the medium had required that they be reasonably
well educated, have some knowledge of chemistry, and enough time and
money to master the skills involved in taking pictures. [14]

She sees the 'early' amateur as a "gentleman amateur" "who had an
understanding of science, a curiosity about the chemical properties of

photography, and an interest in photography as a craft phenomenon." She believes these amateurs were part of the means by which photography was "invented" and "developed." These privileged amateurs created the first photographic society and the first photographic journal, both of which were given to sharing fundamental observations of historical, cultural and technical concerns. In contrast to late Victorian amateur photographers, "early amateurs would not have made a distinction between pictorial photography and those aspects involving chemistry and optics, and they were not hostile to scientists."[15]

Seiberling believes that "the distinction of the early amateurs" was the achievement of "an integration of their various fields of endeavor." They worked with the cultural resources available to them in such a way that they provided a starting point for the rapid expansion of photography in many directions. Unlike the participants in later amateur movements, these men and women were not self conscious and had little interest in making names for themselves as artists or scientists.

> ... they saw photography as an extension of many different strands of activity. It was their sense of connection with many things that was one of the sources of their success in providing a starting point for disparate directions. [16]

Seiberling observed that the late-Victorian amateurs who followed them developed into three distinct types, each with a narrower concern than the early amateur. The largest and most diffuse category is analogous to those who take snapshots with automatic cameras today. This amateur has no interest in understanding photography, no artistic aspirations, and no concern with craft (process). As has already been observed, this type of amateur grew rapidly in numbers after the introduction of the roll film camera and automatic processing. They had no interest in contributing to the medium and simply used it.

Seiberling's second group of late-Victorian amateurs "took photography seriously, belonged to societies, exhibited photographs, and read journals. They followed the lead of the earlier generation of amateurs, but were less flexible, and less interested in photography as an art-science." The change in this group was noted in 1898 by T. N. Armstrong.

> In the early days of wet colodion working . . . it was by no means uncommon for ardent workers to prepare their own sensitive plates . . . but with the modern dry plate, few

workers, if indeed any, now ever think of preparing a batch of emulsion and coating their own plates. Nor is the reason for this change of procedure difficult to understand, when such uniformly excellent plates are so conveniently obtained at prices that place them within the reach of all . . . The production of sensitive dry plates may, therefore, be said to have passed out of the hands of photographers generally into those of the special manufacturers, and all being considered, the change has been beneficial to photography. There is no doubt a degree of pleasure and satisfaction derivable from a thorough knowledge and ability . . . to make one's own sensitive plates and materials, that is unknown to those later day workers who seldom study photography from its scientific point of view, but, on the other hand merely work by rule of thumb, producing effects without the knowledge of how such marvellous changes have been brought about by the various manipulations they have been the means of producing.[17]

Seiberling's smallest group of late Victorian amateurs consists of photographers who worked against established conventions and institutions, both artistic and scientific. In general these amateurs were concerned with the extension of photography as an art and wished to separate it from what they viewed as the stifling concerns of scientific development and the distracting concern with profit. She considers Julia Margaret Cameron, Lady Hawarden, P. H. Emerson, and H. P. Robinson as descendants of the early Victorians. The pictorial movement can be regarded as a late nineteenth century manifestation of one the directions, the art-plus-craft interest the group of early amateurs fostered.[18]

As this history progresses we shall have opportunity to observe if Seiberling's last two categories remain distinct—that is whether later amateurs divide quite so distinctly into a group pursuing the development of process and technique on the one hand, and another group pursuing the development of art. We shall examine if the overlap continues which links the development of process with the fostering of artistic vision. Whatever the case, Seiberling's description of the origins of amateurism and the character of early workers will help us to better understand the motives and activities of succeeding generations of photographic amateurs, including those who developed the kallitype.

V: The Pictorial Amateur

While convenient and dependable manufactured film was available by the early 1880's, a suitable paper on which to make artistic prints was not. The kallitype was born in a period when amateurs were intensely dissatisfied with available printing materials and were determined to find a more satisfactory paper. The vigorous competition among commercial silver papers and other papers was documented in the previous chapter along with the difficulties commercial kallitype paper experienced in the marketplace. We have also noted the variety of "non-silver" print media which competed for the favor of amateur printers. Among this group of processes were platinum, gum, ozotype, carbon, carbro, bromoil, and gravure. Pat Fuller, in an unpublished **Catalog for An Exhibition of Control Processes** held at the International Museum of Photography, discussed amateur interest in "control processes" and the significance of the growth in popularity of such printing processes.[19] Regarding the motivations of amateurs who pursued control processes, she wrote,

> To the photographers associated with the Pictorialist movement, the recognition of photography as a fine art could only come through demonstration of the medium's capacity to serve as a vehicle for personal expression, to give evidence of the individuality of the artist and to serve his temperament and sensibility. Seeking to escape identification with the rapidly increasing ranks of amateur "snapshooters," and to disprove the criticism that photography was a mechanical process which did not admit selection or modification by the artist, they sought ways to invest their pictures with greater evidence of artistic handling. They turned away from modern technological advances—sharper and faster lenses, gelatin silver emulsions and the casual depletion of the hand camera—in favor of reviving older techniques which promised greater opportunity for transmuting the photographic record of the negative into a picture displaying the familiar characteristics of traditionally accepted media. This demanded first, the modification of tones and values in the print, . . . and [second], the suppression of superfluous detail indiscriminately recorded by the negative. To these ends, the platinum and carbon printing processes, already favored by the earlier generation of pictorial photographers over silver emulsion for their permanence, rich

tonal scale, atmospheric qualities, and surface finish, began to
be explored and modified for unique individual effects. [20]

Fuller provides some details on the development of the various
control processes, introduced during the pictorial movement. She
begins with the revival of the gum-bichromate process in 1894, "which
permitted the photographer to isolate important features, to eliminate
unwanted elements, and to achieve a variety of surfaces and textures
associated with the graphic arts." She sketches the development of the
various platinum processes between 1870 and 1890 and the modifications
available in platinum printing—gum platinum, glycerin development,
and the various color modifications brought about by chemical additives.
She also sketches the development of carbon printing, Artigue, and
Fresson, and the development of later non transfer methods based
on the hardening properties of gelatine infused with bichromate—oil
pigment and bromoil printing. She concludes that the latter processes,
introduced after the turn of the century, offered "such control that the
final prints were at times difficult to identify as photographic in origin
and were hailed by the vanguard as at last providing the Pictorialist
worker with complete freedom of expression."[21] Fuller indicates that the
control processes received heated opposition and vocal criticism. The
manual alteration of the "truthful" image produced by the camera drew
the complaint, stated at its worst, of "faking." Opponents charged that
altering the print by hand manipulation was a misuse of the photographic
medium and a distortion of reality. They loudly condemned hand worked
prints made to resemble the appearance of traditional media—drawing,
lithography, etching, and painting.[22] T. N. Armstrong also criticized the
misrepresentation of reality in his *Guide to Practical Photography*,
written in 1898. The truthfulness of photography, he wrote

> . . . is unquestionable, and in many instances of priceless
> value, notwithstanding the sneers of the high art devotee, who
> seldom rests satisfied with nature as The Maker of the Universe
> has chosen to lay it at the feet of mankind, but must perforce
> out of his own puny imaginings add thereto or remodel and
> clothe many of our fairest scenes to suit his individual taste
> and fancy. [23]

By the late 1920's the esthetic of hand control and self expression
found competition from a new realistic esthetic. About this time

pictorialism faced a challenge from a "new" esthetic (actually a revision of an old one) based on objective camera vision and the straight, unmanipulated silver print. The period when pictorialism prospered, a period during which the values of self expression and the hand controlled print dominated photographic esthetics, were the years in which amateurs were most active in working with kallitype. As we have seen in the early claims for the process, the kallitype offered a number of choices which could be taken advantage of for artistic expression. The kallitype could be made in almost any color—brown, black, sepia, green, blue, or red. The contrast of the print was subject to control by the printer. And the image could be printed on any fine art paper. In the chapters that follow, writer after writer will proclaim the pictorial advantages available to the artistic amateur who used this process. It is notable that interest in the kallitype waned when the new esthetic of the unmanipulated photograph became popular.

VI: Social History and the Amateur

While this history does not aim at an extensive description of social and economic influences on photographic printing, a few paragraphs may sketch some broad relationships. During the last quarter of the nineteenth century when photographic practice became convenient, life in general became easier for many in England and America. At this time more people lived in the cities and life in the city, with the introduction of electricity, became less demanding. An educated middle class with surplus time, money, and energy grew. Individuals in the middle class looked for something worthwhile to do with their free time. Among other pastimes—sports, biking, reading, etc.—the new pastime of photography was often chosen.

The choice is not surprising. After all, photography was a recreation that returned a number of satisfactions in exchange for the required development of skill and the acquisition of knowledge. The promised rewards had to do with the pleasures derived from the exercise of artistic choice, imagination and personal expression. Photography became an activity designed for a class of people not content with a sedentary type of leisure, such as reading, on the one hand, or a leisure that is totally dependent on physical satisfactions, like sport, on the other. Welling provides evidence that educated and prominent people were attracted to photography.[24] One need only note the identities of those who flocked to camera clubs and photographic societies to see how frequently the photographers were professional people—doctors, professors, businessmen, lawyers, etc. The historian is tempted to conclude that photography as it presented itself at this time was the ideal post-enlightenment, post-puritan, democratic recreation. It combined work, education, science, and art and allowed one and all to pursue them, regardless of special talent, training, or position. This point of view was expressed by several writers of the time.

> By the mid twentieth century photography has become deservedly popular with all classes of society. Nor is this to be wondered at, when its widespread advantages are taken into consideration. As a hobby, especially for those desirous of producing tangible results from the expenditure of their spare time and means, it has perhaps no equal among pastimes, offering as it does, at one and the same time, a healthy exercise for both body and mind. Its usefulness enters largely into almost every phase of life, so much so, that photography has become a household necessity, and the aid it now lends to commerce, industry, science and the fine arts is universally recognized as indispensable . . . [25]

The kallitype process was an ideal process for the involvement of the new middle class photographer. It was a process that challenged the mind, the sensibility, and the hand. It required some skill and some knowledge of chemistry—but not too much. It held out the possibility of making art to those who did not possess skills or training in a traditional artistic discipline. It was a marvelous opportunity for a class of people seeking an intellectual and artistic challenge. And there always remained the possibility of personal fame or gain.

Since its inception, photography had a history of being expanded by people seeking such challenges. Da Guerre, Fox Talbot, Hunt, Maddox—so many of the great early contributors to photography had been amateurs of a sort. As Seiberling has shown, photography had developed in mid-Victorian England a tradition of the gentleman photographer, that is, the individual of ample means who had good basic education and who chose to explore photography scientifically, technically, and esthetically. Such men were responsible for the development of a good deal of basic photographic process. A generation later, more than a few kallitype investigators saw themselves as playing a role that connected them to such early models as Phillip Henry Delamotte, Hugh Welch

Diamond, and Thomas Damant Eaton, men who contributed much to the early development of the photographic medium.[26] Like them, the later kallitypists were delighted to use what educational background they had to explore a process and participate in a great democratic extension of craft and art.

VII.   Kallitype and Photographic Publications

While the private part of their activity occurred in their darkrooms, the stage on which the amateurs played their public part was often, the photographic periodical, and occasionally, the book. In the absence of original kallitype prints and the personal documents of the photographers who made kallitypes, we must rely on what exists in photographic publications. The exalted salons no longer exist and most, if not all, of the photographs they exhibited have disappeared with time. The records of amateur exhibitions, contained in salon catalogs which have been preserved, cannot always be trusted to record accurately whether prints were platinum prints or their lookalike, kallitypes. Nor can the scholar tell from the description, "silver print" what kind of a silver print an amateur entered, since that description could refer to kallitype prints as readily as to other types of silver print—salt prints, bromide prints, etc. We concluded earlier in this chapter that at this time the historian can only write the history of the kallitype on the basis of evidence drawn from published articles. This necessity will continue until more primary sources such as biographies, private papers, and actual kallitype prints surface. The discovery of further material will certainly round out the story, but until it appears, there is justification for basing a study now on the many available publications about kallitypes.

The study of kallitype is facilitated by the fact that its appearance was accompanied by a phenomenal rise in amateur photographic publications. Welling reports that in one year, 1889, the year the kallitype was patented, three "prime new photographic publications" were initiated "aimed at what was presumed to be a permanent new market of amateur readers."[27] The number of photographic periodicals expanded to as many as twelve in the decade from 1895-1905.

The number of photographic periodicals provided more than adequate space on all of the topics that interested the growing numbers of amateurs. They published all kinds of material relating to artful printing processes. They printed formularies which spelled out how amateurs could mix the solutions needed for the variety of processes in use. They described how to compound the basic chemicals necessary to make the solutions. A number of periodicals provided introductory accounts of photochemistry and even of basic chemistry. ***PHOTO-MINIATURE,*** a monthly magazine was committed to providing inquiring amateurs, not only the information necessary for the individual preparation of a wide range of photographic processes, but also published full monographs on scientific topics related to photography.[28] Photographic journals published detailed descriptions of how to work various processes and the theory underlying them repeatedly, to assist photographers in the discovery and mastery of photography's possibilities. Finally, as capabilities grew for lithographic reproduction of photographs, magazines published reproductions of amateur prints so other workers could "see" the character and quality of prints made with the various processes.

Regarding the kallitype, the journals printed verbatim reports of the patents as well as critical reports of the success and failure of the process when tried by editorial staff or outside experts. They reported announcements of the sale of the product and further announcements of improved versions placed on the market. They reported the demonstrations of kallitypy given in camera clubs. They printed descriptions of how to compound the solutions for the process and detailed the manipulation. They printed amateur questions that arose from difficulties in getting "promised results." And they printed answers to inquiries written by amateurs who "knew" or thought they could "explain problems encountered by less knowing tyros." Finally, photographic magazines published detailed comparisons of process with process and drew conclusions about their relative ease and value. Formularies spelled out how amateurs could mix the solutions needed for the variety of processes in use.

The journals permitted experimentally minded amateurs to theorize about a process; write a historical account of the development of a process; report the results of their personal process "investigations"; and express criticism of the process efforts of others. Amateurs could become process "stars" and publish their expertise now on one process and later on others. Men like Martin Hammond of Philadelphia, James Thomson, of Boston, and James Brooke in England wrote accounts of how to work as many as ten different individually prepared papers. In time, being an "experimenter" became a kind of role that very advanced amateurs could play, and the photographic press, discovering that reader interest in kallitype was perennial and pervasive, provided the good players a stage and a bright spotlight. James Thomson, one of the star amateur writers published 28 articles on the kallitype and allied subjects in journals on both sides of the Atlantic between 1903 and 1923. That is, 28 have been located. There may have been more.

From 1895 to 1925 amateur photographic publications were locked in combat for readers.[29] They sought out and cultivated those workers who made attractive prints and who could write about them. They published many process descriptions which worked well, and some which didn't. They awarded prizes for the best-written articles sent in by amateurs.[30] G. W. Frederick, one of several medical doctors who contributed significantly to innovative and clear publications on the kallitype, won a prize from *Camera* for the best written amateur essay on photography. On the basis of thorough research, Frederick described how to solve all of the problems associated with making kallitypes. His article was mentioned in a number of later articles written by other amateurs.

The American photographic press during the last decade of the Nineteenth Century and the first quarter of the Twentieth Century published a good deal on kallitype. One British kallitypist complained, "it appears to have published everything ever written."[31] The extent of the interest in kallitype can be gauged by the editor's, report that *Photo-Miniature* sold 5000 copies of its first issue on kallitype.[32] Tennant, the editor of *Photo-Miniature* went on to report that that issue also led to the formation of an amateur portfolio exchange group dedicated exclusively to sharing kallitypes.

Fortunately for the historian, publications for most of the years between 1895 and 1925 contain considerable kallitype material. After 1925 rather fewer articles appeared. The number of articles for the years after 1925 show increases and decreases, as if interest in kallitypy was responsive to an assortment of influences like the cost of platinum, the favor of salon judges, or market saturation with the topic.[33] This writer wonders if the quantity of publication on kallitype follows a kind of cyclical pattern—a period of waxing interest followed by a period in which interest wanes. Scenarios come to mind. The cycle begins when a worker "discovers" the kallitype process by a chance reading of an article on the subject. Becoming interested, he, or she,[34] makes some prints and publishes a report of what happened. More often than not the article discusses the problems encountered when trying the "found" process and the solution worked out by the present writer. When this article is published, another reads the account and tries its process. The second worker finds new problems and writes about them in a

'Letter to the Editor'. Other letters or articles are written to resolve the problems raised in the letter. The suggested chemicals or manipulation are tried. Then another worker writes. If the recommended process works, he confirms the last recommendation. If the recommendations don't work, he reports the failure and raises again the question of how to do the kallitype. A flurry of interest and activity occurs. Then the discussion of the Kallitype quiets down for a while—until the whole cycle begins again.

VIII. Historically Important Writers on Kallitype

Some writers on kallitype had more impact on the history of the process than others. The names of Bennett, Frederick, Hall, Thomson, and Blumann appear frequently in the American literature on kallitypy, while the names of Burton, Brooke, and Smith appear often in the English. As we shall learn, these amateurs all contributed importantly to kallitypy by formulating basic and sound approaches. Each in his own way contributed. Some contributed understanding of the chemistry employed, some contributed innovations in manipulation, while others, like Blumann, contributed an irresistible enthusiasm for the process.

The published approaches of these innovators appealed to other amateurs writers who often, after experimenting with what they read, proceeded to re-write what they read, sometimes openly announcing their source. The "historical figures" in amateur kallitype history are the writers whose methods were good enough to be adopted by a number of followers over an extended period of time. The kallitypists who persist in history are workers who present such clear, attractive, and effective descriptions of making kallitypes that their methods continue to be used and republished for succeeding generations of reader workers. We shall try to discover why their approaches are passed on. Ironically, one writer received a great deal of notice, even though users regularly complained that his methods did not work well. Perhaps the quality of the writing, the spirit of experimentation, and the beauty of the published prints moved readers to admiration as often as the effectiveness of a particular approach to a processs.

The historically important writers on kallitype, those listed above, have several characteristics in common. Most made a study of the past approaches to the process. All did basic chemical experiments on the process. And, all designed a relatively distinct way of doing kallitype. With whatever understanding of the chemistry of the process they could muster, they studied the influence each chemical had on the final print, determined successful working techniques, and communicated an effective sequence of processing steps. By testing they found optimum quantities for each variable in the sensitizer, developer, fixer, clearing bath and toner. Frequently they varied each solution with one additive after another to see if they could maximize a favorable effect. All of the innovative workers followed the process where it took them, continuing their experiments till they arrived at an approach which they thought worked well and made good prints.

In contrast to the best original writers, the following survey of amateur writing on kallitype will make abundantly clear, there were many writers who were intelligent, who had good technical backgrounds, but who, when they became involved with the kallitype, were content to publicize a process that had been put together by someone else. In short, some writers merely passed on the formula and methods others had created. It is amazing how many writers on kallitype were "rewriters," writers who re-packaged a process worked out by those who went before. It is also notable how frequently they gave little or no credit to their source. Regardless, "they also served . . ."

IX.   The "Progress" of the Kallitype Process

The documentation in the following chapters on amateur kallitype writing will make clear that the succession of amateur writers did not evolve a single upward progress to the creation of a perfect kallitype print process. The sequence from Bennett to Blumann is not comparable to the sequence that begins with the Wright brothers' "pipes and canvas" and leads to the space shuttle. Sad to say, that kind of straight line development did not happen. In kallitype history the processes written about later were not progressively better than the ones published earlier. In fact, if one looks for a kallitype recipe in a contemporary encyclopedia of photography, [35] he is as likely to find Bennett's approach (one of the earliest written) as Thomson's, in spite of the fact that 30 years of experimentation occurred between them. Approaches published later cannot be counted on to work better. Late writing may be less scientifically sound; it may produce prints of inferior quality, or the later recommended process may not work at all.

This lack of progress toward a perfect process may be viewed as the curse of the kallitype. The reader quite naturally may expect that after 90 years of active investigation, examination of a process ought to converge on a single well-working method. This did not happen during the 50 years of pictorial concerns with kallitype process. Unfortunately, the practical application of a process is where the problems arise. Each new worker who tried to master the kallitype had to understand, not only the principles of the process; he also had to achieve control of a specific set of the chemicals he worked with. During the time when many of the articles were written, photographic chemicals varied a good deal in purity and photographic quality. The same is true of paper stocks. Accomplished kallitype workers frequently report that even their own successful methods occasionally failed to work, for some unaccountable reason. Often the reason for failure is later discovered to be a chemical that had deteriorated but had given no indication of change. Sometimes the problem was a newly purchased chemical, one whose quality was inferior to the chemical it replaced that enabled a process to work well. Problems were compounded when workers far removed from the writer used chemicals which possessed quite different properties from those the author used when he established his "successful procedure."

Another problem that slowed the upward progress of kallitype was flawed communication. More than one writer discovered the article publishing his carefully done research contained basic typographical errors or omitted critical components which altered his meaning or falsified vital information. The problems of chemical variation and erroneous publication of data help to explain the kallitype's erratic development.

There is a positive side to the kallitype's resistance to amateur mastery. The new worker, then as now, was often delighted to find that unsolved problems existed. The true enemy to the continued enjoyment of a photographic process, it turns out, is not its difficulty or the gaps in what is known about it, but rather the boredom that sets in when the process no longer offers a challenge. Curious amateurs, as opposed to the mere consumers who want only convenience, quickly lost interest in a process that was too easy. They despised the "fatal facility" of the commercial print processes that were readily controlled. They preferred processes that required ingenuity and effort to master. Even today amateurs delight in the problems the kallitype presents. They are excited by the prospect of resolving difficulties and figuring out their own way of doing the process. Like it or not, in the history of the kallitype, later is not always better, any more than it is in the history of politics or cooking

Another preliminary conclusion: though there has been unceasing kallitype activity through the years, the popularity of the process seems to continue, as far as can be told, at about the same level. No one knows the number of kallitype workers at different times in the history of the kallitype. Nevertheless, since the invention of the process there has always been a sufficient number to keep the process going. The sizeable numbers that were attracted to kallitype during the pictorial period is understandable when one considers the romantic aura attached to the revolutionary expansion that occurred to photography when new films, cameras and print processes were developing and when new esthetic possibilities beckoned. What is surprising is not the interest in kallitype in 1890, but the interest in the process that exists today in 2012, when there are so many interesting and challenging things to explore.

Considering recent times, there has been a surprising amount of activity in kallitype since 1950. A number of books have appeared since then which report process descriptions. The Society for Photography Education scheduled professional lectures to discuss it and published a booklet of photo processes. In colleges and high schools teachers teach the process. And entrepreneurs sell kallitype kits.

Part of current interest in the "old processes" may be due to the fact that photography has been academicized, at least in the United States. In this country thousands of students in some 3000 colleges "take photography," and many of them are exposed to the existence of processes like kallitype in advanced courses. High school students do projects in the easy processes of blue print and brown print. Out of these inputs, enough interested individuals appear to keep the alternative processes alive in the age of digital photography.

One change is worth noting however. It is much easier to be a kallitypist today than formerly. There are available numerous accounts of the process, most of which will produce an acceptable, if not distinguished print. Kits are sold with directions which guarantee results without requiring the individual to re-invent the process. Also, fresh, reliable photographic chemicals are readily available from several suppliers. Finally inexpensive laboratory apparatus is available to the worker who wishes better control of his process.

But we are getting ahead of the story. The following chapters will show that events relating to kallitype rarely follow a plan. One year the kallitype thrives and the next it appears to be dying. Happily the facts show it never dies. Like the phoenix it continues to arise. What keeps it interesting, historically, is that each time new amateurs commit themselves to a serious investigation, the kallitype became a slightly different process. Later incarnations have always differed from earlier in slight ways. Thomson's process was not rewarmed Bennett. The reader may become impatient as he reads the following chapters, wishing for a linear historical evolution in which a single perfect system is generated by successive events. Unfortunately, the history of the kallitype doesn't converge on single perfect process.

What this history of amateurs working kallitype shows most clearly is that from the efforts of successive workers a number of quite distinct methodologies evolved. One line of development, that of Nicol—Bennett—Burton, led in the direction of extreme simplicity in the preparation of the sensitizer and developer. Another line of development, as exemplified by Thomson, included many ingredients and more processing steps. Frederick and Hall, who saw the kallitype as a variable process, developed an approach that did not confine the kallitype to a single invarying set of solutions and manipulations. They approached kallitype with a set of stock solutions which could be variously compounded to produce quite different prints.

Almost every historically important worker published his investigation of the kallitype with high claims for full understanding and control. After periods of study of varying duration each investigator believed he had mastery of the process within his grasp and published what he thought was *the* way to do the kallitype. In response to every published "one true way," other workers reported that the published methods failed, in part or totally, for them. This happened to all the major formulators of the process throughout the history of the kallitype. We shall see there are reasons why methods that worked well for one investigator may not always work for another.

Part of the difficulty the kallitype had in getting standardized may be attributable to typographical errors. Printing errors were not infrequent at the turn of the century, as a close reader will discover. On occasion, quantities were type-set grossly in error (1 oz. for 1 gram) and chemicals were misnamed, e.g. ferrous for ferric. Grains were sometimes substituted for grams. Examples of errors abound. Thomson wrote an article on making kallitypes that was published in the ***Photo Era.*** It contained a critical error regarding the quantity of a chemical. He wrote to get the error corrected. Before the correction could be made, a British publication reprinted the article. In their type-setting they repeated the first error and in addition made another error.[36] Eventually corrections were made, but in the interim readers tried the figures they were given.

The present reader, considering a formula to apply, should be warned that even when formulas are printed correctly, results may diverge from the writer's. One moral to be drawn from our historical survey of amateur experience with kallitype is that a worker cannot put complete faith in any published account of a kallitype process, because the prints produced were dependent on the status of the chemicals and the precise processing procedure which two people, author and reader use. The history of kallitype process tells us one thing for certain, if nothing else: a published kallitype process should be seen only as a starting point from which further personal experimentation must proceed.

X.   Conclusion

We conclude these introductory remarks on amateur contributions to kallitype by serving notice that the following chapters will proceed by a chronological presentation of the published material on the kallitype the writer has been able to find after a diligent search. Minor articles will be briefly summarized and those considered major will be reported in detail. The aim will be to describe each major approach to kallitype with sufficient detail so that an experienced reader could work the process. Succeeding chapters shall detail for each major version of Kallitype, the specifics of the sensitizer, developer, fixer, clearing bath, etc., and also any special manipulation that is advocated. It is hoped that this approach will enable the historically oriented reader to understand the variety and sequence of the approaches that have developed during the history of the process. The processes will be presented in chronological order, so that the reader may follow any evolution of the process. The processes will be so presented that the historically minded reader will be able to recognize the sources of currently published kallitype methods.

Finally, the processes will be so described that the present day worker can compare sources and perhaps produce a personal approach to the process. The harried kallitypist of today may find interest and comfort in seeing his methods in a historical perspective. He may learn that he succeeds or fails in good company.

Notes: Chapter IV. Kallitype And The Amateur: Introduction

1. James M. Reilly, *The Albumen and Salted Paper Book,* The History and Practice of Photographic Printing, Light Impressions Co., Rochester New York, 1980. This book is one example of a number published on the history of photographic process. Similar studies have recently been published on the gum and platinum print processes.
2. Josef Maria Eder, *History of Photography*, translated by Edward Epstean, Dover Publications, Inc., New York, N. Y., 1972. Eder provides basic historical information, often gained from those involved in the research, invention, and development of many photographic processes.
3. Beaumont Newhall, *Latent Image, The Discovery of Photography*, Anchor Books, Doubleday and Co., Garden City, New York, 1967.
4. For Reilly, see above.
    David Scopick, *The Gum Bichromate Book*, Rochester, N.Y., Light Impressions, 1978.
    John Haffey and Tom Shillea, *The Platinum Print,* Graphic Arts Research Center, Rochester Institute of Technogy, Rochester, N. Y., 1979.
5. *The Encyclopedia of Photography*, ed. by Willard D. Morgan, Greystone Press, New York, 1971, p. 1906. This encyclopedia is typical of most. It reprints formulas and fails to credit them. The present author has traced the formulas to O. P. Bennett and W. K. Burton, as a later chapter of this book explains.
    The Cassel's Encyclopedia of Photography, Bernard D. Jones, editor, 1911, republished in facsimile by the Arno Press, New York Times Co., New York, N. Y., 1974, pp. 314 ff., reprints the Bennett-Burton Formula without crediting either of these two writers. It does give credit to James Thomson for his formula. It appears that the former process may not have been attributed because the authors were not known.
6. Brian Coe and Mark Haworth Booth, *A Guide to Early Photographic Processes,* Hartwood Press in association with the Victoria and Albert Museum, London, 1983. The book illustrates 21 early photographic print processes and defines 17 processes in the glossary without mention of the kallitype. Nevertheless it claims to "provide a guide to the recognition of the major photographic processes from 1840 to

1914 and to illustrate the finest examples of each process." p. 7 See also:

William Welling, ***Photography in America, The Formative Years, 1839-1900***, Thos. Y. Crowell Co, New York, N. Y., 1978. Though this otherwise excellent book gives extensive coverage on the evolution of photographic processes and to the amateur movement, it completely ignores kallitype as a process. Many other histories which contain glossaries or directories of processes fail to mention kallitype also.

7. A 1978 visit to the International Museum of Photography for the express purpose of examining vintage kallitypes revealed that the print collection did not include one kallitype print for professional study. When asked if the collection had "an obligation" to acquire either vintage or current kallitypes, the curator, gave an ambiguous but essentially negative reply. The visitor was left wondering how a historical photographic collection could rationalize such a selective responsibility.

8. In 1978 the following Museums were visited and the indexes of their print collections were searched for kallitypes.

> The Art Institute of Chicago
> The Boston Museum of Art
> The Philadelphia Museum of Art
> The International Museum of Photography, Rochester New York.
> The Smithsonian Museum
> The Museum of Science and Technology, Washington, D. C.

Only one kallitype was found, a gum-kallitype print. It turned up in the Smithsonian Collection. What proved interesting was that discussion with curators with the exception of the Smithsonian Institution, suggested they collected Art, not examples of print processes. Art proved to be prints made by photographers with great reputations in the world of Art. There was no particular responsibility for preserving examples of prints involved in an important development of the medium or the general culture.

9. My evidence for the number is indirect, but plausible. It is based on an editorial report made by John Tenant in ***PHOTO-MINIATURE***, Vol. xvi, No. 185, Jan. 1922, "Kallitype and Allied Processes," p. 210. Tenant writes:

"The Blue Print and its Variations, . . . set a goodly number of American Amateurs experimenting with the 'variations' including kallitype. This begat such popularity for the method . . . that the photographic journals re-echoed with Kallitype for some years thereafter. Thus, in 1903, Henry Hall published a monograph on the method in No. 47 of this series, of which over five thousand copies were sold.

I'm estimating that kallitypes were made by at least one thousand of the five thousand who bought the description of the process.

10. *The American Annual of Photography*, ed., Percy Y. Howe, published by *The American Annual of Photography*, Inc., New York, N. Y., 1921. Reproductions of Thomson's kallitypes on pages 62 and 67 compare favorably with the other reproductions in this well printed book.
11. Robert Taft, *Photography and the American Scene*, "A Social History 1839-1889," Macmillan Co., N. Y. 1938, pp. 374-376.
12. Gisele Freund, **Photography and Society**, David R. Godine, Boston, 1980, p. 201.
13. Henry J. Newton, *American Photography Bulletin*, vol. 20, March 9, 1889, pp. 147f. Newton's remarks were quoted in Welling, op. cit., p. 160.
14. Grace Seiberling, *Amateur Photography and the Mid-Victorian Imagination*, The University of Chicago Press, Chicago, 1986, p. 107.
15. Ibid. p. 115.
16. Ibid. p. 116.
17. T. N. Armstrong, *Guide to Practical Photography*, London, Dawbarn and Ward Ltd., 1898, p. 20.
18. Seiberling, ibid. pp. 106-116.
19. Pat Fuller, *Control Processes, an exhibition at IMP/GEH*, p. 1. I do not know whether this informative document was published, or whether the exhibition was ever held. The document I refer to was a working copy, unpaginated and contained corrections. I thank the curator of documents at the International Museum of Photography for permission to use this working copy.
20. Ibid. pp. 2-18.
21. Ibid. p. 2.

22. Ibid.

23. T. N. Armstrong, Ibid. p. 1

24. William Welling, *Photography In America 1839-1900*, Thomas Y. Crowell, New York, 1978, p. 297.

25. T. N. Armstrong, Ibid.

26. Seiberling, Ibid. pp. 106-116.

27. Welling, Ibid, p. 335. See also Robert S. Sennett, *The Nineteenth-Century Photographic Press,* **A Study Guide**, Garland Publishing, Inc., New York and London, 1987. A sample of Sennett's writing on American photographic periodicals follows:

> "Considered as a whole, American photographic journals were more commercial, more ephemeral, and more ambitious than their European counterparts—qualities which could be found in Americans themselves. In America, photography was a social and a business activity; in Europe, a science and a profession. These qualities are subtly and admirably reflected in the press." p. 11.

28. In the first 40 issues of *PHOTO-MINIATURE*, seven processes are treated at monograph length: three volumes deal with chemical notions, chemicals, and chemical manipulation. Two volumes deal with Kallitype. See Index contained on p. 2, *PHOTO-MINIATURE*, vol. iv; no. 47; Feb. 1903.

29. Welling, ibid. p. 403-4.

30. *The Camera* awarded prizes in a "Literary Competition" for best essay on a photographic subject. See Vol. V, #9, Sept. 1901. G. W. Frederick won in the "No. 8" Competition with "The Successful Employment of Kallitype."

31. H. Wild, "Kallitype," *Photography*, Oct. 26, 1899, p. 733.

32. John Tennant, *PHOTO-MINIATURE*, Vol. xvi, No. 185, Jan. 1922, "Kallitype and Allied Processes," p. 210.

33. James Thomson, *Camera and Darkroom*, 1918, p. 223-4. "In the presence of a marked advance in the price of the metal platinum, we must also expect a rise in platinum papers. It therefore behooves the amateur to look around for something to take its place. In this exigency, one need look no further than kallitype."

34. Known female kallitypists of the time were: Margaret Walpole, Eleanor W. Willard, and A. Leonara Kellogg. Each wrote an article on Kallitype. See Bibliography.
35. **Encylopedias of Photography**, footnote #5, above.
36. James Thomson**, British Journal of Photography**, "The Kallitype Process—To The Editors," January 31, 1908, p. 93.

"Gentlemen, An article by me published in the October, 1906, issue of *the Photo Era*, with the title "The Kallitype Process," you were good enough to reprint in a later issue of your journal. In the article as originally published was a regrettable error, which, of course, was carried along to your reprint. The fact that my process has still further exploitation in the latest "**B.J.**" *Almanac* prompts me to thus communicate with you in the effort to correct the mistake. The original typographical error I sought to correct in a later number of the *Photo Era*, but in making the correction still another blunder was made, therefore, I thought it about time to give it up.

The mistake occurs in prescribing a drachm of oxalic acid in the diluted developer where a grain is what should been.

Chapter V

The Amateur and the Kallitype

I.   Nicol's And Bennett's Approaches to K II.

It is difficult to characterize the moment when the kallitype ceased to be a commercial process and became a process "in the domain of amateurs." As mentioned, there was no public statement which severed the cord of proprietorship. There was only the appearance of first one, then another, and then more articles written by amateurs telling other amateurs that here was a wonderful process to experiment with and use. Like squatters, the amateurs took possession of the kallitype.

In the Nineteenth Century commercial photographers were not as readily distinguishable from amateurs as they are now. At that time amateurs explored processes and not infrequently became manufacturers. Amateurs were the agents of discovery and change more often than commercial operators. The literature of the late 19th century abounds with discussion of the contributions to photography made by amateurs. These included the invention of the wet plate process, the gelatin dry plate, the use of bromide to make faster negatives and on and on. Nicol, himself, though a professor of chemistry, had been an amateur. Some amateurs of that time felt a responsibility for the exploration and the extension of photography. The modern clear cut differentiation of function between the manufacturer (the source of research and production of the product) and the amateur (the user and consumer of the product who expects to do little more than push the button) had not yet developed. As discussed in the previous chapter, many early amateurs regarded gathering knowledge, experimenting, and preparing materials as the normal activities of a photographer.[1]

This being the case we are not surprised to find amateurs moving into the territory vacated by the proprietors of the kallitype like weeds taking over an abandoned town. A vocal group of amateurs had explored the kallitype from the beginning. Even when it was a commercial paper, they offered their advice to puzzled readers in correspondence columns in competition with the voice of the Birmingham Photographic Co. When the company discontinued providing information, it was not long before the amateurs exercised a self declared right and moved in.

We begin our account of amateur activity with Kallitype II in May 1894, the date O. Prescott Bennett, an American physician, published "his" method of making and processing kallitype paper in *The Photo-Beacon*. Bennett's article contains no reference to the rightful owners of the territory he invades, no bow in Nicol's or Lewis' direction to show respect for their proprietary rights or expertise. Bennett simply offers his approach to the process as if it were the inalienable right of an amateur to improve another's process and offer a personal variation for public use. He states his motivation as follows.

> While perusing photographic literature in order to find some printing process by which an amateur might with ease and with little expense prepare his own paper . . . I came across several articles on the use of the iron salts in photographic printing processes[2]

Bennett does not say, in fact he appears to avoid saying, what published material he had read. It is likely he read Nicol's patent for the K II process which **the *British Journal of Photography*** printed. Nicol's patent spells out in excessive detail the chemicals and quantities for each step of the kallitype II process and, as we shall see, Bennett's formulas are quite similar to Nicol's. That Bennett's source was Nicol's patent is also suggested by Bennett's statement in his 1894 article.

> My idea in presenting this paper is not that I claim any originality in the use of the process, or the amount of each chemical used, but that I hope to bring this inexpensive and very artistic process more fully to the attention of the amateur and professional.
> It is certainly the ideal paper for amateurs . . . [3]

From these remarks it is apparent that Bennett did not develop an original K II process. Examination of his processing steps and the quantities of the chemicals he recommended reveals a close similarity

to those of Nicol's patent. A study of Table #I, below, which compares
Bennett's method with Nicol's makes clear the solutions and the quantities
of the two processes are for all practical purposes the same (compare
the first two columns.) Bennett changed Nicol's weights and measures
from metric to avoirdupois, the system more familiar to the American
amateurs he addressed. In the table I have recalculated Bennett's
avoirdupois figures (right column) into metric measures (central column)
and adjusted them to a common base of 100 cc of solution, so that his
quantities and Nicol's may be compared (first two columns).

Bennett also reduced Nicol's quantities to amateur sized amounts.
This can be seen by comparing Bennett's two ounces of sensitizer to
Nicol's 100 cc—closer to four ounces. Later revisionists would go the
final step and reduce the quantity of sensitizer in the formula to an even
smaller volume, one ounce, about right for an evening's work.

Table 5-1

Comparison of Nicol's and Bennett's K II Process

	Nicol pat. 7312	Bennett (adj) Metric	Bennett Avoirdupois
Sensitizer			
Gelatin	0	.5 gram	5 grain
Water	100 cc	100. cc	2 oz.
Ferric Oxalate	15 gram	14.5 gram	150 grain
Silver Nitrate	3 gram	2.9 gram	30 grain
Potas. Oxalate	0 gram	2.9 gram	30 grain
Developer			
Water	100 cc	100 cc	16 oz.
Borax	7 gram	6.3 gram	1 oz.
Rochelle Salt	10 gram	9.4 gram	1.5 oz.
Restrainer			
Potas. Dichromate 5 % solution	.1 to .4 cc	.2 to .4 cc	40 grn/1 oz

Devel. Time	15-30 min	15-30 min	15-30 min
Water Rinse	none	2 rinses	2 rinses
Fixer			
Water	1 liter	1 liter	1 quart
Ammonia (.880)	3 cc.	7 cc.	2 dram
No. of changes	2	2	2
Clearing Bath	none	none	none
Water Wash	30 min	5-6 changes	5-6 changes

The first two columns make abundantly clear that Bennett's quantities are Nicol's quantities, translated loosely from metric to avoirdupois. However, Bennett did make some changes. His sensitizer differed in a number of respects from Nicol's: it contained some gelatin "to keep the image on the surface" and some Potassium Oxalate "to assure good blacks." These and the rinse after development were the most notable differences in the two methods. It turns out that the first "amateur" kallitype process, as Bennett was honest enough to admit, is really only a slight alteration of the process provided by the inventor himself.

If Bennett's process is so similar to Nicol's process, we may ask if Bennett provided any service and what credit is due him?

Bennett's greatest service may have been to fill a void left by the vacating manufacturer and inventor. He suggested a way amateurs could proceed without them. He initiated the process of the amateurization of the Kallitype. He placed before amateurs all the data required to make the paper on their own and gave them assurance they could successfully work the process themselves without any dependence on a manufacturer. He also spelled out in simple, unambiguous detail the manipulation of the process, which was much less clearly described, probably intentionally so, in the patent. Finally he made known the "reasons why photographers should try kallitype paper."

> Amateurs can prepare their own paper with little trouble and with little expense, there being no cracking, curling, expensive solutions, tedious washing, etc. as noticed with other papers.[4]

Elsewhere he wrote,

> It is certainly the ideal paper for amateurs, for, as I said
> before, it is inexpensive, prints can be made and finished in
> a very short time, solutions are always ready, and above all,
> good prints can be made from almost any negative . . . [5]

Bennett also provided some excellent original suggestions on
working the K II process. One practical example shall be provided—his
approach to the control of contrast. Contrast control in kallitype printing
is effected by the use of a restrainer, which is added in small quantities
to the developer. The restrainer allows the printer to print deeper blacks.
The dichromate holds back the development of the light areas of the print
which would become too dark from the prolonged exposure required for
deeply printed shadow areas. Bennett warned his readers that proper use
of the restraining solution required "a lot of judgment." His approach
to its use illustrates the practical bent of his mind and the clarity of his
presentation.

> For a negative that is very hard, with strong contrasts, little . . .
> [restraining solution] is necessary. For very weak, flat negatives, a dram
> of it, or even more, may be necessary to produce pure whites. A good
> plan to secure a suitable developer is to prepare three solutions with
> different strengths of restrainer:

> First, to four ounces developer add ten drops of restrainer, for a very
> strong (contrasty) negative;
> Second, to four ounces of developer add thirty drops restrainer, for
> medium negatives;
> Third, to four ounces developer add sixty drops restrainer, for flat
> negatives.

> If you are in doubt as to the proper amount of restrainer to use, cut a
> print into three pieces, and develop one piece in each solution, and the
> result will indicate to you the proper amount. The use of the restrainer
> produces brilliancy, and any amount of contrast can be secured by using
> a sufficient quantity of it, but an excess destroys the half-tones. [6]

Bennett also provided a check list of what might go wrong in processing kallitype and made suggestions about how problems could be corrected. He emphasized the importance of following directions carefully, of protecting the coated paper from moisture, and of getting "the right chemicals for mixing up the solutions." He ended with the following closing statement.

Do not think you cannot use this process, if the first trial does not give satisfactory results. Have a little patience, read over directions again and again, and you will be well paid for all your trouble.[7]

To sum up Bennett's contribution: he found a process for a paper which was already commercially dead. He saw it could answer the need for a pictorial print paper and believed that amateurs could make the paper and control it. He showed amateurs they could make a lovely print paper that, short of platinotype, would meet their esthetic expectations and at the same time cut their costs. He leap-frogged the barrier of forbidding legal and technical language in which Nicol had buried the process in patent # 7312. Bennett gave Nicol's process to amateurs in a simple and usable form. Finally, Bennett's encouraging manner gave amateurs, inexperienced in paper preparation, the confidence they needed to get started.

Regarding the question whether Bennett's process worked, the evidence of those who reported trying it is generally positive, although there were some some negative responses. The editor of **The Photo Beacon** found prints Bennett sent him "excellent." Three years later, Dr. Henry Stiefel's evaluation was: "the process is accurately described and gives perfect results." W. K. Burton, to whom we shall turn next, responded favorably but made specific criticisms of Bennett's sensitizer.

II.   W. K. Burton's Refinement of Bennett'sApproach.

W. K. Burton, a well known English "amateur" photographer read Bennett's article and tried the approach while working as a Professor of Sanitary Engineering at the Imperial College in Tokyo, Japan.[8] Burton used Bennett's description as the basis of his own thorough study of the Kallitype II process. He reported his experimental approach and presented his findings at a demonstration of K II to the Photographic Society of Japan on Dec. 13, 1894.

There can be no question that Burton had an excellent knowledge of photography and that he was a well trained investigator in photographic matters. By this time, he had published several books and numerous articles on photography, on subjects ranging from bromine-emulsion chemistry to photographic techniques in the graphic arts. In addition to the kallitype, Burton introduced the platinotype and the collotype processes to Japanese photographers.

At the meeting of the Photographic Society of Japan, at which Burton demonstrated the kallitype, two other print papers were demonstrated. Mr. K. Arito showed some solio prints that were made in London. The interesting point of his demonstration was that he used this paper not as a print-out paper, which it was designed to be, but as a paper subject to short exposure and development. The other paper demonstrated that day was made by the American Aristo Paper Co., and was called "Aristo-Platino." This paper possessed a matt surface and an image color which resembled "as nearly as possible that of a platinotype." This new developing-out paper required careful exposure control, since no image printed out during exposure.

In this meeting at which three quite different papers currently involved in the 'war of process" were on the agenda, Burton explained the advantages of Kallitype: ease of working, cheapness, provision of a visible indication of exposure, a matt surface, and the production of a contrast-controlled black image-even from a thin negative. Burton did not state the reason for his involvement with an individually prepared kallitype paper. We are not told whether his motivation was to interest the Japanese in a commercial kallitype paper, or whether he was interested in advancing the cause of a print material that would be individually prepared by amateurs.[9]

Burton's description of the Kallitype process was published in the English journal, *Photography*, on January 24, 1895. The article was reprinted in *Anthony's International Annual* for 1895 and in the March 1895 issue of *The Photographic Times*, both published in New York and also in *the Photo-Beacon*, published in Chicago. Such interest on two continents in a process demonstration given on a third continent, distant Asia, can be taken as a sign that amateur curiosity about an individually prepared kallitype paper was alive, whatever its status in the commercial market. It may also indicate the level of interest in a Burton contribution to kallitype.

The K II process Burton demonstrated was, as he put it, "a recent modification of an old process by Mr. O. P. Bennett." Burton's article reports his modification of Bennett's solutions. Table II compares the two approaches.

Table II
Comparison of the K II Process of Bennett and Burton

Sensitizer	Bennett	Burton
Water	2 oz.	1 oz.
Gelatin	5 grains	0
Ferric Oxalate	150 grains	75 grains
Potas. Oxalate	30 grains	0
Silver Nitrate	30 grains	30 grains
Developer for Black	Bennett	Burton
Boiling Water	16 oz	10 oz.
Powdered Borax	1 oz	Saturated Sol.
Rochelle Salt	1.5 oz.	1 oz.
Restrainer		
Potas. Dichromate	40 grain	1% sol.
Water	1 oz.	
	Use 10-60 drops	Use 7-8 minims
	of Restr. to 8 oz.	of 1% sol. to each oz
	of developer.	of developer.
Wash	Change water 2 times	10 min. wash
Fixer		
Ammonia .880	2 drams	1 oz.
Water	1 quart	10 oz.
	10 minutes	10minutes
Final Wash	5 or 6 changes	1/2 hour water
	of water.	wash.[10]

Burton wrote a companion article in the January 24 issue of *Photography* entitled "The Kalitype Process" in which he detailed the simple but decisive experiments he used to test Bennett's method. Bennett had used gelatin and potassium oxalate in his sensitizer and had used just enough silver nitrate to barely avoid "solarization," the reversed printing of shadow detail.[11] Burton reported that in his tests of Bennett's process, the use of gelatin made no detectable difference in the print. His test was to submit a kallitype coated on a single piece of paper with one half the sheet sized with gelatin and one half not sized to the developer. His result showed no detectable difference in the prints. Also, Burton found that potassium oxalate did not "guarantee . . . good blacks," as Bennett had insisted, but "was a distinct disadvantage in their production."

Burton found Bennett's quantity of silver, "stinting." His test prints indicated Bennett's 30 grains of silver was too small a quantity in proportion to the amount of ferric oxalate used, 150 grains, resulting "in a most unpleasant appearance in the shadows when printing from negatives other than thin." Burton recommended a 100% increase in the ratio of silver nitrate to ferric oxalate. He achieved this by using the same amount of silver Bennett used, 30 grains of silver nitrate, with one half the amount of ferric oxalate used by Bennett, 75 grains of ferric oxalate. The reduction of ferric oxalate by one half effectively doubled the proportion of silver to ferric oxalate used by Bennett. Burton also reduced the amount of sensitizer from 2 oz. to 1 oz., thus making a smaller and more practical volume of solution, about the right amount for an amateur printing session. [12]

Burton's revision of Bennett's process, hereafter shall be called "the Bennett-Burton approach to kallitype II" to give both men the credit due them for adapting Nicol's process for amateur use. It might seem that the process should be simply called the Burton process, since he wrote the final version. But without Bennett there would have been no process for Burton to revise. Nicol's name might well precede the other two men's names to indicate his initial contribution to all kallitype formulas. That seems unreasonable, for the reason of brevity. By naming the process after the two revisionists, my intention is to honor Bennett and Burton for their contributions to the amateur preparation and use of kallitype paper.

The Bennett-Burton sensitizer was reprinted in a number of periodicals, in various annuals, formularies, and articles, soon after

Burton's publication. Rarely was either investigator given credit for his contribution. The Bennett-Burton formula was reprinted throughout the late nineteenth century in formularies and continues to be reprinted in photographic reference books to the present time.

For purposes of identification, let us note again the formulas of the Bennett-Burton sensitizer and developer.

Bennett-Burton Kallitype II Sensitizer

Water	1 oz.
Ferric Oxalate,	75 grains
Silver Nitrate	30 grains
Oxalic Acid	5 grams (Optional—used when more acid is needed)

This is the kallitype II sensitizer that is most often reproduced in currently published photographic reference books. It is the classic kallitype II individually prepared sensitizer, the one that has been reprinted with the greatest frequency in the greatest number of books and periodicals and over the longest time. It is a simple sensitizer, requiring only two chemicals, ferric oxalate and silver nitrate. They are used in the particular ratio of 75:30 grains of ferric oxalate to grains of silver nitrate in one ounce of water.

The Bennett-Burton developer as it is found in many reprintings is more variable. Almost always the developing solution is made with Borax and Rochelle Salt. The quantity of borax is varied to influence print color. The amount of Rochelle Salt normally used is 3/4 ounces to 10 ounces of water, or 1 ounce to 16 ounces of water. The amount of borax used is 1 ounce in 10 ounces of water, or 1.5 ounces in 16 ounces of water. These mixtures produce a "black toned print." Reducing the amount of borax makes the print more brown. When no borax is used in the developer, the print has a sepia image color. Proper contrast for printing a normal negative is achieved by adding 4 to 15 drams of a 1 per cent solution of potassium dichromate to the developer, regardless of the ratio of Rochelle Salt to Borax.[13] The following formula can be taken as a representative Bennett-Burton developer.

Bennett-Burton Developer for Black Tones

Hot Water	16 oz.
Borax	1 oz
Rochelle Salt	1.5 oz.

Reducing the quantity of borax produces a progressively brown print color. Increasing the borax makes the color blacker.

Contrast can be altered by the addition of from .5 to 5 ml. of a 1% solution of potassium dichromate to the developer. The amount depends on the contrast of the negative—less contrast requires more dichromate.

The reader will find that the above Bennett Burton formulas for the kallitype or slight variants from them are the kallitype formulas most often found in reference books.

In the remainder of his article "On the Kalitype Process," Burton discussed several other details of the Kallitype II process—the chemical theory of the kallitype and the management of the printing. He discussed the fundamental chemistry of the kallitype sensitizer, developer, and fixer. He supported the ammonia fixer, patented by Nicol, because "it left no sulfur in the print to later degrade the image." He reported how he tested the effectiveness of fixing prints in ammonia by exposing fully processed prints for extended periods of time to direct sunlight. He found that if they did not change, they proved they were fixed. This finding means that, ammonia had dissolved away any remaining non-image forming light sensitive silver salts. It is important to note that Burton also tried "hypo," as a fixer instead of ammonia. When he did so, he found the prints "suffered some degradation," or loss of tone. He therefore discouraged the use of hypo as a fixer for kallitype, as had Nicol before him.

Most of Burton's experiments were based on the simple method of cutting similarly prepared pieces of sensitized paper in half and submitting one half to an experimental condition and the other half to the standard routine. It should be noted that by publishing how he arrived at his findings, Burton taught amateurs how to experiment with photographic variables in their own darkrooms and test for favorable outcomes. With the simple comparative procedures Burton introduced, amateurs could examine any procedure to see if it worked better than some alternative procedure. Burton ended his discussion of experiments with an explanation of how the amateur could make his own supply of ferric oxalate, a chemical which was then somewhat difficult to locate. He recommended that the amateur make his own to assure a fresh, reliable supply.[13]

In passing, it may be noted that when Burton gave his demonstration of kallitype, the chairman of the Photographic Society of Japan was C.D. West, also an amateur kallitypist. Burton reported that West used a Kallitype II developer made with Sodium Acetate instead of the Borax-Rochelle Salt developer recommended by both Nicol and Bennett. Burton reported that West's Sodium Acetate developer gave a cold bluish black image color, quite distinct from the browns and brownish black of the Borax-Rochelle developer. However, there was a drawback; the sodium acetate developer did not clear away the non-image-forming iron well as well as the borax Rochelle salt developer, regardless of how long the print was left in the developing bath. Burton reported West's observation that prints developed in sodium acetate developer required the use of a clearing bath. West recommended a saturated solution of potassium oxalate should be used after the developer and before the fixing bath to clear away the yellow stain, which indicated the presence of undissolved ferric or ferrous oxalate.[14]

In a letter to the **Photo-Beacon** published November 1895, Bennett responded to the attention Burton gave his process. He considered it "an honor to be criticized by so eminent an authority on matters photographic." He felt "highly gratified to know my article has been the cause of awakening so much interest in this process." Bennett also responded to Burton's criticisms. He defended his use of gelatin in the sensitizer on the grounds that 1) it facilitated spreading the sensitizer; 2) it kept the image on the surface of the paper, and 3) it prevented the 'sunken in look' he got without it. Bennett defended his use of potassium oxalate in the sensitizer by simply repeating his early statement that he could not get blacks without using it. 4) Bennett replied to Burton's charge of "stinting on silver" in the sensitizer. He claimed that he used the correct amount of silver—all that the ferrous oxalate could reduce.[15] Burton made no reply.

From the vantage point of the present, Bennett's and Burton's articles on Kallitype did more than their authors could have imagined. Their methods were popularized in their time by widespread reprintings in formularies of photo-annuals and other reference books.[16] As generations of publishers reprinted their formulas they became classic formulas perpetuated without author identification even to the present day. This widespread publication of the unidentified Bennett and Burton Kallitype II formulas may be responsible for the notion the process belonged to the public. Compared with other kallitype approaches, developed by later innovators, the Bennett-Burton approach has remained a simple, memorable, and user-friendly process. Their simple approach did not intimidate amateur photographers who had never tried to make their own paper and processing solutions. Yet it worked as well, if not better than more complex methods, as many experienced kallitypists would soon attest.

The interchange between Bennett and Burton influenced more than a recipe for a process. The exchange publicized a model of and a rationale for amateur experimentation. Initiated by two workers a continent apart, their approaches showed amateurs how to produce a paper with excellent working characteristics and print qualities that matched the then current pictorial ideal—black tones and matt surface.[17] Their contribution came at a time when the manufactured kallitype paper was no longer made or sold. They launched the kallitype as a project by and for amateurs. They helped hesitant amateurs to feel confident about preparing their own paper by showing the simple chemical formulas and manipulation the process requires. After Burton, amateurs performed experiments in home laboratories and described their successes and failures in publications and at meetings of photographic societies all over the world. We shall see that amateur experimenters lacking formal training (both Bennett and Burton had received extensive scientific education in preparation for their respective professions, medicine and engineering,) could acquire a chemical understanding of sorts through self-teaching manuals on photographic process that publishers began to provide.[18] Educated in this fashion, amateurs could play a lesser but still satisfying role in process implementation. If they did not all originate processes, they could at least do the processes that others with more background had passed on to them.

Anyone could try the simple process Bennett and Burton left, and all kinds of amateurs did. After Bennett and Burton, the kallitype became a part of the public domain of photography to be employed by anyone with the inclination to do so. An expanding brotherhood of amateur experimenters who tried the process grew. After Bennett and

Burton's creative revision, the fate of the kallitype fell to the hands of an amorphous lot of amateurs and semi-professionals who came from all walks of life to possess it and make of it what they could.

In the remainder of this chapter we shall recount the aftermath that followed Bennett and Burton's contribution. The response to their activity will complete the first part of the history of the amateur and the kallitype.

III.   Aftermath of the Bennett-Burton Revisions

Wishing to add my little to the practical side of photography, I outline briefly what I have found to be the best procedure in the printing room on some points. [19]

This quote states with brevity the optimistic credo of the experimental amateur photographer. It suggests the hope that drove the amateur activity already reported and the hope that would drive the activity for years to come.

But not all experimentation with the Bennett-Burton process produced positive results, as we learn from an article on kallitype by Dr. Julius Schnauss. Schnauss wrote "I have made extensive experiments with these complicated formulae, but alas, without desired results." Dr. Schnauss, who was reaching advanced years in 1896, (he was 68 when he wrote this article) complained that "through all these manifold improvements, the process has become very complicated and confusing, so that but few careful operators will be able to use it with success."[20] Schnauss was no helpless, untrained tyro, exposed for the first time to the rigors of hands on photography. He had established the earliest photographic school in Germany, and throughout a long life had achieved success as an authority on graphic reproduction techniques, including gravure. [21] He also wrote a number of well-received books on photography during the years 1860 to 1886.

The 1895 issue of *Anthony's International Annual* also indicated the complications that arose when individuals prepared kallitype sensitizers.

. . . several questions have been raised by workers of the above process who seem to have had trouble in mixing the ferric oxalate and silver, owing to the precipitation of the latter salt. With reference to this difficulty we would say that this trouble may very readily come from using impure ferric oxalate or equally from the use of ferrous oxalate instead of ferric, as the ferric oxalate forms no precipitate with silver, while the ferrous oxalate does. We believe that if neutral ferric oxalate is used in this formula, precipitaton of the silver salts will not follow. [22]

Unfortunately, the editor denies us the identity of the contributor of this interesting but brief comment. The troublesome precipitation of the silver that occurs when mixing the kallitype sensitizer was treated at greater length in a book on kallitype process by this writer. There are authorities who claim the precipitation is inevitable and that it can only be delayed, not prevented.

The next major publication on the kallitype occurred in a book, published in 1896, entitled *Plates and Papers, How made and Used*, written by H. C. Stiefel, Ph. D. The subtitle makes clear the book is "based on practical experience in the Factory and Studio." Stiefel was an English chemist who manufactured and sold dry plates in Germany before coming to America. He wrote about his experiences on more than one occasion.

Along with accounts of albumen, gelatin silver, carbon, platinum, and other papers, Stiefel's book contains a chapter on the kallitype process. The eight page description keeps the promise of the subtitle, by carefully detailing the manipulative aspects of the process. Stiefel indicates his process was taken from Bennett's account in *the Photo-Beacon*.

Stiefel's point of view on the ease or difficulty of making kallitype prints is curiously ambivalent. Sometimes, he praised the process for "working perfectly" and being "exceedingly simple." At other times, he warned the reader that at each step "care must be used." He wrote, for example, "Care must be used in coating the paper with the sensitizing solution . . ." "Printing (exposure) requires the most skill of any of the operations." The use of the restraining solution, "requires a good deal of judgment . . ." His final conclusion comes as no surprise. "When this process is carried out *carefully*, (my italics) most beautiful results can be obtained." [23] We shall find the question of the ease or difficulty of doing the process is one that is raised by many writers as we follow the history of the kallitype.

Beside the Bennett-Burton approach described above, Stiefel's chapter presents two other approaches to Kallitype. The first is a formula he found published in the *The Photographic Times*.

Photographic Times Kallitype II Sensitizer

Water   17 oz.
Ferric Oxalate   231 grains

When dissolved, add part for part of the following

Water   17 oz.
Nitrate of Silver   46 grains

Photographic Time's Kallitype developer

Water   34 oz.
Rochelle Salts   154 grains
Borax   107 grains

Add from a saturated solution of Potassium Bichromate, two to four drops. The directions for use of these solutions are conventional.[24]

At first glance the above quantities do not look like Bennett's quantities. Yet the proportions of chemicals are the same as his; only the amounts have been changed. The reader should note the 5:1 ratio of Ferric Oxalate to Silver Nitrate and compare it to Bennett's original formula: Ferric oxalate 150, Silver Nitrate 30, and Potassium Oxalate 30. The ratio is the same. The developer components in the above developer are also in the same proportion as Bennett's original developer: 1.5 parts of Rochelle Salt to 1 part of Borax. The practice of changing the quantities without changing the fundamental proportions is a common one in kallitype history. By doing this some writers made it appear they provide a new approach, when in fact they have only multiplied or divided another's numbers by a common factor. This was not Stiefel's intention. He apparently was unaware that he was giving the same Bennett formula twice.

Stiefel's chapter also reprinted a set of directions for making ink drawings from kallitypes. Making line cuts from photographs was a common practice in newspaper work in those days. We have described the method used in an earlier chapter. It involved making a kallitype print, drawing on the print the outlines of the image with waterproof ink, and then bleaching away the image. The ink drawing that remained was suitable for line reproduction in newspapers. Stiefel probably took his information on this process from the announcements of "Vici paper," a commercial kallitype paper. The advertisements for Vici were printed for a number of years in the *American Annual of Photography Almanac*. [25]

A copy of that advertisement is reproduced on page 183.

In the formulary section of the 1895 *American Annual Almanac*, full directions were given for the pictorial use of Vici paper. These included the compounding of developers for black and sepia tones, the use of bichromate for contrast control, and the use of a platinum toner.[26] It is interesting to find in Stiefel's chapter on kallitype, directions for working both commercial and individually prepared kallitype II paper. It is unfortunate that Stiefel's account included very few details about his personal experience with the process.

It may be worthwhile to include in our account of the Bennett-Burton aftermath two magazine publications which shed some light on amateur photography of the time. The first piece reported a talk made to the Royal Camera Club by Mr. H. C. L. Bloxam which was entitled "The Chemistry of Some of the More Common Processes of Photography."[27] It is illuminating to discover that this erudite paper which examined in detail the photo-chemistry of the iron processes was read orally to a group of photographers which included amateurs. This fact suggests the quality of activities which the best amateur photographic societies scheduled at that time. Incidentally, Bloxam discovered in his research that ferric oxalate, the key ingredient in the kallitype sensitizer, must be kept absolutely dry and protected from light if it is to be kept from turning ferrous. In the history of the kallitype, ferric oxalate contaminated with ferrous is one of the most frequent causes of failure, as we shall see.

In 1897 the first monthly installment of an *Encyclopaedic Dictionary of Photography* was published by the *Photo Times* Co. The Encyclopedia reported on materials, processes, and chemicals, etc., used in photography. Unfortunately the entry on kallitype contained misinformation. It reported that kallitype paper was still sold, (it was no longer sold in England in 1897, but it was sold in America). It reported that the formula for making the paper was not published (it was). The article on kallitype provided directions for developing K I paper, which had not been sold for at least 5 years. This article indicates how misinformation was circulated on kallitype and other processes. Unsuspecting amateurs, believing that such sources were authoritative must have been confused.[28]

Another misleading article was written for the 1898 *Photography Annual* by R. Childe Bayley, who was just beginning a long and illustrious career as a writer and editor on photography. In the article Bayley provided brief instructions and formulas for developing the no longer available Kallitype I paper. Bayley apparently tried some form of kallitype printing, probably K I.

Twenty years later, Bayley again wrote on Kallitype in his very popular and charmingly written book, *The Complete Photographer*. Unfortunately what he has to say about the process is not very complimentary.

> Kallitype is another printing process or rather was, for little is heard of it now and a batch of Kallitype prints turned out of a drawer the other day bore no sign to distinguish the front of the paper from the back. The image, which once had been vigorous enough, had folded its tent like the arab and had quietly stolen away . . . [29]

A powerful statement, indeed, one that raised many questions about the permanence of kallitype prints down through the years. How seriously Bailey's remarks should be taken is a good question. The permanence of kallitype shall be discussed in detail in a later section of this book. There we shall discuss whether the fading of a kallitype image proves the process is at fault or whether it proves only that the worker may have performed the process poorly. Unfortunately Bayley did not provide sufficient information on his process for a judgment to be made either way.

Back to 1895, when the Bennett-Burton approach appeared again in the first publication of W. J. Brooke, one of the more innovative and capable English kallitypists. Brooke's article, entitled "Kalitype Printing," appeared in the December 5, 1895 issue of *the Photography Magazine*. In it Brooke reported his successful use of a Bennett-Burton sensitizer and developer. Brooke also reported his trials of West's Sodium Acetate developer. It will be recalled that Burton mentioned this developer had been explored by C. D. West, in Japan. Brooke's formula for a Sodium Acetate developer follows:

Brooke's Sodium Acetate Developer For black tones

Saturated solution of soda acetate	2 ounces (fluid)
Water	2 ounces

A saturated solution of acetate is generally recommended, but I prefer a diluted bath, as results are quite as good as in the stronger one, and development is more gradual.

Brooke advised that after the print is developed in the sodium acetate developer, the print should be placed in a Potassium Oxalate bath to clear the yellow iron stain from the paper.

Brooke's Potassium Oxalate Clearing Bath

Potassium Oxalate	2 ounces.
Water	64 oz.

The Oxalate bath tends to remove the iron and prevent its subsequent precipitation onto the paper in the ammonia bath, which would cause yellowing of the whites. If this, however, should occur, clear in

Brooke's Citric Acid Clearing Bath,

Citric acid	1 oz.
Water	10 oz.

The use of the Sodium Acetate developer entailed some risk. Brooke reports that the acetate developer fails to accomplish what the Borax Rochelle Salt developer does, clear away the sensitizer iron not used to form the image. Brooke recommended using two successive clearing baths, to accomplish this task. It appears that the first bath, the potassium oxalate clearing solution, was not able to accomplish its purpose and had to be followed by a second, a citric acid bath. The two bath approach to clearing was for him, hardly an efficient solution to the problem and he recommended that a better one would have to be found in time.

Brooke also did some research on the influence of the paper stock on the color of the final print. When using a Borax Rochelle Salt developer, Brooke reported,

> I have found that with Whatman's papers, it is difficult to get a very brown image. But with cartridge paper, most beautiful sepias can be obtained, colours which are quite impossible with any toning bath I know of.

. . . on cartridge paper, engraving blacks cannot be
obtained even with the acetate bath, the results then being a
warm black . . . [30]

This is one of the earliest observations of the variation of print color
caused by 1) the paper stock and 2) the developer as it reacts with the
chemicals in or on the paper stock.

In 1898 Brooke wrote a long letter to *Photography* in response to
a critical query on the permanence of the Kallitype. In it he responded
to claims that kallitype prints were less permanent than other silver
prints.

Many and various are the complaints laid against this
process as to its impermanency, but I have prints made six
years ago, and exposed to all trying conditions, and still they
are as fresh as when made.[31]

Brooke again recommended the Bennett-Burton formulas for
sensitizer and developer. He also recommended an alternative sodium
acetate developer (now a simple 5% solution) and a sodium citrate
clearing bath. The letter contained one notable departure. Most
surprising, in view of his previous support of ammonia, is Brooke's
sudden recommendation of "hypo" as a fixer for kallitype.

Brooke's Hypo Fixer

Hypo (Sodium Thiosulfate)	2 oz.
Ammonia (.880)	1/2 oz.
Water	16 oz.

The reader will recall that Nicol had made the use of ammonia as
a fixer a crucial point in his original kallitype patent. Nicol had argued
that hypo inevitably left a sulfur residue which under certain conditions
attacked the silver metal that formed the image. Nicol believed that hypo
should be avoided completely if a really permanent silver print was to
be achieved. Seven years after Nicol's last communication, Brooke
wrote "the ammonia bath is, in my opinion, the cause of kallitype print
fading." This position, in view of his past use of ammonia fixing,—his
ammonia-developed prints had remained without deterioration for five

to six years—is puzzling, to say the least. This writer is inclined to ask, if his prints had not faded or deteriorated after being fixed in ammonia, why change? Brooke did not answer that question. But from this point on, the use of hypo continued to grow until it became the universal kallitype fixer. The hypo was generally used in a weak solution, something less than 5%. To this solution a few drops of strong ammonia were generally added to keep the solution alkaline.

A year later, in 1899, W. J. Brooke wrote a series of three articles on individually prepared iron-sensitized papers for *Photography*. The first two articles discussed the cyanotype and the platinotype, and the last dealt with Kallitype II—"this much neglected and abused method of printing . . . that well deserves trial at the hands of most amateurs." The article covered the same ground as the letter summarized above. It repeated the Bennett-Burton formulas and made essentially the same points about manipulation.

In the 1899 article Brooke supported his position against the use of hypo for the kallitype by reference to Haddon and Grundy's well-known and much respected paper on hypo and print fading.[32] Brooke reported these authors "affirm that ammonia is not good for effectively removing silver salt and find that nothing beats ordinary hypo." Brooke also recommended that a sodium chloride bath (1 part in 250 water) be used before fixing to prevent the loss of print detail in the hypo fixer. The reader will recall that Burton had rejected hypo for just this loss of density.

Brooke also presented some relatively new information on the "considerable effect . . . the . . . paper (size) has in determining the colour of the finished print." He reported his experiment on a print-out version of kallitype (first given in the 1895 article), and gave a new account of how to use the glycerine brush developing method on Kallitype II. [33]

These articles make it clear that Brooke was one of the most knowledgeable, experienced, and innovative kallitypists in England. His elaborations of the Bennett-Burton approach make profitable reading for the kallitypist today.

The October 26, 1899 issue of *Photography* reported a kallitype demonstration given to the Tunbridge Wells Amateur Photographic Association by Mr. H. Wild. Wild's observations on the popularity of the kallitype in England are historically interesting. According to him, the kallitype

"was never much used in England, and was practically
dead. It had a better reception in America . . . and was still there
under various names, such as "Mirrotype," "Verotype," etc.

Mr. Wild offered little that departed in concept or practice from the
most standard Bennett-Burton approach. However, following Burton's
lead, he experimented with the proportion of ferric oxalate to silver
nitrate in the sensitizer. He tried a range of variation that went from
a low of 75 grains of ferric oxalate with 20 grains of silver nitrate to
a high of 150 grains of ferric oxalate with 80 grains of silver nitrate.
He concluded that 120 grains of ferric oxalate with 50 grains of silver
nitrate gave the best results, a ratio of 2.4 to 1. This is a slightly lower
proportion of silver than Burton recommended—75 grains of ferric
oxalate to 30 grains of silver nitrate—to eliminate solarized shadows.[34]

Wild rejected Brooke's innovative use of hypo and ammonia and
continued recommending ammonia alone for the fixing bath.

In 1899 the second revised edition of Alfred Brothers' *Photography Its
History, Processes, Approaches and Materials* appeared. Unfortunately,
the book presented the kallitype as it was in 1891, providing only "the
instructions given by the patents" for K I and KII. Brothers did not
include anything on the individual preparation of Kallitype II paper.[35]

In the year 1900 two important publications appeared which added
the weight of reputation and wide publication to the increasingly popular
Bennett-Burton approach. The first was the publication of *Ferric And
Heliographic Processes,* written by George E. Brown, a most prestigious
and authoritative writer and editor.[36] Brown's book, a treatise on all iron
photographic processes, summed up in one detailed chapter most of what
was then known about kallitype. It discussed the theory of iron silver
printing, provided a full gamut of formulas for sensitizer and developer
solutions, and detailed the procedures for making prints. It discussed
the effects of all known developers, the colors they produce, and the
rate of their exhaustion; "(10 oz. will develop 5 or 6 dozen one/half
plates)." The book also discussed both the ammonia and hypo fixers;
the print-out kallitype; kallitype bleaching, and kallitype toning in gold,
platinum and uranium. Brown also discussed Kallitype permanence at
length, emphasizing proper sizing, "thorough treatment with alkaline
and neutral solvents of iron salts, and the benefits of toning with gold
or platinum. Brown's book met and still meets all expectations of the
experimental photographer working in kallitype.

The formulas for sensitizer and developer given by Brown are those of Bennett-Burton, but Brown, an Englishman, not surprisingly attributes them only to Burton, his countryman. Brown added his authority to fixing with hypo, and reprinted the fixer formula which Brooke had introduced the previous year. Brown also reprinted Brooke's individually prepared kallitype III formula, Brooke's bleach and redevelop method, the Vici method of bleaching and inking a photograph to make a line drawing, and last, a typical method of platinum toning the kallitype, with citric acid and chloroplatinic acid. Most of these matters have been or will be detailed elsewhere in this book. Taken as a whole, Brown's book admirably summed up what was known about the kallitype at the turn of the century. What Brown presented on the kallitype was basically the creation of Bennett and Burton to which Brown added some interesting modifications by Brooke.

The other landmark publication of 1900 was the *Photo-Miniature*, "a Magazine of Photographic Information," edited by John Tennant.[37] This publication was as important as Brown's book, but for quite different reasons. *The PHOTO-MINIATURE* had by Volume I Number 7 already acquired a reputation for authoritative and complete information on photographic process, in part because of a fine monograph on individual preparation of platinotype. "Platinotype Processes," was one of Tennant's most successful process issues, selling 6000 copies. The Volume I Number 10 issue was entitled "The Blue Print and Its Variations," but "it ran short" so Tennant included a brief treatment of the Kallitype. He gave the justification that "kallitype, like cyanotype and platinotype were allied forms of the same kind of iron printing, sharing a common theory and method." Tennant admits "I quote a clear account of the later method given by C. E. Brown" [sic]. The initials should have read G. E. Brown. Tennant's account of kallitype was a direct summary of Brown's data, but a commendably accurate one. The process Tenant described is straight Bennett-Burton with Brooke's additions. At the end of his monograph, Tennant appended a bibliography which contains five items. Among them, Brown's *Ferric and Heliographic Processes*, just published in London, receives the comment, "It will, I believe, prove perhaps the most useful work on the subject."

This writer feels somewhat troubled about Tennant's booklet, finding it a direct rewrite of Brown's chapter which had been in publication in England only one year and which was soon to be released in America. Rewriting and republishing recently published material was a common, and apparently accepted practice of the time, especially with publications originating in England and the United States. Tennant at least credited the author to whom he was indebted. And, if Tennant added little in the way of new information in his first publication on kallitype, at least he was instrumental in publicizing accurate and up to date information on the process. In his defense, it may be said that his publication expanded widely the number of amateurs who were familiar with kallitype. We shall be hearing from Tennant again, since he published two further monographs on Kallitype.

The PHOTO-MINIATURE was published monthly from April 1899 to 1932, with the purpose of informing the experimental amateur on a wide range of photographic topics. Particularly popular issues dealt with photo chemistry, optics, the various iron printing processes, including several issues on blue print, platinotype, and kallitype; and the other exotic print processes—carbon, oil, bromoil, and gum printing etc. In the introduction of the first volume on "the Blue Print," Tennant stated one of the reasons his magazine continued to provide information about print processes:

> One of the chief reasons urging me to deal with the blue print in the *PHOTO-MINIATURE* . . . is that it introduces us to a very interesting group of photographic printing processes based on the light sensitiveness of certain salts of iron . . . When I speak of the iron printing processes as interesting from the amateur's point of view, I refer not only to the character of the resulting prints, but chiefly to their interest as processes, offering in their manipulation innumerable opportunities for getting a practical knowledge of many things which will be found profitable in general photographic work.[38]

IV.   Summary & Conclusions

With this quotation we shall end the first segment of the history of amateur preparation of kallitype II. The kallitype II process was invented by Nicol in 1991, and marketed until 1894 by the Birmingham Photographic Co. When commercial kallitype paper went off the market, the process was abandoned by both inventor and manufacturer. It was co-opted and modified by a sequence of amateur experimenters, most notably O. P. Bennett, Walter Burton, and W. J. Brooke. Others, such as George E. Brown and John Tennant, while not adding greatly to the formulary of the kallitype established by Bennet and Burton, contributed immensely to amateur awareness of the kallitype process. Compared to other kallitype processes, the main virtue of the Bennett-Burton method, beyond the fact it worked well, was its ease and simplicity for the beginner. The process required few chemicals and simple quantities and involved no complicated steps. Small wonder the Bennett-Burton approach has remained the most frequently recommended approach to the kallitype to this day.

This chapter also introduced the amateur photographer's role as process explorer and do it yourself provider of photographic materials.

Notes: Chapter V.
The Bennett-Burton Approach to Kallitype II

1. Robert Taft, *Photography and the American Scene*, p. 207-209.
2. O. Prescott Bennett, "Kalitype Paper," *THE PHOTO-BEACON*, May 1894, p. 170.
3. Ibid. p. 171.
4. Ibid.
5. Ibid.
6. Ibid. p. 172
7. Ibid. p. 173.
8. Josef Eder, *History of Photography*, p. 714. Eder believed that "photography was the very basis of the appointment."
9. W. K. Burton, "The Photographic Society of Japan," *Photography*, Jan. 7th, 1895, p. 62.
10. Ibid. p. 51 and 52. The Bennett data is taken from "Kallitype Paper," *PHOTO-BEACON*, May, 1894.
11. O. P. Bennett, Ibid. p. 170.
12. W. K. Burton, Ibid. p. 62.
13. W. K. Burton, "The Kalitype Process," *Photography*, Jan 24, 1895, p. 51-2.
14. Ibid p. 51-2.
15. O. P. Bennett, Letter to the Editor, *PHOTO-BEACON*, November, 1895.
16. *American Annual of Photography*, Formulary Section, 1896. pp. 307-308 reprinted Kallitype formulas in the public domain. Such formularies commonly printed the Bennett-Burton formulas and were a major source for popularizing this method.
17. Many writers refer to the combination of black tone and matt surface of the platinotype as the preferable print look of the time. This is the look that replaced the albumen purple-brown and glossy look.
18. *PHOTO-MINIATURE* during the years 1890-1892 produced three monographs on chemicals, chemical notions, and chemical manipulation to provide background for amateur experimentation and print production. Other texts on photo-chemistry and photo-process appeared at this time.
19. A. J. Hoffman, "How I Run my Printing Room." *Photo-Beacon*, Feb. 1895, p. 60.

20. Julius Schnaus, "On Kallitype," *American Annual of Photography*, 1895, p.186.
21. Eder, Ibid., p. 685.
22. Editorial note, no author given, "Kallitype Printing Process," *Anthony's International Annual,* 1895, p. 268.
23. Henry C. Stiefel, Plates and Papers, How Made and Used, Percy Lund & Co., London, 1896, p. 165.
24. Ibid. p. 169-70.
25. Advertisement, *American Annual of Photography*, 1895, p. 139.
26. Ibid., Formulary Section, p. 307.
27. H.C.L. Bloxam, "The Chemistry of Some of the More Common Processes of Photography, *Journal of the Camera Club*, #126, December 1896, pp. 196-200.
28. Photo Times Co. *Encyclopaedic Dictionary of Photography*, NewYork,1897.
29. R. Child Bayley, *The Complete Photographer*, Frederick Stokes & Co. New York, October, 1926. The first edition of the book was printed in 1906.
30. W. J. Brooke, "Kalitype Printing" *Photography*, Dec. 5, 1895, p. 778-9.
31. W. J. Brooke, "Reply to Query # 1583, *Photography*, Dec. 15, 1898.
32. A. Haddon, "On The Cause of the Fading of Albumen Prints," *The Journal of the Camera Club,* Vol. VIII, # 96, May, 1894, pp. 112-4.
33. W. J. Brooke, "Kalitype" *Photography*, April 20, 1899, p. 261.
34. H. Wild, "Kallitype," *Photography*, Oct 26, 1899, p. 733.
35. Alfred Brothers, *Photography: Its History, Processes, Approaches, and Materials*, 2nd revised edition, London, Chas. Griffin and Co.,1899, pp. 117-9.
36. George E. Brown, *Ferric and Heliographic Processes*, Dawbarn and Ward Ltd., London, 1899, pp. 28-36.
37. John Tennant, *PHOTO-MINIATURE*, Vol. I, January 1900, No. 10, "The Blue Print and Its Variations," p. 481-514.
38. Ibid. p. 483.

Chapter VI

Tailoring the Kallitype II

I.   Frederick and Hall Approach

> "I find no better or concise instructions on the Kallitype process than those given by Henry Hall in the *PHOTO-MINIATURE* No. 47, and the formula by G. W. Frederick, M.D. in *The Camera Magazine*, 1903,[sic] and would advise beginners to obtain one or both of these and study carefully."
>
> A. Leonara Kellogg [1]

From 1895 to 1901, except for the publication of Nicol's patented method, the Bennett-Burton approach was the only kallitype II process available for amateurs to use. The few other methods that had appeared in formularies were given little attention, perhaps because they contained little explanation or encouragement. In 1901 a new approach was offered by G. W. Frederick, who was, like Bennett, a physician. Two years later Henrry Hall added a complementary approach.

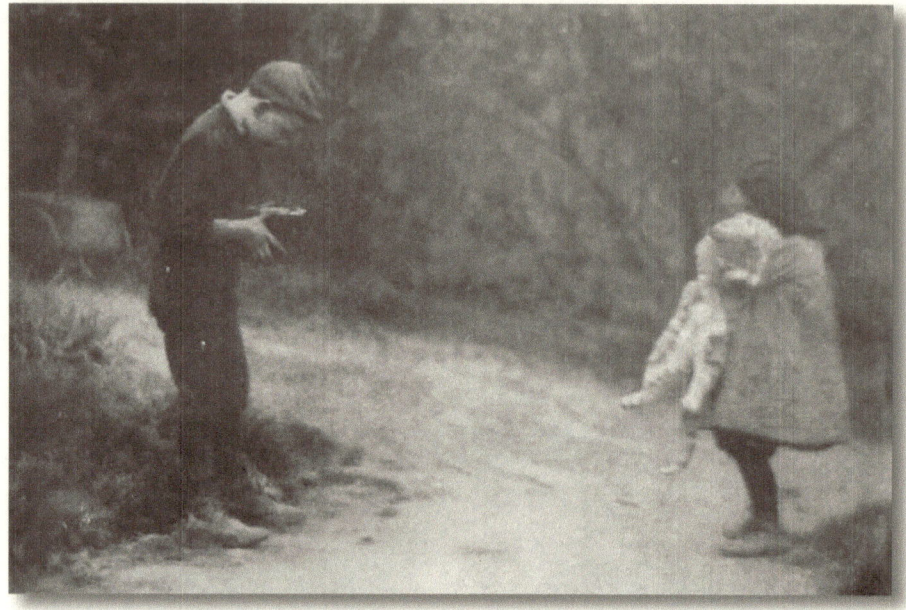

Figure 6. Henry Hall Kallitype, "Pussy's Portrait"

Frederick rethought the problem of making the kallitype and arrived
arrived at an approach that had several advantages over the Bennett Burton
approach and, as some were quick to point out, a few disadvantages.
Frederick wrote an essay entitled "The Successful Employment of
Kallitype," and submitted it to the "Literary Competition Number 8" held
by *The Camera*. It won second prize.[2] When his study of the kallitype
was published in 1901, Frederick had, by his own account, six years
of experiments behind him. His expertise is evident in the informed
and innovative process he described. Two reproductions of Frederick's
kallitypes accompanied his essay. Both were rural landscapes of wooded
streams.

Frederick reported the aim that drove his study was the creation
of a workable process for "the average amateur." He believed that the
kallitype print process met all the requirements of amateurs and that "no
good reason against its more frequent employment will exist, when its
proper manipulation is more thoroughly understood." Frederick felt the
Kallitype II process was both easy to learn and to do. He wrote

the process is at all times thoroughly under the control of the operator, and so simple in its manipulation that its advantages are accessible to the veriest tyro. [3]

Frederick began his essay with a list of "the main obstacles to successful employment of the kallitype" which he collected from a study of worker complaints. He believed these problems had to be overcome before the process could achieve wide acceptance. The obstacles were:

First. The varying qualities of the different makes of ferric oxalate, and even those from the same maker as obtained from time to time.
Second. The deposition of basic oxide of iron, producing an objectionable brown [or yellow] color.
Third. Lack of permanence and keeping qualities.[4]

There has been repeated references to these problems in previous chapters. Some writers believed the problems were endemic to the kallitype and unsolvable. Frederick was convinced that not only were the problems solvable, but that he had the solutions.

Frederick solved the first problem, securing a dependable supply of ferric oxalate, by making his own. In his essay he recommended amateurs to do so and provided a clear and detailed description of how to make a 20% solution of ferric oxalate. The method he described was the common method taught at that time in the schools. The steps in the process were:

a) "dissolve a quantity" of iron wire in nitric acid with the aid of heat
b) after cooling, add strong ammonia to the resulting solution to produce iron hydroxide.
c) After carefully washing the hydroxide in water, filter off the water
d) Dissolve the iron hydroxide with oxalic acid.

If the worker started with Frederick's recommended quantities of material and worked carefully losing no material during the process, he would end with a 20% solution of ferric oxalate in water, which is the desired solution for making kallitypes.[5] Frederick's method of making ferric oxalate, while time consuming and messy, is effective. It produces a clear, lime-green solution that makes an excellent kallitype sensitizer.

Frederick's solutions to the other two problems depend on specific steps in his approach and are detailed in the body of his article. In brief, the yellow stain on the print, is resolved by the use of an effective clearing bath (see below). The third problem, print impermanence, is resolved by the use of a proper fixer, hypo, with or without gold toning. After dealing with the historical objections to the kallitype in the beginning of his article, Frederick launched into a detailed description of his approach to kallitype II printing.

Frederick's sensitizer, on first inspection, bears little resemblance to Bennett's. It requires more chemicals and unfamiliar quantities and proportions. His manner of compounding the sensitizing solution also makes a significant departure from early methods.

A comparison of Frederick's solutions with those of Bennett and Burton reveals several changes. In Bennett's process, the bichromate, the chemical that controls contrast, was added to the developer. In Frederick's approach, the bichromate is found in the sensitizer. The gelatin that Burton removed from the sensitizer, because he could see no effect, Frederick added back to the sensitizer to achieve a more even and flawless coating. He wanted to eliminate the brush marks that occurred during application of the sensitizer to the paper. Frederick also included some potassium oxalate in the sensitizer. This chemical is a solvent of ferrous oxalate, the chemical formed by the action of light on ferric oxalate. Potassium oxalate was placed in the developer to facilitate the conversion of the iron image into a silver one. Frederick also suggested use of a controlled water supply, either distilled or rain water. The changes Frederick made in the sensitizer made it more complex but he felt it provided greater control than the simpler Bennett and Burton sensitizer.

Frederick's K II Sensitizing Solution

Ferric Oxalate Solution (20% sol.)	1 oz.
Potassium Oxalate (neutral)	20 grains
Potassium Bichromate (5% sol.)	18 drops
Gelatin	2 grains
Nitrate of Silver (crystals	30 grains
Distilled or pure rain water	1 ounce

Frederick advised that the sensitizer be made up in the following manner. The potassium oxalate and bichromate are added in that order to the ferric oxalate solution and stirred with low heat, if necessary, until dissolved. The gelatin is dissolved in 1/2 oz. of the water with the aid of heat and is added. The silver nitrate is dissolved in the remaining 1/2 oz. of water and added to the mixture slowly, with stirring.[6]

When appropriate conversions of the quantities of the main compounds are made, it will be seen that the ratios of the quantities of ferric oxalate and silver nitrate in Frederick's sensitizer resemble those in Bennett's process.

Frederick's approach to compounding the developer was more innovative. He made his developer from a set of specially prepared stock solutions, labelled "A" "B" and "C." By varying the quantities of the stock solutions, he compounded a variety of developers that produced a range of colors and contrasts on the prints developed in them. The use of stock solutions permitted the worker to tailor the developer to the requirements of the negative and the intention of the printmaker. Another benefit of the stock solution method was the ease and convenience which it brought to the preparation of the developer for a given printing session. Compounding the developer from stock solutions reduced the time and trouble of weighing dry chemicals. The liquid stock solutions, once mixed, kept well, and their availability on the darkroom shelf made compounding an individualized developer quick and convenient.

Frederick kept his stock developer solutions in separate bottles which he labelled "A," "B," and "C." The developer stock solutions in each bottle were compounded as follows.

Frederick's Stock Developer Solutions

Stock Solution A
    Hot rain water   1 qt.
    Powdered Borax   1 1/2 oz.
    Rochelle Salt   2 1/2 oz.

Stock Solution B
    Rochelle Salt   2 1/4 oz.
    Rain Water   1 qt.

Stock Solution C
    Potassium bichromate   25 grains
    Water   1 oz. [7]

Frederick provided several examples of mixing developers to achieve particular effects in the print:.

> To make a black toned print from a normal negative, compound the developer by mixing four ounces of solution A with 3 drops of solution C.

> For a sepia toned print from a normal negative, compound the developer by mixing four ounces of solution B and adding 1/2 dram of solution C.

For prints from negatives which vary in "hardness," Frederick followed Bennett in recommending three developer solutions kept in bottles marked hard, medium and soft, the first containing normal solution, the second containing five drops more dichromate, and the third containing 10 drops more dichromate. In this way a trial print cut into three could be quickly processed in solutions that produced three contrast levels. The photographer could then develop a "keeper" print in the developing solution that worked best.[8]

The print developer could also be adjusted to produce prints of different colors from the same sensitized paper. For blue-black tones, "rivalling platinum," Frederick recommended:

Developer for Blue Black Tones

Acetate of Soda   1/2 oz.
Water   4 oz.
Dichromate   Add as needed[9]

Interestingly, Frederick believed "the deposit of iron" is caused by extended development, which he advised the printmaker to avoid.

> . . . as usually described, you are instructed to place the print in developer and allow it to remain until cleared, say ten or fifteen minutes. This is wrong, for you not only obtain a

sunken-in-appearance, but in this way you get the deposit of basic oxide of iron that is so objectionable.[10]

Frederick recommended prints be removed from the developer as soon as they reach "sufficient strength," about 2 minutes. They should then be placed in a "clearing bath" to remove the yellow stain.

Frederick's Clearing Bath

Rochelle Salts	2 1/2 oz.
Rain Water	10 oz. [11]

Frederick's directions for the use of this bath are not clear. He gave no indication of the time required for clearing. Apparently he intended prints to remain in the clearing bath until the yellow iron stain was no longer visible.

Frederick recommended a fixer made of one ounce of plain hypo crystals dissolved in 20 ounces of water. Brooke had suggested this fixer and Brown's book had popularized it. Instead of the plain fixer, Frederick preferred a combined fixing and gold toning bath for general use.

Frederick's Combined Fixing and Gold Toning Bath

Hypo	1 oz.
Gold Chloride	1 grain
Water	8 oz. [12]

Frederick's combined bath "rapidly toned while it fixed, producing rich tones ranging from purplish-browns to blue-black, depending on the developer and paper employed." He also recommended a conventional platinum toner and a uranium toner. He reported he sometimes used a uranium toning bath, after gold toning, to produce "red tones of wonderful richness providing the original prints have been printed somewhat deeper than usual." Frederick reported, somewhat in response to Burton, that "there is no change in color . . . and neither is there any loss of strength" during fixing in the hypo after toning in any of the above toners.[13] The heavy metal toners present in the fixer prevented the bleaching of the silver image that hypo normally caused.

Frederick's article on the individual preparation of kallitype II prints was a detailed and thorough explanation. It gave amateurs of the time a new way of making individual prints, one offering efficient compounding of solutions, convenient print variation and control. Since this was a time when amateur photographers valued individual artistry in printmaking, (see the discussion of pictorial printing in the last chapter, and the further discussion to come) it is no wonder that Frederick's essay won a prize. It was clear that he was not just another occasional printer who had tried the process once or twice. Rather he was a committed experimenter who over a long period had carefully tested every nuance of a complex approach to kallitype. It is not surprizing that his article was reprinted the following year in the English periodical, *Photography*. The authorship of the reprint, was erroneously attributed to a Dr. Herbert Mitchell.[14] The mis-attribution may cause unwary scholars some confusion. A careful comparison of the article attributed to Mitchell with Frederick's article in *Camera* reveals they are the same.

Frederick's article provided the basis for Henry Hall's full-scale, monograph on kallitype II that was published in *PHOTO-MINIATURE* **#47**. We shall shortly review this work in detail.

Before examining Hall's contribution to KII, we shall report a few responses to the Bennett-Burton process and quote a few publications that throw added light on the photographic milieu during the first decade of the twentieth century.

1902 was a good year for the kallitype. *Photo Era*, now in its fifth year, (of the "era of photography") began publishing material on kallitype. In the March "Answers to Correspondents," "B.C.R." explained what the name "Kallitype" meant. He wrote the word is a variant spelling on the name Talbot gave to his process, "Calotype," which meant "beautiful picture."[15] B.C.R wrote a second letter to explain how the process worked in August 1902, *Photo Era*. In response to inquiries, he published a brief set of "recipes" for kallitypes I and II. The KII recipes were obviously derived from Bennett-Burton, perhaps from Brown's book, but neither are credited.

Also in the August *Photo Era*, under "Foreign Abstracts," a detailed discussion appeared of Haddon's scientific study of the problem of washing hypo from photographic paper.[16] The article reported Haddon's investigation of the length of time different types of silver prints had to be fixed and washed in order to completely remove unexposed silver salts and hypo. A monumental study, the work affirmed the idea that hypo was not innately a problem in print deterioration and made clear the conditions under which complete and safe fixing and washing were possible. This research dispersed the superstitious cloud that had damaged the reputation of hypo as a fixer for kallitype. After Haddon's study, hypo could be used without prejudice on any image formed of silver, provided the quantity was limited and the wash sufficient.

The same publication sounded a call for entries to the Third Chicago Salon and announced that the exhibition would be held at the Art Institute. The editor warned that "the standard of admission will be one of art, rather than technical excellence . . ."

> Only such works as give evidence of individual artistic feeling expressed in accordance with the canons of fine arts, will be accepted by the jury of selection. Dexterity of technique in the mechanical and chemical processes of photography will be considered, but will be completely subordinate to the composite of imaginative, creative, and technical quality which is the essential of the fine arts."[18]

The editor of **Photo Era**, exhibiting due political caution, commented: "There is no mistaking language of this sort." Excerpts such as this suggest that new photographic standards which transcended perfection of process were beginning to be implemented with increasing authority. The esthetic of Art was confronting the esthetic of Science and Technique.

Finally, as if he had been carefully preparing readers for it all along, B. C. Roloff published an article entitled "The Kallitype Process," in the May 1902 issue of the *Photographic Times*. Roloff was a good writer and a well-informed kallitypist. His approach to the Kallitype II is an interesting combination of the methods of Bennett and Frederick. Roloff found Bennett's sensitizer to be "most practical and satisfactory, and at the same time the most simple to prepare." He recommended the Bennett-Burton classic sensitizer : 75/30/1—grains of ferric oxalate, grains of silver nitrate, and ounces of water. Roloff also described in detail Frederick's modified sensitizer, which he found "very satisfactory." Being comprehensive, Roloff also described a "ferric citro oxalate sensitiser," an early version of a sensitizer for what is usually called a "sepia or brown print" paper. The presentation of such a trio of sensitizers became the common fare of kallitype articles of the future.

Roloff's approach to the KII developer is neither a simple solution, such as Bennett's developer, nor is it compounded from a set of stock solutions like Frederick's developer. Roloff offered a set of six different developers, one for every problem the printer might wish to solve. He provided three different developers for achieving different print colors and three more for different contrasts.

Roloff's approach—one requiring six developers to be mixed, bottled, and stored, each with a separate function, seems a bit impractical when compared with the simpler approaches of Bennett or Frederick. First, compared to Bennett's, it is far more complex. Second, while it might be seen as a logical extension of the stock solution approach Frederick introduced, it fails to resolve an obvious problem: what does the printer do if he wants a developer for a certain print color and also needs to adjust contrast? Roloff's system doesn't accommodate adjusting the developer for two conditions at the same time.

After this well written but not entirely successful attempt to bridge the contrasting approaches of Bennett and Frederick, Roloff did not write again on kallitype. A few of his photographs were reproduced in the popular journals of the time. [19]

II.   Henry Hall—More tailoring—The Stock Solution Approach

A new comprehensive description of the "tailored approach" to working kallitype appeared in a monograph published in the February, 1903 issue of **PHOTO-MINIATURE**. The monograph was written by Henry Hall, an amateur, who lived in New York. Hall's long article, really a booklet, entitled "The Kallitype Process," dealt exhaustively with every aspect of the K II process.

The monograph opened with an introduction and short history of the kallitype, ably written by John Tennant, the editor of **PHOTO-MINIATURE**. Tenant's remarks clarified the reason for the devotion of the entire issue to the subject of kallitype. They also made clear the kind of reader for whom the monograph was intended. Tennant reported that while "kallitype has made it around the world," "kallitype paper is not obtainable commercially at present" but is now "prepared by the individual worker." Tennant believed the kallitype process has "no appeal to the photographer who confines himself to printing papers obtainable from all dealers." It appeals "with particular force to those who have a little leisure and a liking for experiment," "individuality," and "beauty."

> In the home preparation of the paper with the opportunity this affords for the exercise of individuality, in the simplicity of the method, and in the beauty of the results lies the charm of Kallitype for those who are familiar with its virtues and its vagaries. [20]

Hall's approach to the kallitype is clearly targeted toward pictorial photographers who made artistic prints for salon competition. This is most evident in the later sections of the work which treat matters of specific concern to the competitive pictorialist: spotting, flattening, glossing, mounting, and matting—all of which are discussed from the point of view of exhibition.

At the end of Hall's monograph, Tennant appended a brief presentation on the brown print, which he called "a very simple Kallitype method." The account of the brown print was "written" by Tennant, but it would be more accurate to say Tennant closely rewrote G. E. Brown's discussion of the process given in *Ferric and Heliographic Processes*. A later chapter will recount the history of the brown print.

The major portion of the monograph was written by Henry Hall.

According to Tennant, Hall "has pursued the method with indomitable perseverance, beginning his experiments with the outline of kallitype given in the *PHOTO-MINIATURE*, number 10," published in January 1900.

Hall began his text with a forthright statement on the value of Kallitype to the amateur.

> The amateur photographer who has mastered the preliminaries of negative-making and printing will find an ideal printing method in Kallitype . . . . and when he has learned to turn out a desirable print with ordinary printing-out, or developing papers, the Kallitype process opens before him a field wide enough for all the skill and selective taste he has acquired, and deep enough for all his chemical lore, with the added attraction that, if the process is taken up

with a determination to secure better than average results,
his knowledge of plates and papers will be augmented in a
thoroughly gratifying degree. [21]

In this paragraph Hall clearly defines kallitype as a process that will
challenge and extend the experience and skill of the advanced amateur
photographer, a photographer who masters process in order serve Art.

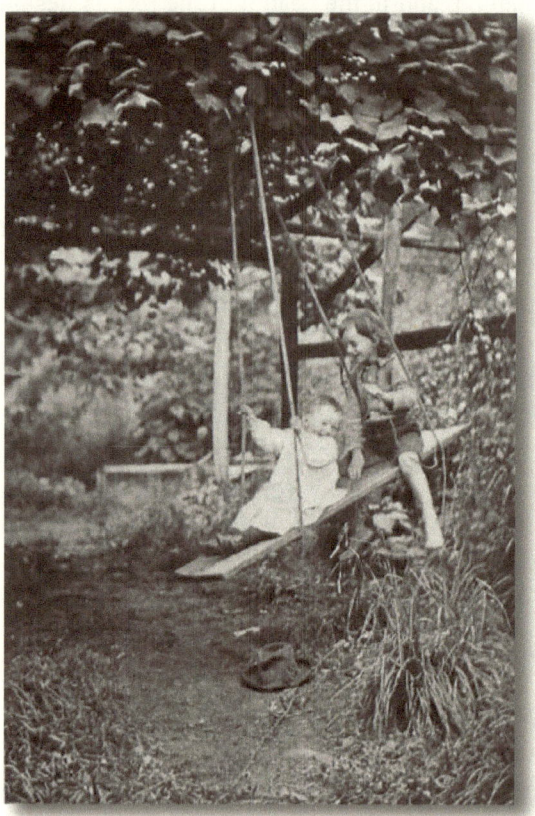

Figure 7. Henry Hall Kallitype, "Swinging"

Hall reported his early experience with the kallitype. Reading a
glowing account of the process, he became convinced that here, at last,
"was the Eldorado of the amateur." His

. . . initiation to the mysteries of Kallitype began with
much enthusiasm and a confidence which seemed warranted
by previous success in photographic manipulation.

> A week later my enthusiasm had lost a good deal of
> its glow, while my confidence had resolved itself into an
> abnormally large interrogation point. [22]

After his early failure—for most writers, failure was a characteristic invitation to explore the mysteries of working the kallitype—Hall decided he needed more information on the process. He reports he "missed Dr. Frederick's very clear and detailed account," and found little help until he discovered Abney and Clark's book on *Platinotype*.[23] In it he learned valuable lessons about chemicals, testing chemicals, and the manipulation of individually prepared paper. Hall reports "it was not long before Kallitype gave me more satisfactory results than any paper on the market." Hall's personal experience—the particulars of his failures and successes—pervade the work, even when his writing becomes topical and objective. His odyssey of learning how to do the process has charm that will appeal to kallitypists at whatever level. Hall is a captivating writer and a knowledgeable teacher.

In the body of the monograph Hall discussed in detail the full range of topics pertinent to working the kallitype: the choice of paper, sizing paper, compounding the sensitizer, applying the sensitizer, developing prints, storing coated paper, developing with sodium acetate for black tones, developing with borax and Rochelle salts for brown tones, clearing, fixing, rinsing, washing, printing (exposing), printing hard and soft negatives, special development practices, drying, spotting, modifying the print by toning and bleaching, and finally varnishing and glossing the finished print. On each topic he is informative, specific, practical, detailed, and thorough.

Because of the limitations of space, I shall not review every topic he treated, but rather will limit discussion of his approach to the sensitizer and the developer, both of which are innovative. Those wishing more detail on the topics not covered are encouraged to consult the original work, which is available in libraries and from dealers in antique photographic books.

Hall openly announced that his approach to the sensitizer is a modification of Frederick's. He adopted Frederick's idea of working from a set of stock solutions. From the stock solutions, he compounded working solutions, which he varied to meet the needs of his negative or his concept of what the print should look like. Hall recommended preparation of four stock solutions, labelled "A," "B,""C," and "D," from which he mixed his sensitizer.

Hall's Stock Solutions for Making K II Sensitizers

A Solution
Ferric Oxalate   1 oz.
Distilled Water   5 oz.
Gum Arabic   48 grains

B Solution
Potass. Ferric Oxalate   1/2 oz.
Water   8 oz.

C Solution
Oxalic Acid   1/2 oz.
Water   4 oz.
Ammonia .880   100 minims.

D. Solution
Potassium Bichromate   120 grains
Water   4 oz.[24]

The "A" solution is clearly the ferric oxalate, the light sensitive iron compound. It has added to it some gum arabic to facilitate even coating of the sensitizer and to prevent the image from sinking into the fibers of the paper.

The "B" solution is a variant iron solution and is used, apparently, to create a buffer of oxalate ion in the ferric oxalate sensitizer. Hall said it made the sensitizer a bit less responsive. It produced prints with a bit less contrast while maintaining rich blacks.

Hall is not very definite about the function of the "C" solution. A mixture of an acid and an alkali, it appears to be another buffer, a solution intended to maintain the pH balance of the sensitizer from deviations introduced by the addition of other chemicals that are more or less alkaline.

The "D" Solution is the familiar "restrainer," or contrast agent. It is used to increase the contrast of the print and to clear the whites, especially on prints made with thin negatives.

When Hall wanted to print, he went to his bottles of stock solution and measured the amount of solution he needed from each. By combining the solutions, he quickly and conveniently compounded his sensitizer. He designed the sensitizer so that it was a) appropriate to the needs of his negatives, b) appropriate to the subject matter, and c) suitable for the desired "look" of the print.

His sensitizer for average negatives is:

Hall Sensitizer for Average Negatives

Sol. 1. Mix the following
      "A" solution   1 oz.
      "B" solution   1/2 oz.
      "C" solution   30 minims.
      "D" solution   4 drops.

Step 2. Add 6 grains of silver nitrate to 120 minims of solution 1.

With stock solutions ready and with appropriate graduates, little time was required to measure the amounts, pour them into a mixing vessel, and stir till a uniform solution was achieved.

It should be noted that in "step 2" the silver nitrate is added from a supply of dry chemical. In the sensitizer being compounded above, Hall recommended adding 6 grains of silver nitrate to 120 minims of the "solution 1" that is mixed.

Hall recommended the following sensitizer to print contrasty negatives. Combine the following amounts of stock solution:

Hall's Sensitizer for Contrasty Negatives

Step 1. Mix the following
      "A" solution   1 oz.
      "B" "   3/4 oz. to 1 oz.
      "C" "   35 minims
      "D" "   none.

Step 2. To 120 minims of the solution made in step 1, add 7 grains of silver nitrate.

Reduced contrast in the print is achieved by using the less active potassium ferric oxalate in place of the ferric oxalate, and by the omission of any potassium dichromate, the contrast producing agent.

Hall does not give specific quantities for compounding a sensitizer for "soft" negatives, that is, negatives with too little contrast. He suggests an unspecified reduction of "C" and double the quantity of "D."[25]

While it may not appear so from a first reading, Hall's stock solution approach to compounding the kallitype sensitizer does have a number of advantages. Experienced workers reported it was a definite advance in the convenience, flexibility, and speed of working kallitype. Hall extended Frederick's method to a point of complete flexibility. Together the two methods enlarge the concept of kallitype method from unvarying simple solutions to complex solutions that permit adjustment to the needs and purposes of any given printing situation.

To give both men their due, I shall hereafter refer to the stock solution approach to kallitype process as the Frederick-Hall method. Their stock solution approach is the most convenient and flexible approach to compounding often used solutions, since it eliminates the measurement of small amounts of dry chemical. Experience shows that a worker using Hall's approach can make up a fresh sensitizer, one geared to the needs of a particular negative, in less than two or three minutes. Measuring small quantities of dry chemicals and dissolving them would require much more time and trouble.

Hall's approach to the kallitype II developer uses stock solutions, also, but in this application fewer stock solutions are necessary. Hall's favorite print developer was the sodium acetate developer. He believed it produced "the strongest blacks the kallitype is capable of making." His formula was:

<u>Hall's Sodium Acetate Developer</u> (For Black Prints)

Sodium Acetate	1 oz.
Water	7.5 oz.
Sol. "D" (see above)	as required
Tartaric Acid	12 grains

Hall advised the worker to prepare a large quantity of developer stock solution, for example 64 oz. or 1 gallon of Sodium Acetate

solution, mixed in the proportion of 1 oz. Sodium Acetate to each
7.5 oz. of water, that is, 17 ounces of Sodium Acetate in 1 gallon
of hot water. For a given night's work, the photographer would take
from the stock bottle 7.5 ounces of stock and add to it 12 grains of
tartaric acid and the required amount of Solution "D" just before
printing. Solution D is a stock solution compounded by adding 120
grains of Potassium Dichromate to 4 ounces of water. The addition
of 12 grains of Tartaric Acid to the developer produces two results:
a) it makes the color of the resulting print slightly warmer, and b)
makes prints clear better than they do without it. Prints made with
Sodium Acetate developer clear less well than do those made with
the Rochelle Salt developer. Hall wrote that phosphoric acid added
to the sodium acetate developer (50% solution of phosphoric acid: 10
minims, added to 8 oz. of mixed developer) changed the black in the
direction of platinum-like blacks." [26]

Hall also provided a Borax-Rochelle Salt developer, which he found
many workers preferred. He indicated his Borax-Rochelle Salt formula
is actually "Dr. Frederick's most convenient formula," "found published
in *Camera*, September 1901."

Frederick-Hall Borax and Rochelle Salt Developers

"A" BLACK—Hot water, 1 qt.; Powd. Borax 1 1/4 oz.; Rochelle Salt
    2 1/4 oz.

"B" SEPIA—Rain Water, 1 qt.; Rochelle salt, 2 1/4 oz.

"C" BROWN—Mixing "A" and "B" in various proportions gives
    various browns

"D" BLACK—Borax 1 oz., Rochelle Salt 1/2 oz., Ammonium Acetate
    40 grains; water 12 oz. "D" from above 50-150 minims. [27]

The "A" to "D" solutions, above, are recipes for separate developers
which produce the image color listed at the left. Regardless of which
developer is used, Hall advises the use of a clearing bath, even though
his Borax-Rochelle Salt developer usually cleared the iron stain away
by itself, given enough time. Hall strongly recommended the following
clearing bath for prints developed in the acetate developer.

Hall's Clearing Bath for Use after Acetate Developer

Sodium Citrate	1/4 oz.
Citric Acid	20 grains
Water	8 oz.

Hall recommended a different clearing bath for prints which failed to clear yellow stain when developed in a Borax Rochelle Salt developer.

Hall's Clearing Bath for Use After Borax-Roch. Salt Development

Rochelle Salt 2	1/2 oz.
Water	1 qt.
Dichromate sol. (Sol. D)	1/4 oz.[28]

Hall recommended the prints should remain in the appropriate clearing bath for at least a half an hour.

After either clearing bath is used, the prints should be rinsed in water to free them of iron salts before they are placed in the toner or fixer.

Silver fixers generally have the task of dissolving the unused silver salts so they can be washed away, leaving the paper free of silver compounds that remain capable of reacting to contaminants or light which discolor the print. Hall is aware that "some authorities insist Ammonia .880 (60 minims in 10 oz. of water) is the primary kallitype fixer." He writes,

> I have not been able to secure with this bath prints that would maintain the purity of their whites for half a day in strong sunlight [29]

He reports that a 10 minute soak in the following bath will "remove the last traces of unreduced silver."

Hall's Fixer

Hypo	1 oz.
Ammonia .880	120 minims
Water	20 oz.[30]

It will be seen that Hall's fixer, following Brooke and Brown, uses a combination of the known silver solvents, hypo and ammonia.

Hall completed his formulary with extensive instructions for toning and bleaching. He provided formulas for uranium, gold, and platinum toners, all of which were well known. They work. He reprinted the combined toning-fixing bath "given by Dr. Frederick." He also includes a solution for reducing dense prints, that is, prints that have been overexposed and appear too dark.

Hall's Print Reducer

Hydrobromic Acid	35 minims
Water	1 oz.[31]

With this we conclude our presentation of Hall's method, necessarily abbreviated as it must be. For more information, interested readers should study his complete presentation in *PHOTO-MINIATURE* No. 47.

In evaluating Hall's contribution from a historical perspective, we notice his gracious acknowledgement of Dr. Frederick, from whom he received the idea of stock solutions. But Hall's work with Frederick's basic idea goes beyond the slavish copying that so often appears in kallitype writing. Hall's work is valuable in part because of the insight and understanding he brings to working the kallitype but also because of his effectiveness in communicating the many particulars he worked out for each step of the process. Eighty years after its publication, Hall's monograph remains a landmark publication on the manipulation of the kallitype that no worker can ignore. Because of Hall's comprehensive grasp of every aspect of the complex process, and because of his ability to communicate so well what he had learned, Hall's monograph remains a primary source for the student of Kallitype.

III.   After Hall to Early Thomson.

After Hall, innovative and well researched writing on the Kallitype II process, kallitype exploration by amateurs began a twenty year decline. During this period, Frederick and Hall's processes continued to be favorably received, although some criticism was made of their complexity. The flow of articles recommending the Bennett-Burton method for its simplicity continued. During this period most writers on kallitype wrote short derivative articles based on earlier approaches—articles that offer little that extends theory or practice. The articles appear to be the product of casual workers who lack the inspiration, experience, knowledge and drive of the experimenter who searches for and tests new ideas. We shall notice these mediocre efforts and give them their due for keeping the kallitype alive in print. But we must be excused if the tale of their activity is brief.

Interestingly, the next major figure in our history, and one of the exceptions to the mediocrity mentioned above, is James Thomson, a Bostonian. Thomson began his long career of writing on kallitype and related topics with a response to Hall. James Thomson was to become the most prolific writer on kallitype of his time, and he remains so to the present, with little likelihood of being challenged. Thomson wrote over a period of about 20 years and published some 28 articles on kallitype and related subjects. Although he wrote on many aspects of amateur print making, his international reputation rested primarily on his work with the kallitype. He also wrote on salt printing, gum printing, and brown printing. On the topic of the individual preparation of printing papers using iron as a sensitizer, he wrote on cyanotype, uranotype, platinotype and kallitype. On the subject of kallitype, unlike the classic writers, Bennett, Burton, Frederick, and Hall, Thomson, after an early interest in K II, concentrated his efforts on the Kallitype I process and the brown print for the best part of his long career.

Thomson was a writer whose production appeared in profusion, sometimes four or five articles a year. This is a good deal of writing, especially when it is remembered that his articles were concerned with reports of experiments. Many of his articles were reprinted, sometimes by more than one journal, in this country and abroad. Thomson was also a letter writer, who often communicated his thoughts and feelings on a variety of subjects to editors who were pleased to publish them. His letters give the historian a sense of the man which goes beyond the little that is known of most other workers in kallitype, who remain anonymous despite the researcher's best efforts. It may not be an exaggeration to say that the history of the kallitype from 1905 to 1925 is dominated by Thomson's writing and by the writing of those who responded to it.

In spite of Thomson's dominance, the majority of amateurs before 1910 worked with and wrote about the Bennett-Burton approach and the Frederick-Hall approaches. We shall continue to briefly document the writers who periodically recapitulated these approaches, in spite of the fact that, as time went on, they added little that was new. The justification for this documentation is that, while such writing may not be innovative, it reveals that the attraction of the kallitype for the amateur persists and reveals changes in photographic culture.

IV.   James Thomson's early work on K II

Because this chapter is somewhat short and because James Thomson's writings are many, we shall end this chapter with a discussion of his brief involvement with K II. Thomson is one of the few kallitypists who began with a K II process only to abandon it for prolonged experimentation on K I and the brownprint. His work on those processes shall be treated in later chapters.

We first learn of Thomson in a brief article he wrote in the September 1903 issue of *Photo-Beacon*. Thomson indicated he had only recently begun working with kallitype, partly as a result of exposure to Hall's account of the KII process in *PHOTO-MINIATURE*. He wrote:

> Some time ago I invested good dollars in photographic chemicals and started to try for myself the various printing processes I had read about. In short, to prepare my own paper. I can say with truth that I have had the worth of my money in pleasure and knowledge . . . and have learned that there are other enjoyable features connected with photography outside the taking of pictures.
>
> Among other printing methods, I tried Kallitype, from the description given of the process in the *Photo-Miniature.* To those who have tried it, it will be no surprise when I say that it has one serious fault that seems difficult of elimination, and that is the bronzing of the shadows when using a negative of more than ordinary contrast. All the writers seem to ignore this objection to the process . . . [32]

The "bronzing of the shadows" Thomson refers to is a common problem that occurs in all iron printing, whether cyanotype, platinotype, or kallitype. This effect, often called solarization, appears when printing contrasty negatives. What happens is that when the printing is continued long enough to achieve high light detail, the tones of the shadow areas, often reverse, and print lighter, as exposure for the dense high lights is extended. The result is a strange tonal shift of the dark print tones from blacks to browns. In effect a two toned print is created. Thomson believed that this problem was resolvable by chemical intervention, and he began to experiment with various chemicals added to the sensitizer "to overcome this trouble." Like many experimenters who attempted to resolve this difficulty, he reported "sometimes I succeeded, and then, when most confident, mixing up a new lot of sensitizer, I would for some reason fail. Not only fail, but get entirely different colors with the same manipulation from what I had previously obtained." This variability of the Kallitype in which different results mysteriously occur from the "same" circumstances plagued Thomson throughout his career. We shall hear more of it as the years pass.

Early in his association with the kallitype Thomson was determined to publish, whether he had full control of a process or not.

> I have, however, succeeded more than once in getting beautiful results, and in the hope that others may profit by my experience, I have felt constrained to give my formula . . . [33]

Thomson published his first K II sensitizer in 1903.

Thomson's first K II Sensitiser

Silver Nitrate	30 grains
Ferric Oxalate	75 grains
Citrate of Iron and Ammonia	10 grains
Chloride of Copper	9 grains
Distilled Water	1 oz.

We can see immediately that he began with a modified Bennett-Burton Kallitype II sensitizer in the proportions 75:30: 1 and that he added two unexpected ingredients, ferric ammonium citrate and copper chloride. The addition of these two chemicals characterized Thomson's sensitizers over the years. While it is impossible at this late date to ascertain with certainty his reasons for including these two chemicals, one can speculate. It seems likely the ferric ammonium citrate, which is the sensitizer in Van Dyke brown prints, (which see) has been added to make the sensitizer print out a more visible image during exposure, as it does in the brown print. In the brown print the image prints out almost completely from the action of light alone, even with no development while the kallitype, even when well exposed, normally has only a faint suggestion of the complete image.

Thomson sought a kallitype that printed out a fully visible image. He believed such a process would enable the beginner to make a better judgment of when the print had received adequate exposure. Thomson may also have added Ferric Ammonium Citrate to the sensitizer to achieve a longer, flatter scale of tones, since the brown print is generally a less contrasty print media than kallitype. He may also have added ferric ammonium citrate to open up the shadow areas enabling them to print with less bronzing, or as it is often described, with less solarization.

The writer does not recall that Thomson ever explained the effect the copper chloride was to have on the print, and the effect is difficult to infer. One obvious effect of any chloride in the sensitizer would be the precipitation of silver. (Chloride plus silver makes silver chloride which is insoluble in water.) So the copper chloride would precipitate some silver and thus weaken the tones of the print. It could also be that the copper was responsible for the the unusual color of Thomson's print—a salmon brown. [34] When this writer tried this sensitizer, the prints turned out a muddy grey color.

Thomson developed his exposed prints in

Thomson's first K II Developer

Rochelle Salt	25 grains
Borax	45 grains
Water	2 oz.
Potassium Dichromate	4 to 20 drops
(4 grains / 1 oz water)	

This developer is conventional. After fifteen to thirty minutes of development the print is rinsed.

Thomson fixed his K II prints for 10 minutes in a bath made of 1 dram of Ammonia .880 in 16 oz. water. Thomson apparently had not yet been converted to the use of hypo as a fixer.

Thomson's early brief article on Kallitype II was reprinted in England in the Oct. 3, 1903 issue of *Photography*. In Thomson first approach to the the kallitype, we note several characteristics that continued through his 20 years of writing on the Kallitype:

a) he ignored what others were doing. He spent little time trying to work others methods or understand the ideas behind them. He concentrates on what he observes in his own darkroom.

b) he operated from an empirical or pragmatic base, as opposed to a theoretical base. His inclination was to try anything to see if something worthwhile happened. Thomson rarely tried to account for a result with theory, nor was he led by chemical principles.

c) Thomson was a "putter-inner," not "a taker outer." He habitually added chemicals to formulas, to his own, as well as to those of others. He rarely removed a chemical from his formulas, with the result that they appear to have everything in them, as one writer observed, but the darkroom sink. His formulas become long, complicated, and full of materials that sometimes make no sense from the point of view of theory or successful practice.

d) On the positive side Thomson communicated a spirit of intense concern for the kallitype process and an abiding enthusiasm for searching and experimenting.

On occasion the mix of these qualities makes reading Thomson painful. The absence of a theoretical bent makes his writing seem, at its worst, a succession of mindless formulas tried on the basis of whim not reason. But at its best, Thomson's writing has the excitement of a personal odyssey. He began each quest with high expectation which he richly communicated to his reader. Sadly, many of Thomson's explorations ended in disappointment. Frequently readers wrote that they could not get his formulas to work. Eventually the Thomson reader comes to understand and perhaps admire this man, who so often rushed to publish the results of each new set of trials, always with boyish hopes that "he had it." Somehow he managed to face the disappointments that so often came.

V.   Conclusion

In this chapter we have described the "tailored" approach to the preparation of kallitype II solutions. We followed its introduction by G.

W. Frederick and watched Henry Hall extend and perfect the method. Both sought to create an approach to kallitype that emphasized flexibility and convenience and permitted extensive variation of the kallitype print. We mentioned in passing a number of writers who rewrote the approaches these men created. Their approaches to the variability of the kallitype appealed to the strong interest in expressive pictorial printing popular at that time We also introduced the work on Kallitype II of James Thomson, an inveterate, if hapless, American experimenter who explored many kallitype applications, which we shall shortly take up.

In the next chapter we shall discuss the broad esthetic context in which the kallitype existed: pictorialism. We shall be interested in seeing how the esthetic interests of the time helped the kallitype prosper in some ways and troubled it in others. We shall also review the accumulating responses of amateur users to the processes that have already been described. Finally, we shall discuss in detail James Thomson's prolonged efforts to perfect an amateur approach to the seldom explored kallitype I process.

Notes Chapter VI: Frederick and Hall Tailoring the K II

1. A. Leonara Kellogg, "The Kallitype Process," *American Annual of Photography*, 1909, p. 231. A kallitype photo is reproduced with the article.
2. G. W. Frederick, "The Successful Employment of Kallitype," *The Camera*, Vol. V, # 9, Sept. 1901, p. 234.
3. Ibid.
4. Ibid.
5. Ibid., p. 325.
6. Ibid.
7. Ibid., p. 327-8.
8. Ibid., p. 328.
9. Ibid.
10. Ibid.
11. Ibid.
12. Ibid. 328-9.
13. Ibid., p. 330.
14. Dr. Herbert Mitchell, "The Successful Working of the Kallitype Process, *Photography*, January 2, 1902, pp. 14-16. This is the same article G.W. Frederick published in *Camera*, September, 1901. See above.
15. B. C. Roloff, "Answer to Correspondents," *Photo Era*, March 1902, n.p.
16. Foreign Abstracts, no author given, "Kallitype Processes" *Photo Era*, August 1902, n.p.
17. Unsigned editorial comment, **Photo Era**, September, 1902, p. 110.
18. Ibid.
19. B. C. Rolloff, "The Kallitype Process," *The Photographic Times*, Vol. XXXIV, #4, 1902, pp. 169-172.
20. John Tennant, "The Kallitype Process," *The PHOTO-MINIATURE*, Vol. IV, No. 47, February 1903, p. 513.
21. Henry Hall, "The Kallitype Process," *The PHOTO-MINIATURE*, Vol. IV, No. 47, February 1903, p. 515.
22. Ibid.
23. William de Wiveleslie Abney and Lyonel Clark, ***Platinum, Its Preparation and Manipulation***, New York, Anthony and Scovill, 1898.
24. Henry Hall, Ibid, pp. 521-524.

25. Henry Hall, Ibid, p. 524.
26. Ibid. p. 530.
27. Ibid.
28. Ibid.
29. Ibid. p. 532
30. Ibid.
31. Ibid. p. 546.
32. James Thomson, "Modified Kallitype, *Photo-Beacon*, Sept. 1903, p. 274.
33. Ibid.
34. Ibid.

Chapter VII

Pictorialism

I: The Historical Context of the Kallitype 1890-1925

In this chapter we will consider the kallitype as it relates to photographic practice and esthetic values between 1890 and 1930. We have already considered photography in mid-Victorian times when a selection of upper-middle class people began to consider photography as a science to be explored, a commercial prospect to be considered, and as an expressive medium for art. We later viewed photography in the historical context of 1880 when middle class amateurs began to champion artistic interests and values over technical and commercial interests. It is important that we look now at the kallitype in the historical context of the turn of the century. During the 50 years that elapsed between 1880 and 1930, photography developed a new focus in a society that was undergoing fundamental social changes. We will make an effort to see if changing circumstances in society influenced the development of photographic values and ultimately the manner of making photographs.

We noted earlier that photography in the middle of the nineteenth century was an extremely demanding process—one which required a combination of knowledge and skill to control difficult processes. We observed that before 1880 the heavy burdens the wet plate process imposed on the photographer. Because photographic supplies were not then fully commercialized, the photographer had to prepare materials his work required often at the site of picture taking. At that time so much of the photographer's attention and energy went into getting the apparatus and the chemicals to produce a photograph that success was largely defined in terms of flawless execution. That is to say, the operation of image-making was regarded as artistic, if the photographic print described accurately and without flaw what the camera had been employed to record. As a result of the emphasis on solving technical problems, the image on the print, not infrequently was regarded as mechanical and prosaic by sensitive and creative viewers. Believing that plain description was all the medium would ever achieve, critical viewers dismissed photography and its productions as being unworthy of the attention of the cultivated individual. Story describes the common rejection of photography at this stage of its development.

> There was a time when the art world used to sneer at photography. They said it was going to ruin art with its crude stiff facsimiles of objects, like enough to pass for the real thing, but in truth so bald and lifeless, and moreover, in general detail, so distorted and awry as to destroy all beauty, and in the end, to wither the taste of those who put up with its vapid and soulless mimicry. There was some truth in the charge, of course—enough to weight it, in fact, and almost to drive it home; but that was in the early days, ere the light-picturer had time to grow. There are few who would venture to repeat the denunciation now [1898] . . . for, it was soon found that, like the ugly duckling, it was a thing of power and of potential beauty. It came amid muttered curses and maledictions; it stayed to aid and to bless.[1]

We get a further insight into the prevalent negative assessment of photography when we consider the way photographs were exhibited. In the 1880's photography was exhibited in places like modern state fairs. In England exhibits were organized by the Royal Society of Photography, an organization which attempted to promote three as yet undifferentiated concerns: the advancement of the science and technology of photography, the extension and refinement of commercial photography, and the advancement of photography as an art. The order of the list may be taken to describe the priorities of the institution and of most members. George Bernard Shaw, in his customary acid manner, commented on the Society's early exhibitions and its later compensatory activity.

> The Royal Photographic Society mixes up optics and fine art, trade and science in a way that occasionally upsets the

critical digestion. It is divided between two quite incompatible interpretations of the word exhibition, which means sometimes a huge international display of industrial products, with gold medals for future use as advertisements, and sometimes a collection of works of fine art. To complete the muddle, the Royal Photographic Society has been so effectually laughed out of its old actions that photographs are to be esteemed according to certain technical conditions in the negative, that it has now arrived at the conclusion that a pictorial photograph is one in which the focusing and exposure are put wrong on purpose. Consequently, whilst it solemnly medals some of its exhibits as if they were sewing machines, it is afraid to give a medal to any picture which does not look more or less mildewed, lest it should be ridiculed for Philistinism. [2]

Shaw's remarks clearly announce that values were shifting from objectivity and accuracy of reproduction to expressive communication through creative manipulation of the photographic process.

In 1892 a group of artistic members seceded from the Royal Photographic Society to found the Linked Ring Brotherhood, an organization dedicated solely to the advancement of the art of photography. By the time Shaw wrote his comment, (in 1892) new approaches to the exhibition of artistic photographs were already in place. [3]

In 1891 the first international salon for art photography was held in Vienna. This was a high point in the establishment of artistic values and process for photography. It followed closely a similar salon held in Paris that exhibited new artistic efforts in painting.

We have already noted that by the 1880's photographic devices and processes were becoming less problematic, from the point of view of control and use. As a consequence of the ready availability of dry plates, commercial print papers, and smaller, more portable cameras, there was a sudden growth of all types of amateur activity and a proliferation of photographic societies and clubs. The greater ease and convenience of products developed by the growth of photographic technology opened the medium to the photographically naive "amateur"—in fact to everyone.

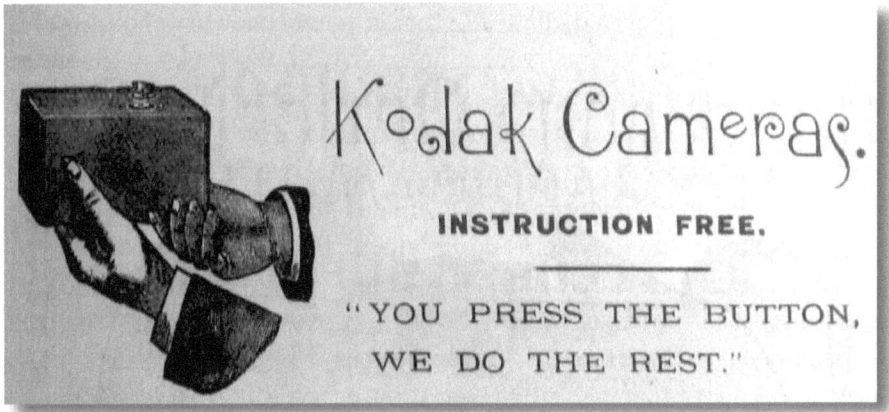

Figure 8. Kodak's Famous Ad. "You press the button"

Kodak's famous "you push the button" slogan coined at this time, attracted many who had no inclination to explore advanced photography or use it as a medium for art. Such photographers wanted only to "take" pictures. The growth of consumer-driven photography led to a phenomenal increase in the production of photographs taken by all kinds of people for all kinds of reasons. Some of the photographs revealed accurate descriptions. Others caught spontaneous, fresh, everyday activity. Others revealed the world in out of focus black and white snapshots. But most photographs were so casual and devoid of visual taste, that photography as a medium lost, rather than gained, respect. Franklin Jordan, commenting on these developments, wrote

> It was not long after photography had been developed into a workable process for making pictures that serious workers felt the need to modify the strictly mechanical results that it gave. As a straight record, it left nothing to be desired. For the realist it was, and is, a complete medium. But what about the man and his audience who were not greatly intrigued by a plain statement of obvious fact, who wanted to say more than the obvious, something a little subtle that would convey his message by suggestion and indirection, that form of expression which has ever delighted painters, writers, and musicians? The undiluted medium was too baldly literal for their niceties. [4]

Earlier we reported that as the convenience of photographic materials, equipment, and processing improved, the ranks of amateurs grew, and the number of processes available for printmaking multiplied. Some of the commercial print processes marketed at this time were easy and required little "control" (silver bromide) while others permitted considerable modification of the image by the printer. Sensitive photographers and viewers complained that prints produced by commercial processes had a 'mass produced look'. Prints produced by artist prepared processes were viewed as possessing an individualized appearance characterized by "distinction and elegance." Individually prepared papers that permitted variation enabled the photographer to make prints that looked uncommon. In the later 80's and 90's a class of "amateur photographers began to employ print processes that aimed at individualized esthetic impact. As a result of such efforts it became clear to the sensitive that photography was not doomed to be prosaic. Rather, photographic art began to be seen as limited only by the imagination and skill of the photographer who worked with processes that permitted individual manipulation.

II.   The Question: Is photography an Art?

After 1890, the question, "Is photography an art?"—a question that for years had usually been answered in the negative—suddenly began to be asked with greater urgency. In many journal articles and at many club meetings photographers asked quite seriously, 'Could photography rise to the status of an art among traditional arts?' They inquired 'What qualities made photographs artistic?' These questions could be answered in the light of the old notion of accurate depiction or they could be answered in terms of some new conception of photography. To answer such questions in the light of the newer photographs being made required a rethinking of the medium. For the emerging *avant garde* photographs often had little to do with accurate depiction. A parallel situation was occurring in painting. The new impressionist painters also were demanding the acceptance of new ways of painting that went beyond the traditional studio disciplines.

As the 90's proceeded, sensitive photographers increasingly made photographs in which their dominant concern was the creative manipulation of the photographic medium for expressive purposes. Writers on photography labored to create new goals for photo artists.

Critics and photographers—those in favor of the newer work and those against it, initiated a dialectical exchange that continued for some fifty years in photographic journals. (Some say the discussion they initiated is still going on today. Those in favor of change tried to establish a new conception of what photography is and should do. They wanted to align photography with the new esthetic values continental artists claimed for traditional mediums like painting and drawing.

By 1890 the key problem of photography was no longer a scientific or technological one. In 1889, one of England's leading research scientists in photographic theory had already lamented the fact that scientists were abandoning the theoretical study of photography.[5] The technological needs of photography, after a series of brilliant innovations, were regarded as presently being met. While complacency about technical progress was by no means justified, it was a time when the needs of photography were already commercialized and reliable products for most needs were available. The time had passed when photographers had to investigate, and produce the products they worked with, a generation earlier. The time had arrived when well made, but hardly exciting, prints could be taken for granted.

The new "buzz" in photography was the cry for art by a class of artists who were concerned about a new esthetic conception of photography. Photographers continued to preserve the "straight" values but they were becoming a minority. As we have seen and shall continue to see, kallitypists often were a part of that minority, believing that the "real" photographer was one who investigated his medium and prepared his solutions and papers to satisfy the demands of excellent craft. But many kallitypists were concerned with the kallitype's possibilities for making "Art" that communicated creativity.

The new class of "artistic" photographers were divided about whether photographers should be "technicians" or creators who expand and liberalize process to achieve more powerful expression. Some amateurs asked why should artistic photographers concern themselves with process investigation, when easily controlled print materials were available. Gillies focussed the basic question. Should the photographer spend his time "looking for pictures or playing with chemicals?"[6] Against this view there were numerous published complaints that the craft of photography was dying with the arrival of convenient but vapid commercial processes. Some felt that contemporary printmaking had been simplified till little remained but "the slavish following of recipes."

After 1880 complaints abound about the newer films and papers that offered the photographer few creative possibilities and little intellectual challenge. The implication was that photography without process decisions and media variability gave the worker little sense of personal involvement and little pleasure of accomplishment.

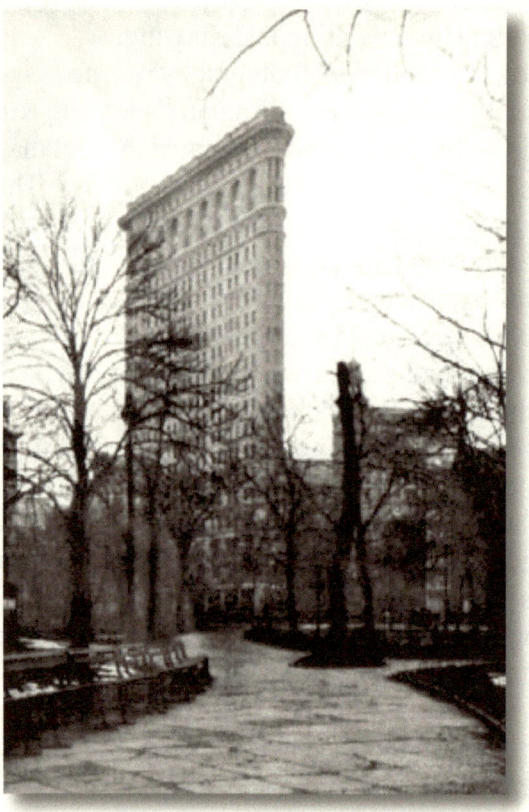

Figure 9. J.M. Mc Corckle, pinhole photo "Flatiron Bldg."

But a problem remained that called for attention: the explosion of dull and banal photographs. The easier the medium became the more people made more artless snapshots.

The result was more critics repeated the idea that photography was limited to artless mechanical productions. As the century turned, photography had its first serious identity crisis. It had been one thing; it now wanted to be something else. It could even be an art. But what would be the identifying characteristics of an art of photography? of the artistic photographer? of the esthetic photograph? If photography were to be an art these difficult questions awaited answers. Here is a resume of some of the questions raised.

Questions about *technique* were raised. Are technically sound negatives essential for making an artistic photograph? Should printing aim at accurate, flawless, depiction? Should photographic prints hide or display the photographer's "hand"?

Questions were asked about the qualities photographers should cultivate. Should the photographer aim to be an excellent technician? Should the photographer concern himself with knowing and controlling the chemical and technological qualities of his medium? Was the artistic photographer responsible for extending the medium through experimentation and investigation? Should the photographer develop, not manipulative skills, but rather sensitivity, imagination, creativity, a knowledge of art traditions?

Questions were raised about the art photograph. Did the art of the photograph depend on the quality of objective representation or on the vision conveyed by a creative, less objective image. Was it acceptable to express ones idea by means that circumvented traditional technique to achieve an emotional response through image manipulation? Is the ideal photograph a technically arresting representation of the objective world or is it an image variously modified, that engenders a moving subjective response? If the ideal photograph is a sensitive, well made individualized print, what are the visual characteristics of successful prints ?

Should art photographers establish new institutions to help photography become a fine art subject, to its own laws. Questions like these were asked and answered many times in the period we are considering. We include such questions to suggest the ideological context in which the kallitype existed. Objectivity was eventually redefined during this time as having expressive possibilities by artists like Stieglitz, Evans, and Strand. They continued to exert a conservative influence on printing as well as a voice for the value of expression.

The new force that began to make things happen was a group of sensitive, articulate, artistic photographers who insisted on a new focus for photography's identity: the individualized handmade print.

The new artistic photograaphers believed that appropriate approaches to selection, processing, and presentation of photography should be developed and they set about creating them. Finally they believed that a new set of ideas on the criticism of photography as an art should be written and disseminated. They set about publishing their ideas in the new photographic journals that recently proliferated.

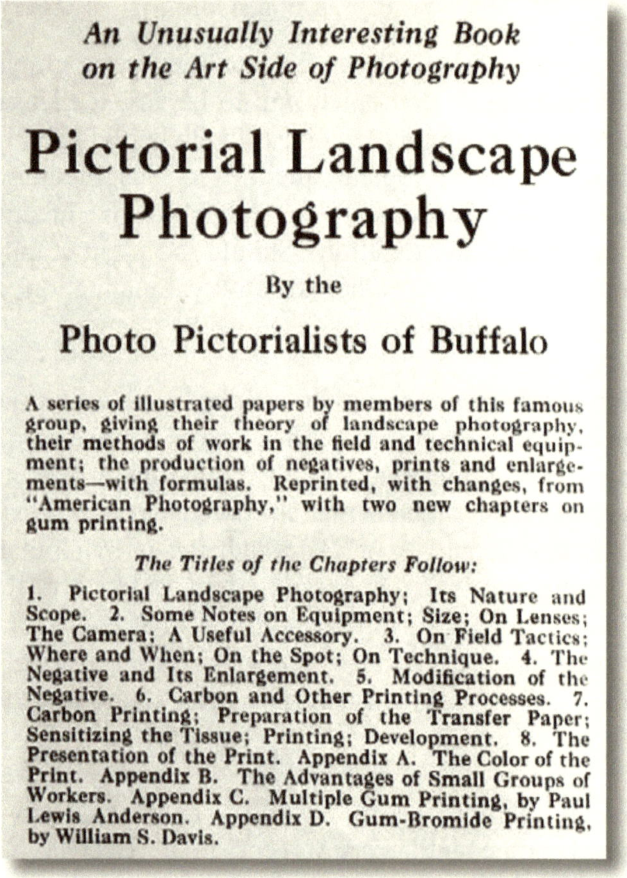

An Unusually Interesting Book on the Art Side of Photography

Pictorial Landscape Photography

By the

Photo Pictorialists of Buffalo

A series of illustrated papers by members of this famous group, giving their theory of landscape photography, their methods of work in the field and technical equipment; the production of negatives, prints and enlargements—with formulas. Reprinted, with changes, from "American Photography," with two new chapters on gum printing.

The Titles of the Chapters Follow:

1. Pictorial Landscape Photography; Its Nature and Scope. 2. Some Notes on Equipment; Size; On Lenses; The Camera; A Useful Accessory. 3. On Field Tactics; Where and When; On the Spot; On Technique. 4. The Negative and Its Enlargement. 5. Modification of the Negative. 6. Carbon and Other Printing Processes. 7. Carbon Printing; Preparation of the Transfer Paper; Sensitizing the Tissue; Printing; Development. 8. The Presentation of the Print. Appendix A. The Color of the Print. Appendix B. The Advantages of Small Groups of Workers. Appendix C. Multiple Gum Printing, by Paul Lewis Anderson. Appendix D. Gum-Bromide Printing, by William S. Davis.

Figure 10. Pictorial Photo Instruction Book

Turning photography into an Art during the period from 1890 to 1920 became a crusade for well known photo artists like Stieglitz, Davison, Demachy and Coburn. They and many other passionate photographers on both sides of the Atlantic undertook a variety of projects to establish photography as an art. These involved the creation of photographic societies, salons, exhibition galleries, and publications to promote the new idea and their new work They created and supported new creative treatments for negative and prints that enabled original expression. And to a great degree they accomplished what they intended. Because of their efforts, photography became something more free, more plastic, more original, and more respected than it had been. In time, a new label was attached to the kind of photography concerned with art: "pictorial." The new prints were called "pictorial photographs." Pictorial prints were made any way that communicated individual vision and feeling.

We will join others in calling the period during which this activity occurred the "pictorial period" of photography. In the next sections we shall make an attempt to describe selected pictorial developments, which bear directly on the history of the kallitype. In coming sections we shall discuss at greater length what pictorialism entailed with respect to the ways of making art, the appearance of the art, and the subject of the art. We shall address these issues as they relate to the kallitype. In particular we shall concentrate on describing pictorialist conceptions about photographic printing that influenced the development of the kallitype.

The real problem of the Kallitype photographer was to decide on which side of fence—art or technique—he or she should camp. Was the kallitypist an agent working toward new values—a force for creative expression through individualized process—or was he a conservative force trying to preserve the old ideal of technical mastery.

III: Technique and Art in the early 1900's

> "Isn't it too bad your photographs are not paintings. If they had been made by hand, they would be Art."
>   Said by a German Painter to Alfred Stieglitz.[7]

In 1903 C. Sordes Ellis, F. I. C., F.C.S., published *An Elementary Chemistry of Photography.* During the time of the Pictorial revolution, works on the chemistry of photography became popular as amateur interest in kallitype and the profusion of alternate printing methods grew. Experimentally minded amateurs wanted to know more about photographic chemicals, their properties and their effects, and their combination in scientifically balanced formulas in order to make more powerful and different expressive prints.

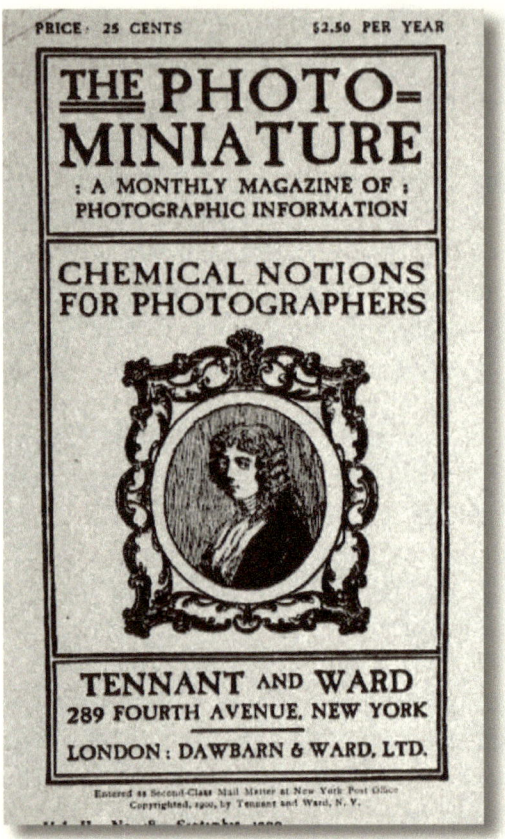

Figure 11. *Photo-Miniature* Monograph,
"Chemical Notions for Photographers"

To do this they wanted to know how to make their own photographic chemicals, especially those which were not readily available from local suppliers, or if available, were not always in prime condition. Ferric oxalate was one of the compounds of great interest to the kallitypist. It was often unavailable or, when it was in supply, its condition and efficacy were often suspect. The chemical manuals of the day explained why compounds deteriorated, the signs of print deterioration, and what happened to the activity of the photographic solution when chemicals went "bad." Photographers also wanted to know about acids and bases, how they influenced the activity of photographic solutions and how they could be measured and controlled. **PHOTO-MINIATURE** published a series of monographs on photochemical topics for the photographer during the pictorial period.

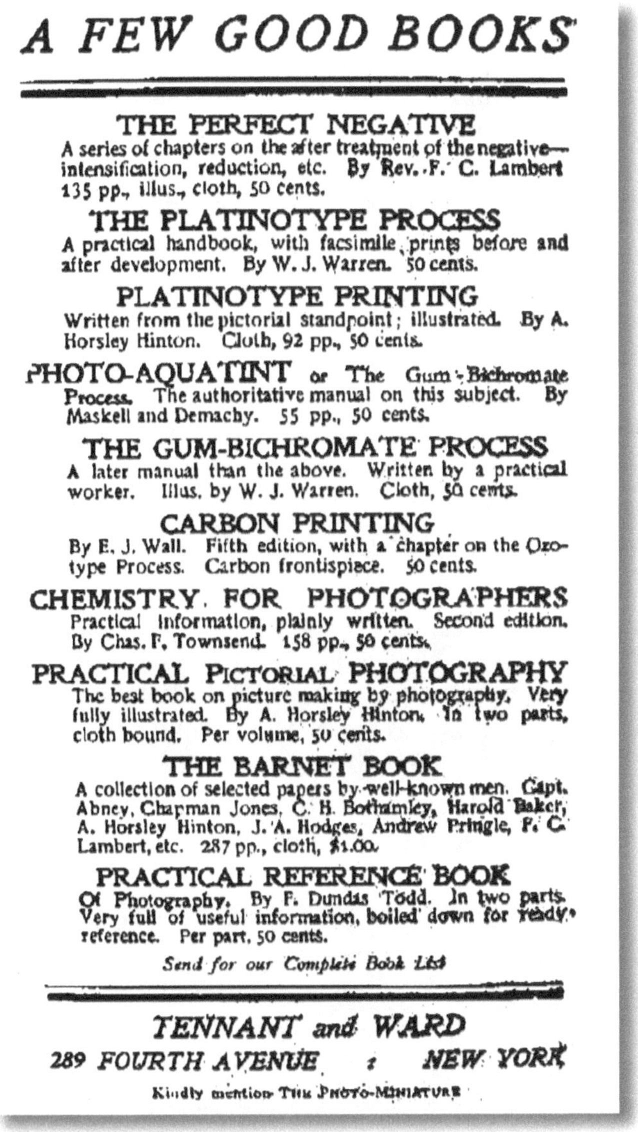

Figure 12. Tennant & Ward Process Monographs

They explained such topics and provided basic chemical information on the use of equipment (scales, glassware, etc.), the mathematics of chemical reactions, and the dangers of common photographic chemicals.[8] As Henry Hall reported in the preceding chapter, photochemical manuals were very useful to kallitypists who wished to understand more about control of the materials used in this "often puzzling" process.

The eighteen nineties were a time when art-oriented photographers valued knowledge of technique. In his critical book, *Principles of Pictorial Photography*, Gillies, emphasized the need for photographer-artists to develop greater awareness of chemical and photographic technique. He added that it was equally important for the technically competent to develop esthetic sensitivity.

> There are a great many pictorial workers who have so successfully evaded the technical side of photography that, unless they turn back and learn something about proper chemicals and equipment, their further progress is barred. Then there are those good technicians who cannot see a picture. Something of both technician and artist is demanded of the real pictorialist, his technique being so perfect that he uses it without thought, free to give his entire attention to . . . making the picture. [9]

Gillies' remarks document the tension that existed between the concern for mastery of photo-chemical technique and the concern for artistic expression at this time. They suggest that art, while much concerned with expression still demands technical mastery. Gillies leaves little doubt which of the two concerns was more valuable from the pictorial point of view.

> The business of the pictorialist is to make pictures, and when he has exposed the negative the work is ninety percent done, the rest being sure knowledge of the medium. If he chooses to dabble in chemical processes, he can very well remember that he is satisfying his curiosity in a chemical and not in a pictorial direction.[10]

IV.   Pictorialism and Photographic Technique.

As early as 1883, The *Philadelphia Photographer* published a statement on "Photography and Art" written by Charlotte Adams. Adams saw new esthetic attitudes were developing and made an attempt to clarify them.

> A new school of photography is arising, notably in New York, the actual art center of the United States . . . This new school of photographic art is not satisfied with merely producing likenesses or literal representations of figures or objects. It accords to its subjects, and seeks to employ in its methods, approximately the same treatment that would be given to a picture in oil or watercolor. Pose, composition, both of ensemble and of detail, warmth or coldness of tone, breadth, force and delicacy of treatment, effects of light and shade, the various qualities of color produced by different handling of black and white, are all considered by the artistic photographer of today. The progress of American photography is keeping pace with that of American art, and the . . . reciprocal influence of these two forms of artistic production, hitherto considered widely distinct, are every day becoming closer . . . The term" photographic" is losing its contemptuous significance as applied to the character of a picture, and on the other hand, the highest praise that can be accorded a photograph is to say of it that it is artistic or pictorial. [11]

By the end of the century, the major trend in artistic photography in America and abroad was a loose collection of principles, practices, values, and process preferences that now go under the name of Pictorialism. We must try to specify some of the photographic qualities that were supported.

In ***Pictorial Photography***, Gillies suggests that "pictorial photography," was concerned with the use of photography for "individual expression."

> . . . It was not until the eighties, when photography had been enormously simplified, when the modern amateur arrived and photographic societies multiplied like flies in summertime, that there was any glimmering of the use of photography as a means of individual expression and what we today call pictorial photography came up over the horizon.[13]

As with most art movements, Pictorialism had as many definitions as writers defining it. Modern Scholars tend to define it within large and abstract structures of thought. Taylor, for example saw it as a conflict between individual expression and esthetic convention.[14] Gillies' description was expressed in terms the artists of the time often used. He saw the pictorial movement as a concerted effort to differentiate the artistic photograph from the snapshot and the record photograph.

> It was the beginning of a new day and rose, here and abroad, in the discontent of an increasing number of serious amateurs with the inanity of the "snapshot" and the coldly mechanical perfection of the record-photograph of the time.[15]

Elsewhere Gillies wrote, the "pictorial" photograph "has a special 'art' value not found in other photographs, they being of a merely reproductive nature." The further problem of identifying and specifying what was artful in a pictorial photograph remained. During the early years of the pictorial movement the issues often seemed simpler than they were later perceived to be, and the answers that satisfied photographers then may now seem too simple and too easy. The writers of the time should not be scorned for doing no more. Some of their questions have not been resolved to this day.

We shall follow Gillies lead and view pictorialism as evident concern with expression through artistic manipulation of the photographic negative and print.

Historically, the task of defining the esthetic values of pictorial artistry and the acceptable qualities and limits of media management fell to the competitions and the salons. Taylor's analysis of British pictorialism suggests the esthetic problem of the pictorialists, as distinct from the political one, was the resolution of the conflicting claims of individualism and convention. The artists wanted to be free from conventional expectations of what a photograph should look like in order to create and express, in freedom, a personal response to their world. The dialectic between convention and individual expression in photography occurred within a larger and essentially similar dialectic that occupied French painting and drawing. That confrontation involved similar problems about individuality, creativity, and expression, on the one hand, and the restraints and dictates of conventional painting discipline on the other. [16] The most obvious painting example of media change was in the impressionists championing of a variety of new brush stroke techniques and approaches to color, which clearly were in opposition to traditional techniques based on objective representation.

The pictorial dilemma over photography's technique involved the acceptance of a number of liberal practices of printmaking. Some of the most common were: the acceptance of individual sensibility as a value, the approval of certain categories of subject matter and treatment, and the approval of media strategies and manipulative technique similar to those used in traditional art media, especially painting and drawing.

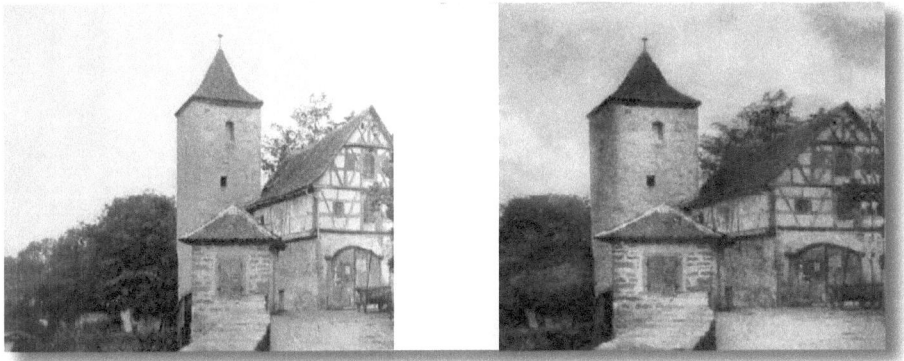

Figure 13. Example of Photomanipulation by Brushing Oil Pigment on print

Comparison of a straight print with a print made with "treatment" by the bromoil print process. Pictorial photographers placed low value on certain practices, most notably the practice of sharp focus, full scale of light to dark values, and "straight" printing. Pictorial print makers endorsed hand manipulation or marking of negative and or print. For the pictorialist, "expression" of the individual's response to the world might be achieved by any

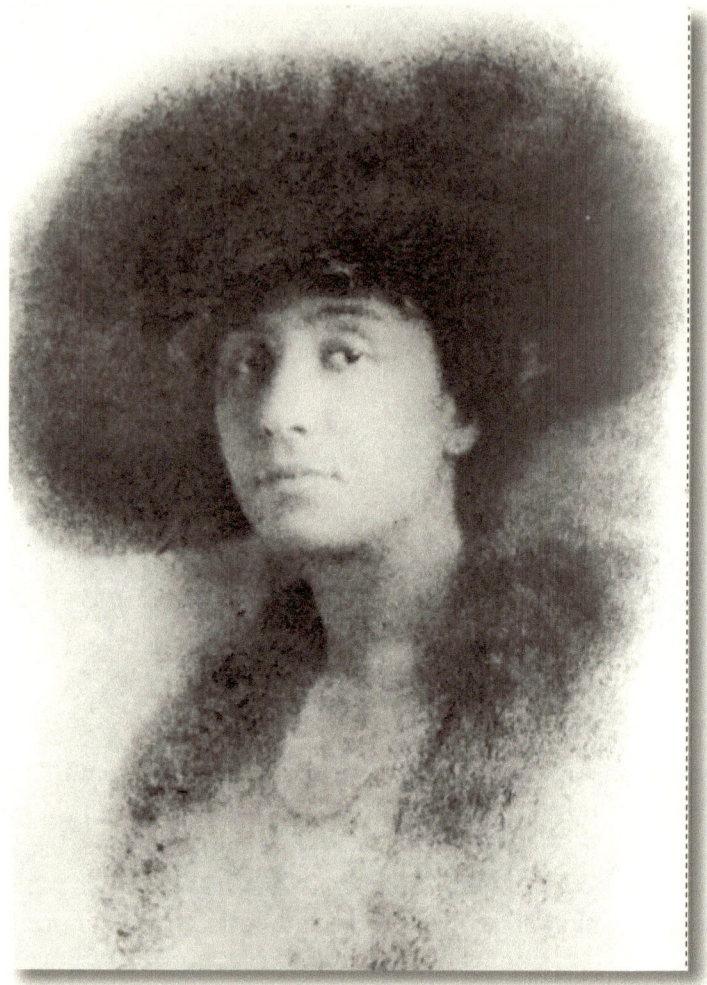

Figure 14. Robert Demachy, Gum Bichromate Print

means available: soft focus lenses, selection of vantage point, composition, the handling of light and shadow, the alteration and exaggeration of contrast, extensive hand manipulation of the negative, the choice of tone or color of the print, manipulation of the print, the choice of textured papers for printing, and refined presentation etc. Any or all of these media options could be used to communicate the maker's unique personal response. They could also clearly signify the photographer's interest in being recognized as a pictorialist. Pictorialist photographers of the time were criticized by traditional critics for their emphasis on a variety of subject matters and out of focus treatments.

Pictorialism encouraged media exploration and the hand-influenced negative and print. Any technique or print "look" was acceptable so long as it was a fresh exploration of media possibilities and expressed something. Pictorialism was an expansive esthetic force in the history of photography. It strongly emphasized the centrality in art of media freedom and originality so long as the liberal treatment enhanced expression.

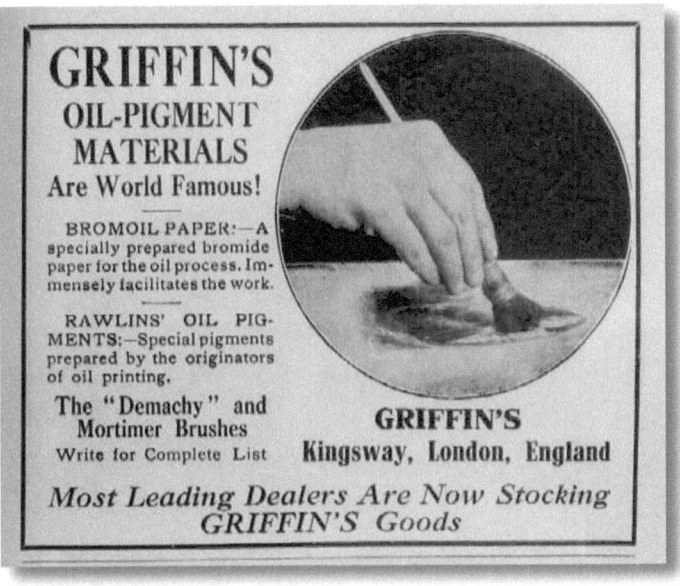

Figure 15. Oil Pigment & brush for painting on Photos and Negative

Of course the media freedoms Pictorialism encouraged could be abused by photographers and quickly they were . . . by photographers who lacked concerns to express or those who had little understanding and taste in their use of materials and process. As Shaw commented earlier, any out of focus muddy image could be and often was defended as pictorial. Nevertheless, it is possible to see the pictorial revolution as the source of the freedom of photographic artists today, when artists commonly employ any means to the end of original expression.

As time passed, the revolutionary aim of the pictorialists became associated with particular manifestations. One manifestation was interests in certain subjects—"scenic" rural nature, the nude, the still life and the portrait—became identified as pictorial subject and pictorial treatment. One manifestation was

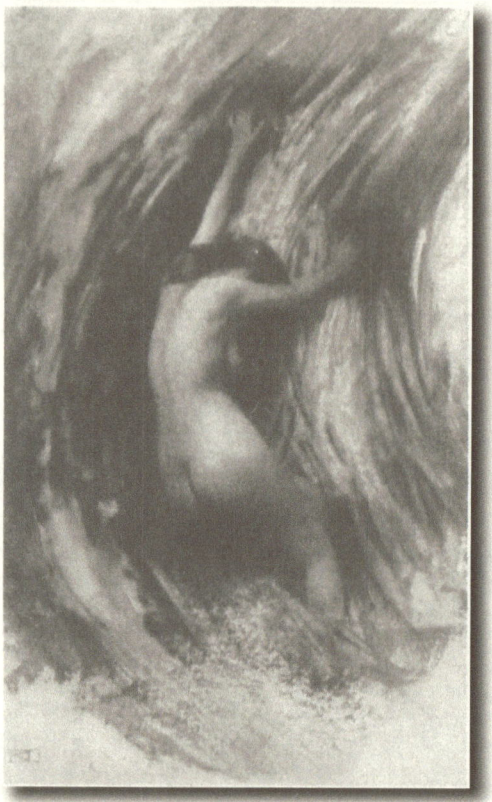

Figure 16. R. Demachy. Brushwork on a photographic print

impressionistic or indistinct "treatment" of subject matter. This was variously accomplished by the focus of the camera or enlarger, by hand manipulation (drawing, blurring, marking, screening etc) of the negative and print. Still another technique was the avoidance of plain development and "straight" printing and commercial papers. Special individually manipulated variable print processes must be included here. For example, gum bichromate, kallitype and platinotype. were commonly recommended as pictorial processes The use of chemical and tonal" variations" at times became stereotypical behaviour. Some variations were merely trendy and tended to be used with as little expressive potential as the objective records pictorialists set out to replace.

Interest in certain subjects and modes of expression often led to a stereotypical "soft look" in the work of many photographers. While the subjects and the treatment are by no means identical from photographer to photographer, common emphases in subject and technique were readily observable.

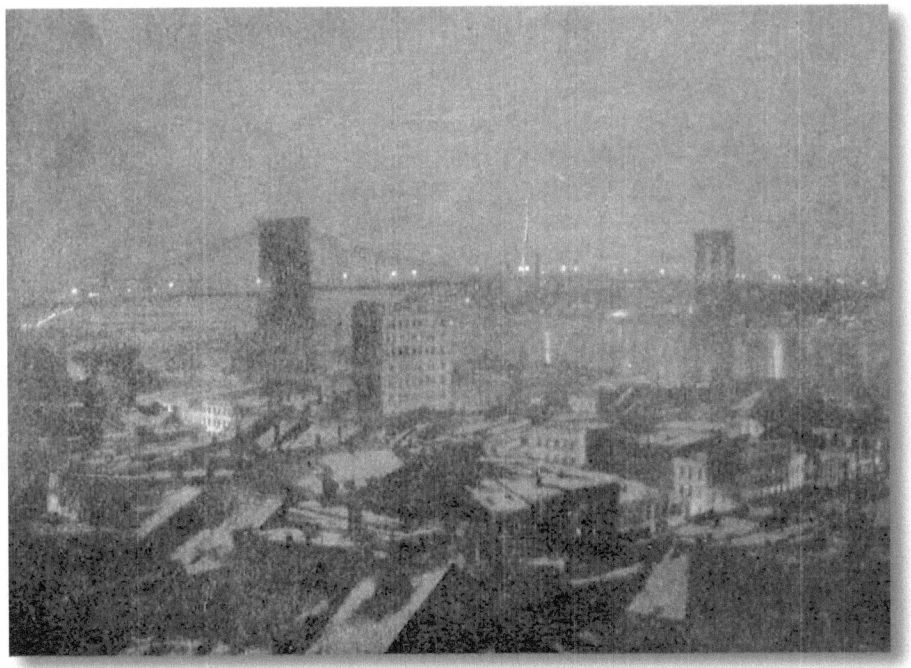

Figure 17. J. F. Strauss, The Bridge. Popular heavy dark tones

Pictorial photographers often exploited the affective qualities of light and shadow, suppressed detail, and emphasized mass and shape in order to communicate a subjective response to a world the camera tended to make all too definite.

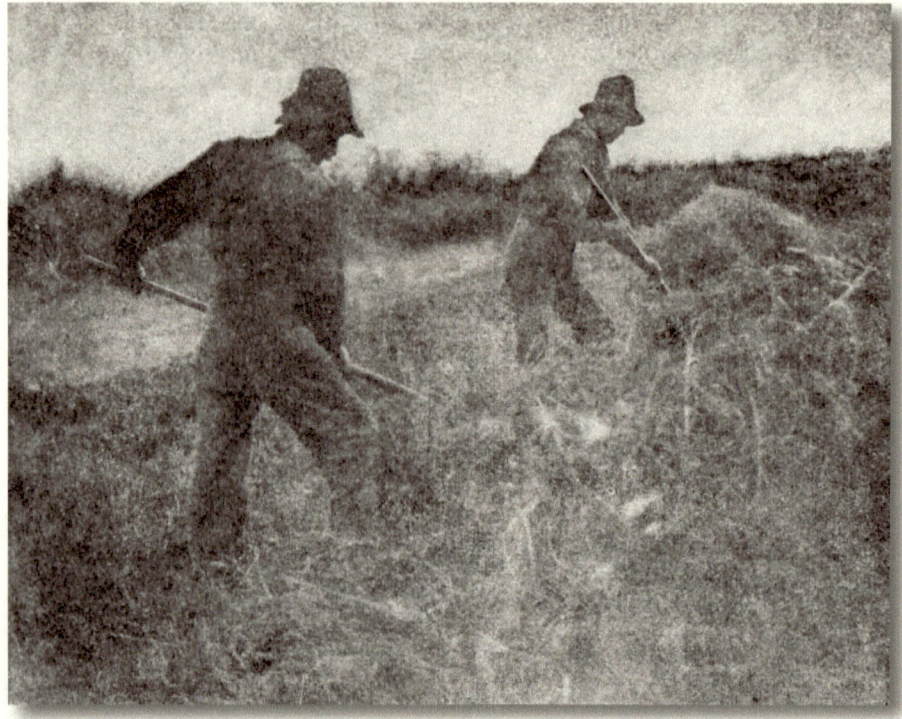

Figure 18. Typical impressionistic "fuzzy" rural image

One critic complained of "the incessant nocturnes" and the too frequent "mid day harvest scenes done in a few tones of dingy brown."[17] Critics allied to the pre-pictorial esthetic values of precise representation, objected that pictorial photographers, disdained the real—"which God had made"—and complained that pictorial artists all too frequently made photographs that denied technical credibility, that is, were unsharp, murky, and plain fuzzy. Traditional critics charged that pictorial photographers were either perverse in their attitude toward photographic discipline or incompetent.

Early in the movement, writers referred to the pictorial look as impressionistic, and traced its influence to impressionistic painters, English and French, especially James MacNeil Whistler. The avant garde of photographic pictorialism felt they were a part of an important crusade which they shared with revolutionary painters of the time. Will Cadby, writing in 1904 observed,

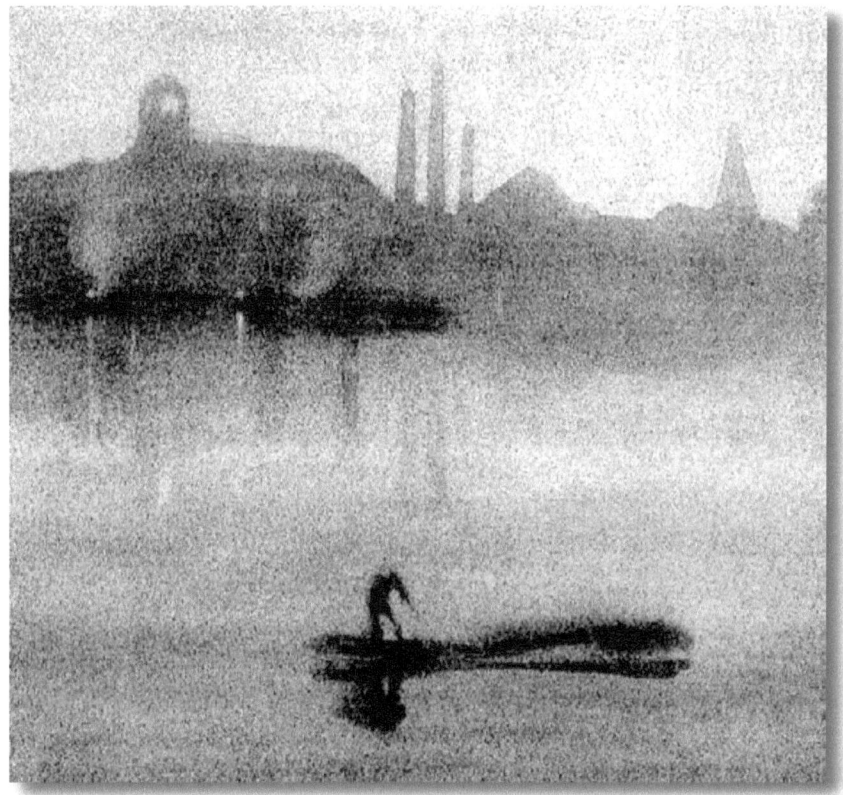

Figure 19. Whistler, impressionistic painting that influenced photographers

> few camera workers [before 1900] could have found in
> Whistler's utterances on art much that they would have had
> the temerity to apply to their own craft, such is not the case
> now. Photographers have become bolder, and take their own
> work more seriously. [18]

Liberated photographers and painters would arise and together would save the arts from the canons of realistic conventional realism.

Certain attitudes about photographic printing evolved out of the pictorial dialectic on principle and practice. From a technical or process point of view,

1) the pictorialists opposed making an automatic record (negative) of what was in front of the lens and

2) the pictorialists, generally opposed the printing of an all too direct print of what the film had recorded.

There were important exceptions, most notably, Frederick Evans, whose straight depictions of cathedral interiors come to mind. But it is notable that he printed on platinum paper, approved by the pictorialist credo. Regardless of their objectivity, few critics regarded Evans prints as bereft of sensibility in view of his moving use of light and shadow, detailed representation and sensitive composition.

Photographers were advised to make images that transformed the real world by the expressive force of individual sensibility, thought, and feeling using manipulative techniques. For example, if a photographer wished his landscapes to be considered artistic, he was advised to compose what nature offered according to known artistic principles of composition, or at least according to his own cultivated sense of organization, before he snapped the shutter. Photographers were advised to learn principles of composition, by studying great master paintings or their explication in the books by competent writers, such as Robinson and Hammond from within photography, or Van Dyke and Claffin, etc., from without.[19] In the printing room photographers were encouraged to further nudge the image away from direct depiction toward subjective interpretation. When printing they might accomplish this by employing one of a variety of "plastic print processes"—especially those which allowed the hand to shape and emphasize, and thereby to express the subjective response in the print. While making the print, photographers were encouraged to employ those printing techniques which aided expression, for example paper negative manipulation, combination printing, metallic toning, etc.

The pictorial interest in expression interacted with the lively interest in photographic processes of the time. The two interests reinforced each other. The pictorial concern supported elegant, hand-controlled printing, and the process movement in printing supported the growth of expression in photography. William Mortenson, in a book on *Pictorial Lighting* published in 1937, taught

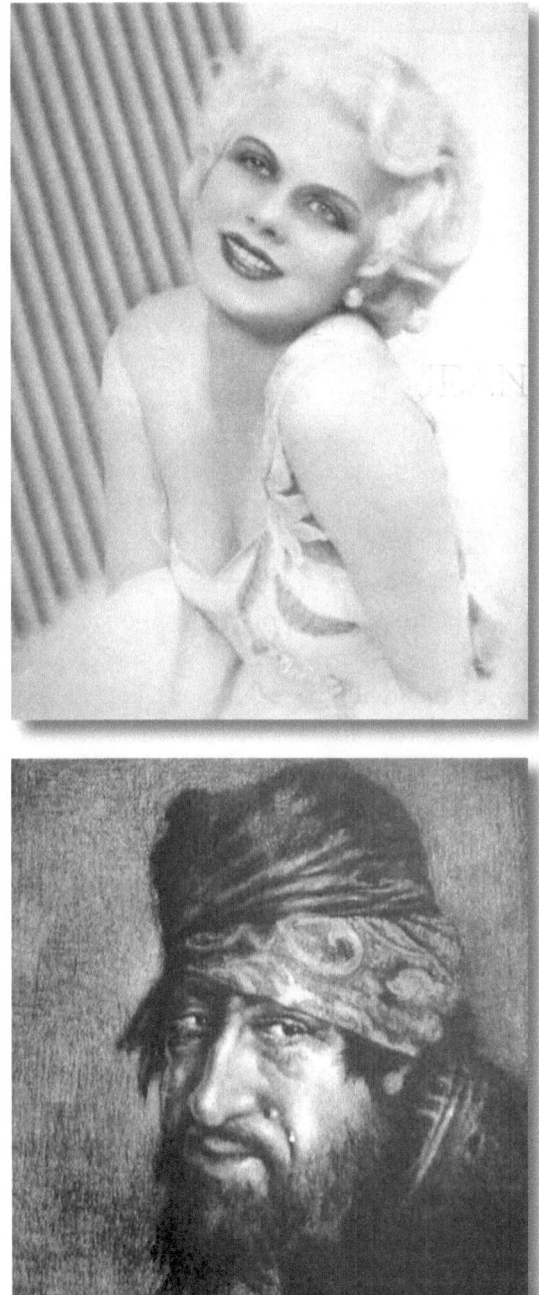

Figure. 20. William Mortenson, hi key and low key Images

a much admired use of pictorial techniques that extensively modified his original negatives and prints to present highly poetic concepts. The print processes that merely reproduced, those which had little capacity for individualized control, such as albumen silver and silver chloride—such media were belittled and ignored by the the pictorialist. Print processes that permitted variation and control of the color and surface qualities of the print, such as kallitype, platinotype, and gum bichromate were seen as offering pictorial opportunities. Those processes that allowed considerable hand manipulation, such as gum bichromate, and carbro, were used by photographer artists with great enthusiasm because they facilitated expression by offering expressive possibilities through brush manipulation. Pictorialists were even more enthusiastic about the very plastic photo media such as platinum glycerine resist or bromoil, in which hand manipulation of the photograph could be so extensive that the photographic identity of the finished print is left in doubt. Wheeler in his manual on printing processes had this to say about the importance of printing control for the pictorial photographer.

> The great photographic exhibitions held yearly by the London Salon and the Royal Photographic Society convey a good many lessons, but none more forcible than the emphatic fact that, putting choice of subject, composition, lighting and technical ability in negative-making all on one side, printing [how the print is made] counts today more than ever it did.[20]

During the pictorial period, photography became defined as an expressive art. For a print to be considered expressive, it helped if it was made on media that was variable. At that time evident hand control was not regarded as the eyesore that it came to be in the purist period that followed pictorialism. The pictorialists delighted in a display of purposeful individual manipulation of print media chosen for their plasticity. They took great delight in the alterations of reality that the print processes of their time provided.

To summarize, pictorial printers endorsed print processes in which surface texture and image color were subject to individual control. They preferred "hand-intensive" print processes, which permitted extensive manipulation of photographic images. DeMachy's gum prints can be said to represent the pictorial ideal. They supported the fantasy that photography is an art like the arts of painting, printmaking, and drawing which utilized the hand and manipulable media. Controllable photographic print media, such as kallitype provided a number of the media options which the pictorial artists required for their expressive imagemaking. At one extreme, making a variable paper was seen as expanding the possibilities for personal creativity, since it permitted choice of the color and texture of the paper the image was printed on. Print process options provided possibilities which artistry could utilize. Of course, some print media gave more options than others. Glossy commercial printing paper, in some minds, the least variable of all print media, provided the fewest opportunities. We are not surprised to learn that the use of a mass produced paper was seen by pictorialists as a radical curtailment of the photographer's expressive potential. As Gillies put it, photographers are

> "victims of materials which are manufactured with no regard to their applicability to picture making. . . . in photography there are few manufacturers of materials who even consider the pictorial use of the materials they sell. It is a sad state of affairs."[21]

V.   Kallitypists, photographic Artists, and the Pictorial Movement

This may be the place to position such leading kallitypists as Bennett, Hall, Thomson, et al. with respect to the pictorial movement. All of the writers we have discussed so far, including Nicol, were active during the pictorial period of photography. In Thomson's writing, as in Hall's, the reader cannot fail to notice a greater confidence about their advocacy of process manipulation than about the artistic quality of their photographs. Both writers made defensive apologies for their limitations with respect to the "rules of art." Hall addresses his audience as "amateurs," lovers of photography. Thomson consistently refers to himself as an "experimenter" and to to his readers as fellow "workers." He rarely uses the term "artist." The brotherhood with which Thomson most often identifies himself is the fellowship of process investigators.

Both Thomson and Hall submitted photographs to competitions, and the record indicates that both received occasional acceptance. Both had images reproduced in the photo-journals of their time. The same is true of Bennett and Frederick. But these workers did not, nor did most kallitypists we shall write about in this book have their kallitypes selected for major prizes by salon judges. And, while all of the kallitypists achieved fame as writers, none was lionized for his prints by well known pictorial writers and critics. Their photographs were admired, there is evidence of that, but the admiration came from editors, whose hunger for copy may have clouded their estimate of the esthetic value of the prints they praised. Their kallitype images, when viewed today, appear no better or worse than other published pictorial photographs of the time. Their photographs, general of landscape or flowers for Thomson, genre studies of children at play for Hall and for Frederick, when reproduced in the dot screen reproduction of the printed magazine, fail to show the charm and loveliness the kallitype process must have lent them. It is unfortunate that the unavailability of the vintage prints prevents a more direct assessment of the esthetic force and value of their work.

The writer has found no evidence that any of the major photographic artists of the pictorial era, Stieglitz, White, Coburn, Strand, etc., knew about Bennett's or Thomson's or Hall's work on kallitype. The names of the kallitypists do not appear in the published correspondence of the photographers who achieved fame during the pictorial period.

Sad to say, none of the major photographic artists of the time left records or prints showing they personally investigated the kallitype. Stieglitz had books in his library that discuss the kallitype, but no evidence that he tried the process has been found. [22] An examination of the catalog of his extensive print collection reveals no prints identified as kallitypes, and none of his circle of photographers were active kallitypists. It is notable that Paul Anderson, a well known latter day pictorialist worked the kallitype. [23]

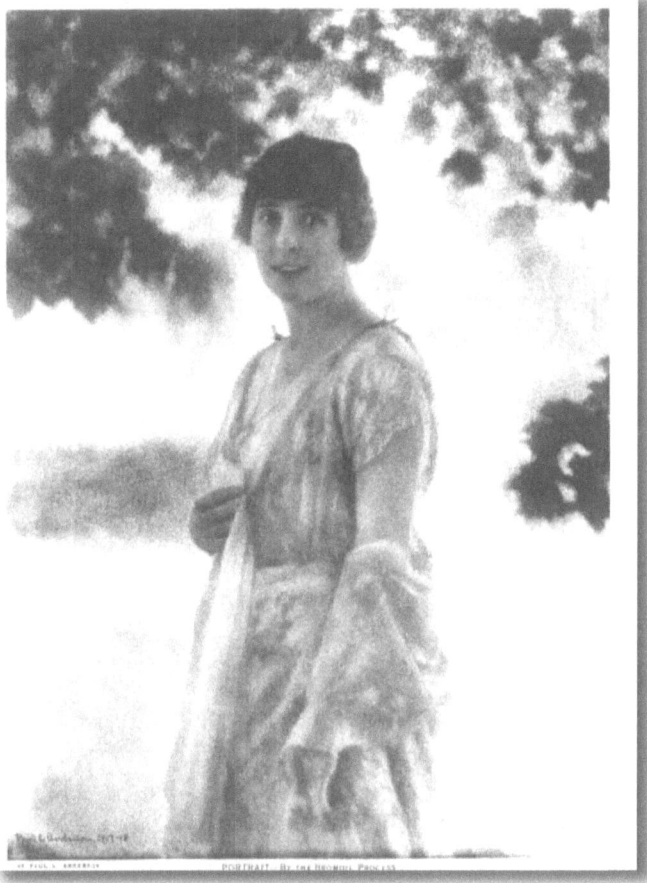

Figure 21. Paul Anderson, manipulated pictorial print

To understand why the "great" pictorial photographers did not print kallitypes, a process of considerable charm in the minds of all who knew it, we must turn our attention to the platinum print, the sister iron process. The platinum print was the darling of pictorial artists. During the entire span of the pictorial period, the journals echo with repeated and unstinted praise of the platinum print. Here is the likely reason why the kallitype had so little success with the artist photographers of the period. The kallitype was not used by the well-known artists of the period because the platinotype was. Artist's of the time thought the platinotype had insuperable advantages. We shall discuss them in the following section.

Before leaving the topic of the kallitype and the pictorialists, it seems appropriate to say a few words on the need for creating a list of the photographers who made kallitype prints during the era of pictorialism. A list of the photographers who worked the kallitype which included the localities in which they worked and the dates when they worked the process would be valuable for assessing kallitypes' contribution to pictorial art.

VI.   Platinum Printing and the Kallitype

Throughout the pictorial period, platinum was the paper of choice for the serious artist. In the May 1911 issue of the *Photo-Miniature*, in the second monograph the magazine published on platinum printing in a decade, Tennant, the editor raised the question "Why platinum printing?" His answer provides a clear indication of the high regard pictorial photographers had for platinum paper.

> Go into the leading portrait studios of any large city and ask what printing-paper they employ for the highest grade work. The answer will be: platinum paper. Go to any one of the famous pictorialists whose exhibition prints excite your envious admiration, and ask him what paper he uses in obtaining the superb qualities which make his prints so desirable. The reply, in most cases, will be: platinum paper. Go to the great reproduction firms which supply the wonderful copies of paintings and works of art bought by our libraries and print collectors, and ask what printing papers they employ. The answer will be: silver, carbon and platinum, the last named being used for the best grade of small reproductions. Here we have the opinion of experts admittedly well qualified to judge as to the qualities desired in the best sort of photographic prints. Why do they all prefer platinum? The answer is not far to seek. In all photography there is nothing more beautiful, or more everlastingly permanent, or more completely satisfying to the cultivated eye than the platinum print. [24]

There are a number of reasons for this preeminence, some real and some fancied. Platinum paper was admired by the scientifically trained photographers, such as Abney, for its long, straight-line response to the negative, what Gillies referred to as "perfect gradation." Platinum was also considered by many to be a "permanent" photographic medium, one impervious to chemical change, and therefore, a suitable medium for recording the serious artist's vision for future generations. Finally, Platinum paper was considered by many to possess the most satisfying surface texture and image color available to printmakers. The matt surface and neutral black color of the platinum print represented the standard of excellence in printing throughout the period when pictorial values dominated photography, 1890 to 1925. Carbon paper was as inert and therefore as permanent as platinum, but it did not have the tactile and attractive matt surface. Platinum had one other important advantage; it was available as a well made ready to expose commercial paper. Last, but not the least of its advantages, it was an easy hand controlled process to work.

No less an expert on process than Paul Anderson tells us that platinum was one of the easier, if not the easiest photographic process of the time to work, whether the commercial or the individually prepared platinum paper was used. [25] In this he was joined by other well informed writers. Tennant wrote that "platinum printing is one of the simplest and most direct of all methods of photographic printing." [26] R. Childe Bayley, who, like Anderson and Tennant, was knowledgeable about many processes, and for whom platinotype was "a favourite," wrote in *The Complete Photographer* that the platinum print "had no particular difficulties" and that the processing "called for no skill or knowledge." Carbon printing, gum printing, and even kallitype were more difficult to control because the sensitive elements reacted, during printing to more undesirable influences than platinum—which, being a noble and less reactive metal, was prone to behave. Bayley, after twenty five years as the editor of *Photography*, continued to be impressed by "the great beauty of the platinum print." and felt that "in spite of the great advances in development papers, there are still many who would agree with me that the best platinum prints have never quite been equaled by those of any other process whatever."[27]

If platinum was universally regarded as the most beautiful paper of the time, it was also the most expensive. The costs of platinum paper varied from three to five times what other print media cost. For the relatively well-off photographers, such as Stieglitz and Coburn, the increased cost was borne without complaint.

For others, cost was obviously a problem, as journal articles repeatedly inform us. It bears repeating that pictorial artists did not have to make their own platinum paper; they could buy it ready made by several manufacturers, among whom the names of Willis, in England, and Willis and Clements, in America, were most

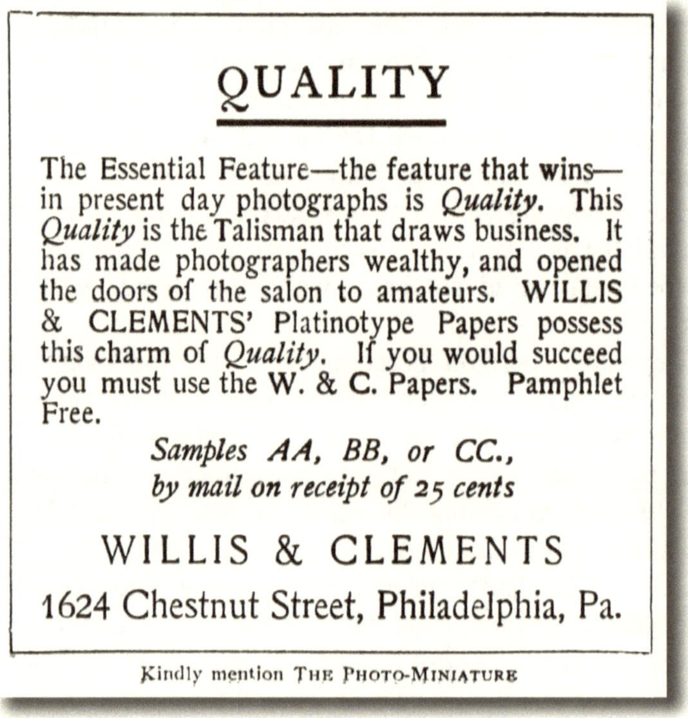

Figure 22. Platinum paper ad emphasizing Quality

reputable. Because of military need for platinum and the short supply of the metal, it became impossible for commercial suppliers to provide the paper, during the first world war, when lack of availability of the metal made continued production of commercial paper unfeasible. During the period of shortage, platinum paper substitutes were made which appeared identical to conventional platinum prints. The substitutes were made with varying proportions of other metals, most often palladium and silver. Such papers had some popularity when true platinum was not available. When the war ended, commercial platinum continued to be made and sold for a brief period, until it became totally uneconomic to continue. Then production was halted permanently. Presently platinum printing can only be done on individually prepared paper. Working solutions, chemicals, papers and instructions are available from suppliers.

There are a number of interesting sidelights on the pictorial use of platinum paper that involve the kallitype. One has to do with the effort of artist-photographers to make platinotypy a more plastic medium. During the pictorial period, artists frequently processed platinum prints with mercury, uranium, or catechu, to achieve artistic effects of color and contrast. Such chemical adventures negated, at least in part, the archival permanence native to the pure platinum print. Platinum prints that display colors other than the neutral black, blue black, or warm black of the unadulterated process, especially those that appear unusually brown or pink, may have been given that appearance by treatment with chemicals that are subject to change. Platinum prints, which were manipulated with additives more than 50 years ago, today display obvious signs of aging, browning and staining. In some, the platinum metal that constitutes the image has not changed, but the paper has deteriorated to an ugly brown appearance from the influence of contaminants. W. H. Smith complained of the practice of mercury-modified platinum to the Royal Photographic Society on two occasions. In his second address, on November, 1915, he warned, "the use of too much mercury meant the impermanence of the print." [28]

During the war and immediately after, critics complained that "platinum" paper was often so salted with silver that the vaunted claim of permanence degenerated to a myth, more fantasy than reality. The later platinum papers were often made of silver in combination with platinum. Thomson felt these prints were often closer to kallitypes with some platinum in them than they were to the real platinum prints of an earlier time.

After the first World War, a class of look-like platinum papers, made with silver, was developed and widely marketed. These simulated platinum papers bore titles that openly attested to their silver content—"Platino Argenta" and "Aristo-Platino," and "Platino-Simili." Thomson reports that Willis and Clements late in 1913 "placed upon the market a combined silver and platinum paper . . . and so far as appearance is concerned, [it is] of platinotype quality." [29] Whether such papers should now be considered 'platinotypes' or 'kallitypes' is a question. The fate of such papers needs study.

Kallitype and platinotype image color can have a variety of print "looks," neutral grey, warm black, and brown, which, when printed on the same paper stock, make prints from the two processes visually indistinguishable. Kallitypes can be made as similar in appearance to platinotypes as a master printer wishes to make them, that is, so similar that it is impossible to tell by visual inspection alone whether a print is one or the other. There are chemical tests that can readily identify if the metal in a print is platinum or silver, but it is unlikely they were used by salon judges or customers. Tennant hints that perhaps they should have been.

> The professional too, has learned to appreciate Kallitype, although for him, put in disguise and under another name, thus proving that Will Shakespeare was right in what he said about 'a rose by any other name.' [30]

Today there are easily administered tests which can quickly reveal whether platinum prints are pure platinum prints or only partial platinum prints with some percentage of silver added. Such tests can be applied with minimal damage to the print. The tests are rarely used, if they are used at all by custodians responsible for museum collections. Non-testing leaves the chemical and process origin of some platinum prints open to serious question. [31]

There can be no doubt that the strong acceptance of the platinum print worked against the acceptance of the kallitype as a preferred medium for artistic work. While advocates of kallitype with some frequency drew attention to the close resemblance of the kallitype and platinotype and the much lower cost of the iron silver process, few well known pictorialists admit to working the iron-silver print. When the kallitype was first placed on the market as a commercial paper in 1891, price and resemblance to platinum prints was a key selling point. Both price and similarity to platinum paper were frequently claimed by writers promoting individually prepared kallitype. While the manipulation of platinum and kallitype are similar, platinum printing is easier and has fewer problems than kallitype. Printers could switch with little difficulty from one process to another. Nevertheless, the argument for kallitype "took" only with the echelon of amateurs a cut below the very best pictorial artists. The best stuck with platinum. At least, so they tell us.

By 1917 new eddies in the flow of amateur photography began to appear. Soon the pictorial concern with the artful image and the hand manipulated print would be threatened by a new esthetic idea and a much less troublesome and cheaper process—straight silver chloride or silver bromide.

VII.   Commentary on artist Prepared Paper

Perhaps the historian may be permitted an occasional editorial. If there were no "lessons" in history, what would be the point of it?

Obviously the relation of print processes and esthetic taste is a subtle and difficult, one, subject to a variety of interpretations. The pictorialists certainly were aware that individually prepared print material could be highly individualized. Anyone who examines artistic pictorial prints sooner or later succumbs to their charm. They possess an affective quality that many of the later objective, formalist, gelatin-silver bromide prints lack. The concern with beautifully made, expressive pictorial print objects surely was a high point in the development of photography.

Of course, there is another side. Concern with photographic print processes can degenerate into process mongering—meaningless, expressionless, empty process play—media without message or massage. But it is difficult to believe that all hand made pictorial prints, including kallitypes, have nothing to offer. One can understand why museums might wish to avoid responsibility for preserving examples of photographic media which have limited continuing cultural importance. It is also possible that curators of photography who refuse to collect well made samples of past pictorial processes may define cultural relevance and their responsibilities too narrowly.

In 1904, addressing the Photographic Convention at Bath, George Davison an excellent pictorial photographer, praised the early photographic investigators for their energy and contribution in bringing about the progress that so expanded printing methods in the 1880's and 90's. But he warned artists to avoid losing sight of their primary responsibility, which was to see and make artistic images. This responsibility should lead artists to avoid, Davison suggested, involvements with printmaking processes that would distract them, or use up time better given to finding images worth recording.[32] It should be noted that Davison, while an excellent pictorial photographer, earned his living as an executive with a major supplier of photographic paper. His argument seems to serve his employer better than the artists he addressed. Or, if it served the artists, it served only one of their several interests.

Davison's position anticipated the concerns of the artists of the Linked Ring, of which he was a founding member. The Linked Ring sought to free the art of photography from a stifling preoccupation with technical excellence. It is worth emphasizing that Davison's position and that of the Linked Ring eventually led to the separation of the art of expressive photography from the exploration and development by the artist of the material on which his art is produced. The advice that the artist should mind his image, not his paper led to the exclusive production of photographic papers by manufacturers and the making of images by artists, who eventually withdrew from any responsibility for the exploration and development of the material they work with. Since manufacturers are not interested in art, but rather in the widest possible application and sale of their print materials and the profits that come from wide sale, their commercial interest has lead them to produce papers which have qualities of limited interest or utility to the artist, as we have seen Gillies claim. And manufacturers have every reason to act in this fashion—that is to produce a highly reliable, high quality, standard product that serves the widest possible market.

But when photographic artists place in direct opposition the search for an image on the one hand and the preparation of media on the other, as Davison above suggests, a short-sighted and ultimately dangerous separation of responsibilities occurs. If the capacity of artists to express is to remain unencumbered by the limitations which paper producers for their own reasons of profit build into the print media they manufacture, then artists must retain sufficient understanding and control of their media to rescue it from producers. This may be one of the lessons that can be drawn from the pictorial concern with individually prepared papers. If photographers wish to cultivate their own garden, to borrow from Candide, they may have to worry a bit about—and occasionally contribute to—the condition of the soil in which their flowers grow.

Notes: Chapter VII Pictorialism

1. Alfred T. Story, *The Story of Photography*, S. S. McClure, Co., New York, 1898, 1908, p.155.
2. G. Bernard Shaw, "Some Criticisms of the Exhibitions," *The Amateur Photographer,* Oct. 16, 1902, p. 307.
3. John Taylor, *Pictorial Photography in Britain*, 1900-1920, Arts Council of Great Britain in Association with the Royal Photographic Society, 1978, p. 75.
4. Franklin I. Jordan, F.R.P.S., *Photographic Control Processes*, Galleon Publishers, Inc., New York, 1937, p. ll.
5. Raphael Meldola, *The Chemistry of Photography*, London, Macmillan Co., 1889, p. vii ff.

> "It is to be regretted that purely scientific chemists have of late years shown a tendency to neglect the *Chemistry of Photography* as being an uninviting branch of their science."

6. John Gillies, *Principles of Pictorial Photography*, New York, Falk Publishing Co.; Reprint edition, Arno Press Inc., 1973, p. 80.
7. William Welling, *Photography in America*, Thos. Y. Crowell Co., New York; 1978, pp. 291.
8. The titles of the photo-chemical issues of *Photo-Miniature* are:

> #18. *Chemical Notions for Photographers*.
> #23. *Photographic Manipulations*.
> #43 *Photographic Chemicals*.
> #101. *Photographic Chemicals*.

> There were similar works on optics. For a list of other works on chemistry published at this time, see the Bibliography.

9. John Gillies, Ibid., p. 73.
10. Ibid., p. 80.
11. Charlotte Adams, "Photography and Art," *Philadelphia Photographer*, Vol. 20, June 1883, pp. 179-183.
12. John Taylor, Ibid. p. 14 ff.
13. Gillies, Ibid., p. 64.
14. Taylor, Ibid., p. 9.

15. Gillies, Ibid., p. 58-9.

16. Taylor, Ibid., p. 13 ff.

17. Taylor, Ibid., p.17. He quotes A. J. Anderson, "A Note On Two Pictures by Annie Brigman," *The Amateur Photographer*, March 8, 1910, p. 236.

18. Will A Cadby, "Whistler and The Gentle Art of Photography," *The Amateur Photographer*, June 21, 1904, p. 436.

19. Henry Peach Robinson, *Pictorial Effect in Photography*, London, Piper and Carter, 1869
    Charles H. Caffin, *Photography as a Fine Art*, New York, Doubleday, Page and Co., 1901
    Arthur Hammond, F.R.P.S. *Pictorial Composition in Photography*, American Photographic Publishing Co., Boston, Mass., 1920

20. Owen Wheeler, *Photographic Printing Processes*, American Photographic Publishing Co., Boston, Mass., 1936, p. xv.

21. Gillies, Ibid., p. 58-9.

22. Weston J. Naef, *The Collection of Alfred Stieglitz*, A Studio Book, The Metropolitan Museum of Art, The Viking Press, New York, 1978, p. 507-519. Books in Stieglitz collection that refer to kallitype include those by W. J. Harrison, **The Chemistry of Photography**, New York, 1892 and Edward L. Wilson, **Cyclopedia of Photography**, New York, 1894. A number of the Annuals and Year Books also include discussions of kallitype printing and kallitype formulas.

23. The International Museum of Photography in Rochester, New York has a collection of Paul Anderson's papers among which is a page of formulas on making kallitype II. The formulas involve a "general purpose" sensitiser, one that could be used as a sensitizer for other iron prints, eg. blue prints, etc., an iron-silver sensitiser, and a Borax-Rochelle Salt developer.

24. John Tennant, *Photo-Miniature*, Vol. X, May 1911, No. 115, pp. 306-7.

25. Paul Anderson, *The Technique Of Pictorial Photography*, Philadelphia, J. B. Lippincott, p. 194.

26. John Tennant, *Photo-Miniature*, Vol. X, May 1911, No. 115, p. 306.

27. R. Childe Bayley, *The Complete Photographer*, New York,

28. W. H. Smith, "Platinotype and Its Modifications," *Photo Era*, February 1916, p. 82.

29. J. Thomson, "A Silver Platinum Printing Paper." *American Photography*, November, 1915, p. 630

30. John Tennant, Introduction, "Kallitype and Allied Processes," *Photo-Miniature* Vol. XVI,

31. Brian Coe and Mark Haworth-Booth, Ibid., p. 27. The authors provide the following information on the identification of platinotypes:

> Platinum prints are generally recognizable by their image colour and plain paper surface. In the 1890's and 1900s manufacturers of gelatin-silver papers tried to match the characteristics of the platinum print with their products. "Platino-matt' prints, if in good condition can resemble the real thing, but usually some image deterioration will be present. If a very small sample from the edge of a print is available, it can be tested by applying a drop of a solution made from 26 grams of mercuric chloride and 5 ml of hydrochloric acid in 1 litre of water. Silver prints will bleach, platinum prints will not. Platinum prints in albums may sometimes be identified by a rust-colored impression of the picture on a facing page, formed by residual iron salts diffusing from the platinum paper.

32. George Davison, "The Position of Artistic Photography," *Journal of the Camera Club*, Vol. VIII, May 1894, p. 114 ff.

Chapter VIII

Thomson and the Experimenting Artist

I.   The Joys and Sorrows of Photo-Experimentation

> "I was possessed of a desire to dabble with chemicals and
> see what would happen." [1]

We shall now return to our account of amateur development of the
kallitype by reporting James Thomson's many concerns with the process.
In the spectrum of pictorial interest in Kallitype, Thomson is an extreme
right champion of process exploration as the following summary of his
work on KI indicates.

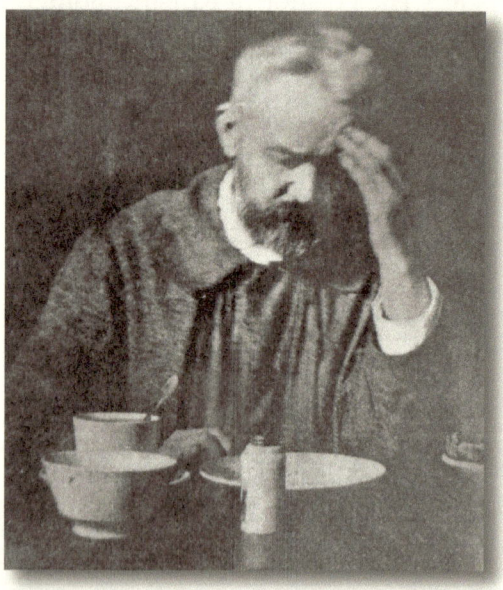

Figure 23. James Thomson kallitype, self portrait

Of considerable interest to this history is an article Thomson wrote on the process-experimenter. More than any other writer on kallitype, Thomson himself personified the process experimenter. From him we can learn a good deal about what amateur experimentation on photographic process was like. In l907 he wrote an article entitled "Experimentation" for **Western Camera Notes**. In it he openly expressed his disdain for those photographers "who do not inquire."

> Photography with some folk consists of but little beyond pushing the button and letting others do the rest. That such people have but small claim to be regarded as photographers goes without saying. Nor can we blame . . . the worker of tried worth and serious purposes, when he looks down on those who would without credentials place themselves on a par with himself. And indeed, what would photography amount to were all of the trifling sort.[2]

He reports a talk with his photographic friends, all amateurs. One made light of the experimenter, saying,

> I have no use for experiments. Let the other fellow have that if he wants it. I am content at the end to come in and get the benefit of his self-imposed drudgery.

If Thomson was not impressed with the trifling amateur, he was not impressed with the typical commercial worker either. He finds

> the professional of old . . . was apt to work by rule, and the result was stagnation . . . he was lacking in essentials.

Thomson ends this discussion with a statement of the benefits that accrue to the amateur as well as to photography from experimenting.

> It is at present the privilege of all to benefit from the exhaustive experimentation of a great many earnest investigators, many of whom are forgotten. Photographic experimentation, even at this late day is extremely interesting. A small investment puts within our grasp a valuable, and in some cases a profitable means of using odd moments.

> The difficulties are not many, and formulas are plentiful
> wherewith one may prepare one's printing paper . . . . The
> iron salts, particularly ferric oxalate—open up for the patient
> experimenter a region of pure delight . . . . the non chemist
> novice is apt in his ignorance to attempt something his more
> learned comrades would not think of trying. Such attempts
> may lead to something of value, for in this, as in other things,
> there are surprises . . . [3]

It appears that Thomson here applies the transcendental philosophy of his fellow Bostonian, Emerson, to experimenting with photography: the amateur, by stretching himself, stretches his medium and all who use it. Thomson on more than one occasion communicates a kind of photographic Self-Reliance that is reminiscent of the Sage of Boston, as any who read much of his writing style will note. While Thomson's ideas on doing independent research are expressed by other kallitypists, his attitudes have a certain poignancy coming at a time when commercial interests were rapidly converting photographic printmaking into a relatively automatic process which all but defied individualization.

II: Thomson and the Kallitype I.

In August 1904, in *the Photo Beacon*, making a surprising tack from his first article on the Bennett-Burton approach to Kallitype II, (1903) and from his article on brown print, published in July 1904, Thomson wrote an article on kallitype I, a process which had been ignored in the more than ten years since it was removed from the market. Thomson's first article on kallitype I is a creative response to Nicol's original description of the process and merits a close look. Thomson worked on the K I process for more than twenty years, publishing revised and "improved" versions with regularity as the years passed. He remains the only twentieth century experimenter who tried to instill new life into the abandoned process. His 1904 article begins with a prologue:

> Being something of a dabbler in the process . . . the desire
> took possession of me to go back to first principles—to see for
> myself what sort of result the man who invented the process
> could have had . . . The prints I got from it [Nicol's Kallitype
> I approach] did not suit me at all . . .

The image was of an inky blue, good enough so far as it went, but lacking vigor. I added something I thought would supply the deficiency, only to get too much contrast, and then began a series of experiments which . . . eventually led to complete success.

. . . Having no desire to keep a good thing to myself, much as is the pleasure that is mine therefrom, I hope . . . to make plain the manner in which I get . . . prints that are in every way as good as those produced by the platinum methods, and in jet black and white from the word go. [4]

Thomson's Kallitype I formula No. 1, for "average pictorial purposes," that is, for black prints from negatives neither flat or contrasty is given below, left. His No. 2 formula, below right, provides "a stronger image," one with more contrast, and a brown black color.

Thomson's KI Sensitizer	Formula No. 1	No. 2
Ferric oxalate	33 grains	30 grains
Citrate of Iron and Ammonia	10 "	6 "
Chloride of Copper	6 "	6 "
Oxalate of Potassium	37 "	30 "
Gum Arabic	10 "	10 "
Distilled Water	1 oz.	1 oz.

The reader will recall that in Nicol's K I process, the iron sensitizer, ferric oxalate, is coated by itself on the paper. Exposure to light behind a negative causes a faint image in iron to be formed. The iron image is converted to a silver image by "development" in a bath that contains, silver nitrate and potassium oxalate. This is a most simple process. One cannot help but notice that Thomson has added to his formula a second iron sensitizer, citrate of iron and ammonia, and some copper chloride, as well. Thomson does not clarify the function of these two chemicals. The gum arabic is a spreading agent commonly used to facilitate even coating.

After the sensitized paper is dried and exposed to sunlight under a negative, the print is immersed in the developer. The same developer is used for prints coated with either of the K I sensitizers.

Thomson's K I Developer (Stock Solution)

Silver nitrate	40 grains
Oxalic acid	10 "
Citric acid	10 "
Phosphate of soda	2 "
Distilled Water	1 oz.[5]

The developer is a "stock solution." For use, add to each ounce of stock solution seven ounces of water. Development should proceed until the image appears in full, which normally takes less than two minutes. A running water wash follows development—five minutes is sufficient. The prints are fixed for five minutes in a solution of 1.5 grains of fixer per ounce of water.

The "No. 2" sensitizer can be used with thinner negatives and is preferable when printing on heavily sized or rough textured paper. Thomson was particularly concerned with creating a paper that reproduced well the gradation of tones on the negative. He felt that these two sensitizers achieved "good gradation and at the same time a vigorous print." His article contained additional details on the employment of these formulas and the manipulation of his paper.

With Thomson it is always difficult to figure out why he included each and all the chemicals in his formulas. Thomson was not much for clarifying such matters with chemical theory so some discussion is in order. K I sensitizers require only ferric oxalate to record a response to light. We ask why does Thomson include ferric ammonium citrate, a second light sensitive compound in his sensitizer? He probably used it, first, to achieve a better print-out of the image during exposure. Ferric ammonium citrate prints out more visibly than does ferric oxalate. Second, he may have added it to produce a print with a longer scale of tones—that is, an image with less contrast. Ferric Ammonium Citrate produces a flatter, less contrasty image than ferric oxalate. The addition of the potassium oxalate is probably intended to make up any oxalate deficiency present in the sensitizer caused by the ferric oxalate turning partly ferrous. Thomson never explicitly revealed why he used copper chloride. The gum acacia assists in the even spreading of the sensitizer and also slightly influences the image color toward a warmer brown.

With regard to the developer, Thomson says it contains two acids to assure the silver salt is all reduced to silver metal when it contacts the dissolved ferrous ion. Thomson says the phosphate of soda, alters the image color from brown toward blue-black. The potassium oxalate in the developer is a solvent of ferrous iron, hence its presence assures complete conversion of the salt to image forming silver metal. Without it the original image in iron would not undergo the reaction that causes an image in silver to form.

The response Thomson received to the publication of his approach to Kallitype I was not what he expected. *The Camera* printed a short humorous article that was written by an amateur printer "trying my hand at making some Kallitype paper the other day." The writing turns out to be a satire on the whole business of making one's own printing paper. After finding an article about making kallitype paper, the writer reports he

> . . . coated some paper, and with my experience with a whitewash brush on the chicken house a few weeks ago, made what I thought was a fine job. You ought to have seen the prints! That particular batch of Kallitype was certainly fitted for some particularly "dink" sort of negative. Even the worst one I had would not give a good print. I hunted up a local authority on the subject and got some advice. He said I should get a softer brush and put on a double coating, going over the paper in an opposite direction the second time. I did this and the prints were lacking in those rusty shadows, except when using the Borax and Rochelle-salt as a developer. Another car-fare, and I found that I had neglected to dissolve the borax in hot water as directed. I was all right again except for a lot of spotted prints that cost me another car-fare to find that they were due to not using enough clearing bath and not keeping the prints in motion for two or three minutes. Cutting down the Borax or adding cigar ashes to the solution, or both, gives a very pleasing warm tone to the prints—I am as yet, not sure which, but will advise later. [6]

This piece, published in August, 1904 could have been in response to any kallitype article, but it seems directed as much toward Thomson as any. It certainly demonstrates power of ridicule.

In October, 1904 Thomson wrote to Todd, the editor of **the Photo-Beacon**, and somewhat sheepishly admitted that his article on Kallitype I may not have been entirely ready for publication.

> . . . Since I wrote the article on Kallitype as published in the August issue of **the Photo-Beacon**, I have discovered some openings that might lead those unacquainted with the peculiarities of working the process to make a failure of it, or perhaps only a partial success. So, in the interest of good work, I have thought well to add a few remarks. That kallitype, of printing methods, is the most elusive of all, every one who has tried it must realize. When one is most confident he has mastered all its intricacies, some variation is certain to crop out. It is as elusive and uncertain as the shell game of the bunco men—you think you can with certainty find the pea when, behold, it is somewhere else. For the past six months, I have been experimenting, and thought I had got it to a reliable basis, but now, I see where one might fail.
>
> The process, as outlined by me, I consider in a state of transition. I gave to my comrades such data as were mine to give to date, with the hope that some other might improve on it. Had I waited until perfected, perhaps my article would never have been written. [7]

Not to be discouraged by so small a matter as partial bad directions and print failure, Thomson further revised his K I approach and published another article in the December 1904 **Photo-Miniature**. Here he tells his readers his aim is "to achieve a more vigorous image in the {printing] frame," and a stronger, browner image after development. Again Thomson's enthusiasm boils over, and he claims "I have succeeded." [8] But again, by January 1905 he receives letters indicating those who have tried his new method were perplexed by the directions and failed to get the predicted results. [9]

Thomson wrote Todd yet another letter on the problems of writing on Kallitype process. This time he discussed Henry Hall's problems, not his own. He reported that "Mr. Tennant, of *the Photo-Miniature* . . . states they have had considerable trouble since publishing their monograph on the subject [of kallitype]. So many failed to get the results promised in the directions." It is to Thomson's credit that he does not attempt to humiliate Hall. Instead, Thomson itemized for Todd the reasons why readers fail when they try kallitype: "variation in sizing and chemicals employed, or unclean manipulation, or bad chemicals." [10] Again Todd published the letter.

In the December, 1904 issue of **Photo-Miniature**, appeared an eight page article, written by Thomson entitled "Postscript to No. 47, Kallitype to Date." *Photo-Miniature No. 47* was the issue that contained Hall's account of making kallitype II by working with stock solutions. Thomson's "Postscript" was hardly a response to what Hall wrote; it made no mention of Hall's formulas, the problems that arose from them, or the resolution of problems associated with his method. Instead, Thomson wrote about his own radically different approach to kallitype I.

He began, "Prints in black and white are possible by other Kallitype methods, but in none are they so easily obtained as in . . . [his own approach]." Thomson advocated and explained his own quite different methodology of making kallitype I paper. He did not allude to the fact that his approaches had already generated their share of problems when readers tried them. Feeling confident again, Thomson said of his new formulas, that they "have been varied in every possible way and proportion, and after a thorough test, I have come to the conclusion that in the shape I am about to present them, there are great possibilities." As in the past, Thomson again turned his back on the difficulties amateurs have when working with formularies and photographic chemicals.

In the "PostScript to # 47" Thomson published two more K I sensitizer formulas, called again "No.1" and "No. 2." They followed the same pattern as his earlier #1 and #2 formulas and they are more complex.

Thomson's revised K I Sensitiser Formulas

	No. 1	No. 2
Ferric Ammonium Citrate	25 grains	28 grains
Ferric Oxalate	15 "	28 "
Copper Chloride	8 "	8 "
Potassium Oxalate	33 "	35 "

Silver Nitrate	15 "	15 "
Oxalic Acid	15 "	l5 "
Gum Acacia	10 "	l0 "
Distilled Water	l oz.	1 oz. [11]

Thomson gave very particular directions on mixing the chemicals and stipulated the formulae would work only if his exact procedure is followed using the exact chemicals named. He wrote there can be "no chemical substitutions" and insisted on the highest purity and freshness of the chemicals.

These sensitizers differ from his earlier formulas by using a greater proportion of ferric ammonium citrate with the ferric oxalate, and in the inclusion of some silver nitrate in the sensitizer. Also the oxalic acid formerly in the developer is now in the sensitizer. Thomson tells us the silver in the sensitizer is intended to maintain the concentration of silver nitrate in the developer at full strength.

Thomson's Revised Stock Developer For No. 1 and No. 2 Sensitizers

Silver Nitrate	40 grains
Citric Acid	l0 "
Sodium Phosphate	2 "
Distilled Water	1 oz. [12]

Mix 1 dram of developer stock solution to one oz. of water. Add 1 grain of oxalic acid for blue black prints; more for brown/black prints. The exposed print is developed for about five minutes in the silver bath and then fixed in a hypo bath (50 grains to one quart of water) for five minutes.

Thomson's aim in these revisions was to re-formulate the K I process so

1) the exposed print will provide the operator a clear indication of complete exposure
2) the shadows would not bronze or solarize, and
3) the print would have a rich black image color.

Thomson reported the No. 2 formula "gave a stronger image," that is one which has a deep brown black color. Inspection reveals that the primary difference between the No. 1 and No. 2 sensitizers is the variation in the quantities of the two ferric chemicals.

Thomson impatiently awaited reader reaction to this revision. He did not have to wait long. In October 1905, he wrote to tell Todd the news. Apparently the response to the "Postscript" is no better than the response to his earlier articles.

> My Dear Sir,—The photographic experimenter along chemical lines has his troubles from adulteration and substitution in the materials he is called upon to employ. The unscrupulous vendor of "something just as good" is responsible for many a slip up, and what makes it all the more exasperating, one may never fathom the cause of failure.
>
> I have no doubt but what I have many a time been unknowingly using water which was not distilled, and that in this fact lay the cause of failure.
>
> I have learned the importance of pure chemicals . . . It may seem unimportant to some, but all bottles should have good corks. Ferric Oxalate especially must be guarded from dampness.

Thomson concluded his lament with concern about novices.

> To the person with more knowledge and experience, failure to make the formula work in accordance with schedule is not so important, for one may know what to do to save the day. But with the subtle chemistry of many printing formulae, it is important that the novice gets the precise quality of chemical or water called for. [13]

Thomson's experience suggests a problem about amateur experimentation in process that may have no solution. If Thomson's chemicals have been "bad" as he reports, how can he tell others how to make kallitypes, since his formulas are based on abnormal chemicals. On the other hand, if his chemicals were good, how can others be successful in working his formulas if their chemicals are "bad"? The experienced worker, as he says, may know how to compensate for the inevitable variations in chemicals, but how can the novice deal with the apparently inevitable differences in chemical quality? Unless good chemicals can be taken for granted, the problem of writing about a chemical process appears unsolvable, unless the reader can be depended upon to have knowledge enough to compensate for abnormalities in chemical performance.

It is also apparent that Thomson was asking for trouble with his formulas based on long lists of chemicals. Simpler, shorter formulas, like those of Bennett and Hall, present fewer occasions for chemicals to perform unpredictably or for a novice to err. It is also clear that the writer of kallitype directions must insist on reader responsibility for some experimentation with the formulas published to make up for any variations in the material used by the author and the reader. With his later formulas Thomson became increasingly circumspect. He sounded warnings about bad chemicals, insisted that experimenters test chemicals for freshness and purity, and encouraged readers to experiment with published formulas until they found an approach that worked with local materials and methods. Eventually it became clear to Thomson that there were limits to the effective communication of photographic processes in amateur journals to workers who know little about what they were trying to do.

We shall return again to Thomson in later chapters.

III.   Summary of Chapter VIII

The last two chapters set out to present a sense of the artistic milieu that encompassed the kallitype process during a period when photographers sought new freedoms for the art of photography. They described the development of pictorialism, a revolutionary movement in which photographers championed new values and freedoms for photographic art. These values emphasized personal expression in prints over accurate objective recording. The result was a new direction in printmaking that stressed printing options that would enhance individual expression. These options led print makers to work with print processs that permitted individual manipulation. Photographers of the period also selected paper processes that provided color, tone, and contrast variation, attractive qualities not offered by the plain papers manufacturers provided. Platinum Paper. not Kallitype became the first choice of the important artists of the period because of its superior visual qualities and permanence. Kallitype success was limited by its variability, its inability to make successive prints identical in visual quality. Kallitype process made prints that were identical in appearance to platinum prints and as platinum became more expensive, during and after World War I, was used to "simulate" platinum prints. It is likely some platinum prints have more silver in them than platinum.

Finally a report of Thomson's over elaborated K I process writing was given. The repeated negative results amateurs had with his processes illustrate the dangers of publishing individual process explorations for other amateurs to try.

Notes Chapter VIII James Thomson

1. James Thomson, "On the Preservation of Iron Salts," *British Journal of Photography*, Jan 27, 1911, p. 62.
2. James Thomson, "Experimentation," *Western Camera Notes*, Oct. 1907, No. 10, p. 251-3.
3. Ibid. p. 253.
4. James Thomson, "Kallitype Again," *Photo-Beacon*, Aug. 1904, p. 240.
5. Ibid., 241.
6. No. author, "About Kallitype," *The Camera*, August, 1904, p. 319.
7. James Thomson, "Letters to the Editor," *Photo-Beacon*, October, 1904, pp. 315-16.
8. James Thomson, "Kallitype to Date," Post Script to No. 47, *The Photo-Miniature*, Vol. LVI, No. 69, December, 1904, pp. 507-516.
9. James Thomson, "Letter to Todd," Collotype, *The Photo-Beacon*, January 1905, pp. 27-8.
10. James Thomson, "A Note On Kallitype", *The Photo-Beacon*, May 1905, p. 146.
11. James Thomson, "Postscript to #47," *Photo-Miniature*, Vol. VI, No. 69, Dec. 1904, p. 509.
12. Ibid.
13. James Thomson, "Pure Chemicals," *Photo-Beacon*, Oct., 1905, p. 293.

Chapter IX

The Brown Print

I.   Introduction to the Brownprint

In July, 1904 James Thomson wrote an article on a process he called the "watertone developed kallitype." He was referring to a variant of an iron silver print process which had been receiving international attention from some of the best photochemists during the last two decades of the nineteenth century. The process Thomson described has been called by various names. It was often referred to as "sepia print paper" because of the light yellow brown color of the image. In time it received other names from those who wrote about it: "brown print," "water developed kallitype," "imitation kallitype," and "Van Dyke brown print." The latter name was derived from a version of the process that produced a dark brown print color which resembled the paint pigment of that name.[1] The process was sometimes called the "American Kallitype," to distinguish it from the Kallitype II process Nicol, the Englishman, invented. The paper in all likelihood received its 'American' name from the efforts of Thomson, who published a number of papers on brown print over an extended period of time and who was the most persistent twentieth century advocate of the brown print. In this book the paper shall be called "brown print," in view of the descriptive quality of that name and because that is the name photographers and writers commonly use today.

The brown print is an interesting approach to photographic printing. Next to the blue print, it is probably the simplest and easiest individually prepared printing paper. The brown print is an iron silver process which first makes an image in iron and then converts the image to a final silver metal image. Almost any mixture of its ingredients, ferric ammonium citrate and silver nitrate will produce an image. The brown print is a contact printing process which requires a full sized negative and a printing frame. Before electricity was available, brown prints were exposed to sunlight. When exposed for a sufficient period, the image prints out almost fully in the printing frame, so judging exposure is relatively easy. After exposure, the prints are developed in plain water. That is the basic process. As we shall see, the basic process went through a number of chemical modifications in the hundred years it has been worked. Variations in chemical formulae and processing have resulted in variations in the color and contrast of the print. The image can be toned.

It has often been said that the brown print and the blue print are—of all individually prepared paper processes—the easiest to learn and the cheapest to work. The hearsay is true. That is the reason why the brown print has been and will continue to be an enjoyable introduction to individually prepared photographic papers. However, though the brown print requires little skill to work, it can produce prints of refinement and delicacy when manipulated by a careful printer.

The brown print has generally been viewed as a version of the kallitype process because it uses iron-silver chemical reactions. While, the two processes are similar in that they are based on the light sensitivity of ferric compounds, they are quite distinct in the sense that they use different iron sensitizers and a different approach to processing the print. In contrast to the kallitype, which usually employs ferric oxalate as the sensitizer, the brown print uses ferric ammonium citrate as the light sensitive compound. As we have seen kallitype II printing generally involves one or two processing baths, usually a developer and a clearing bath, to achieve a desirable image. In contrast, basic versions of the brown print use only water for the total processing of the print! Because the brown print "invented" by Nicol, played an active role in late nineteenth and early twentieth century amateur photography, and because the brown print continues to have an active following of printmakers today, an account of its origin, development, and use merits a place in this history.

In what follows we shall present what we have been able to find on the origins and development of the brown print from an assiduous search of the literature of photography. We shall report the various brown print processes that have been published, their intended purposes, their sources and dates, the composition of the working solutions and the manner of working to produce a print. As we narrate the history of the brown print, we shall select for detailed reporting several brown print processes. Their selection was based on the reputation of the writers who first described them, the importance of the particular processes as historical examples of the brown print process, and their suitability for trial by interested contemporary workers.

II.   The Chemistry and Processing of the Brown print

In most cases the chemical process for making brown print employs a solution of ferric ammonium citrate as the sensitizer to which is added silver nitrate, the final image former. The mixture is coated by brushing or floating the sensitizer onto the printing paper. When the fully dry paper is exposed to light (under a negative the size of the desired print), the ferric compound turns ferrous and the silver nitrate creates a metallic silver image.

Depending on the sensitizer mix, A partially visible brown image is produced by the exposure to light alone. When the exposed paper is immersed in plain water, the image "develops," further. That is, the image becomes fully printed out and deeper in tone. In greater detail, when the print is immersed in water, more silver salt (silver nitrate) comes in contact with the ferrous image and is reduced to silver metal. The complete reaction takes place when the ferrous iron produced by light acting on the sensitized paper is rendered soluble and fully active by the citrate in the ferric ammonium citrate sensitizer. The partial formation of the image during exposure occurs because of minimal amounts of water in the air and the paper which permits a partial chemical reaction to occur while the paper is being exposed. The water bath that follows exposure fully "develops" the image, because it dissolves all of the citrate in the sensitizer, which in turn acts as a solvent for the ferrous iron, permitting it to reduce the silver nitrate to silver metal. Since all of the chemicals that have been coated on the paper are soluble, a final running water wash can be used to remove the residual iron and silver compounds that have not been involved in the formation of the image. Thus, water is used not only as developer, but also as the "fixer" and for the final wash. A further consequence of the water wash, with or without the additional use of hypo, is the deepening of the image color of print to a dark chocolate colored) sepia.

The earliest workers believed there was no need, in theory or in practice, to use any fixer other than plain water. They thought, and with some reason, that all of the products of the chemical reaction occurring in brown printing were soluble in water and therefore removable by water alone. Later workers used a weak hypo solution (3-8%) to more thoroughly "fix" the image. Since the both the iron and the silver salts used in the brown print process remain soluble it is theoretically questionable if a hypo fixer is needed to dissolve away non-image forming silver. Workers following this line of thought insist that the only function of a hypo-fixing bath in brown printing is to darken and enrich the brown of the image.

The normal color of the brown print image is sepia, a yellow brown which is not particularly pleasing. The immersion of the print in a hypo bath changes the sepia color to a chocolate brown color, which most printmakers prefer. A solution of sodium sulfite can be used instead of a hypo bath or, after a hypo bath, to darken and "brown" the image color.

Printers, both early and late, added certain substances to the brown print sensitizer, notably citrates, tartrates, oxalates, and gelatin. They believed the addition of these chemicals increased the deposit of silver, influenced the color of the print, and helped to achieve even coating. A number of workers advocated toning procedures which substitute nobler metals for the silver of the brown print image. Gold toning, for example, produces a second effect of increasing the permanence of the print while altering the image color. Finally, brown print processing usually ends with a running water wash which eliminates any residual chemicals from the fiber of the paper. These are the general details of the various brown print processes in the literature. Individual processes varied in one way or another in these details, as we shall see in the formulas which follow.

Those considering making their own brown prints should be aware that the sensitizer, ferric ammonium citrate, is available in two forms, brown and the green powder. The green is the most desirable because it contains considerably more of the active ferric compound, and as a result prints considerably faster. Both forms of the compound are very hygroscopic. During storage, the dry powder will in a short time draw enough moisture from the air to become as viscous as molasses. While it will still "work" as a sensitizer when it is viscous, it is difficult to weigh or measure. One would think the answer to the problem of keeping the chemical would be to mix up a liquid solution of a known strength. The difficulty with that is the solution grows a mold which thoroughly contaminates it. The best answer to the problem is to keep the powder dry by the use of tight seals and the enclosure of the container in another container which holds some desiccant, such as silica gel.

III.   The Beginnings of the Brown Print.

No record has been found of a claim for the original invention of the brown print, and none is likely to be. While the origin of the idea of the process is known, the origin of the first process is impossible to determine. The historian can find numerous descriptions of proto-brown print processes in the literature, but, without a verifiable claim, such as a patent or publication, there is no assurance that the originator has been found. It is quite likely that the concept of the brown print process grew out of the researches of John Frederick William Herschel, the English scientist, who published the first report of the formation of an "image" by coating ferric ammonium citrate on paper, exposing the paper under objects, and "developing" the image by the application of solutions of potassium ferricyanide, gold chloride, and silver nitrate to the exposed paper. The ferricyanide developer produced a blue image and the print was called a cyanotype. The silver nitrate developer produced a brown image and Herschel called it an "argentotype." Herschel performed his experiments on iron sensitized paper over a period of time and reported them in *Philosophical Transactions* in June, 1842.

Herschel did not follow up on his initial lab observations.[2] They remained little more than laboratory curiosities. Herschel's experiments were widely publicized by Hunt and others who were interested in more definite researches in the direction of a practical process. Blue print paper was rather quickly adapted to the reproduction of technical drawings, but no specific working process resulted from the argentotype for about 45 years. [3] During the middle years of the nineteenth century, many researchers experimented with Herschel's iron-silver reaction, and the knowledge that a silver image could be formed on paper sensitized with ferric ammonium citrate became widespread.[4] Any close reading of the photographic literature published during the two middle quarters of the nineteenth century, will reveal numerous discussions of silver image formation following iron sensitizing. The number of reports of proto-brown print processes leads the present writer to the conclusion that it is futile to look for a single source of a practical printing process involving solutions of ferric ammonium citrate and silver nitrate. The literature supports the conclusion that iron-silver printing was an idea whose time had come in the period between 1850 and 1870, and that the idea came to many.

After 1872 the idea of printing with iron and silver was given greater impetus by the successful manufacture and sale of a commercial iron-platinum paper by Willis. By 1880 platinum printing achieved reliable manufacture, successful marketing, and wide use by artistic amateurs. The likely success of an analogous process based on printing with iron and silver probably occurred to many photographers and chemists.

In view of the wide currency of the idea of iron printing, the search for a single person who invented the brown print seems pointless. We must begin our history of a practical brown print process, if not with Herschel, then with the first useful process that is described in the literature.

When first found as a published process, brown printing appears in a variety of forms. It appeared at about the same time in England and in Germany as a commercial paper offered for sale. In England it was offered as a continuous tone printing paper, while in Germany it was first marketed as a paper for making two-toned (high contrast) copies of engineering drawings. About the same time, a brown print process, in the form of a bottled light-sensitive solution, was sold as a general copy-all photographic process. Photographers could coat the solution on paper to make announcements, menus, postcards, etc. Eventually the process entered the mainstream market for photographic papers. The Eastman Co. sold a version called "Eastman's Sepia Paper" for amateur use between 1900 and 1904. Eventually, as with kallitype, the brown print process was "appropriated" by amateurs as a paper that could be individually prepared for personal use, for postcards and family albums, etc., and with some modification, for artistic use. The pictorial interest in the paper was sustained through the 1920s. By then the process was well known in Europe, England, and America. Articles dealing with "brown print," sepia print," and "van dyke print," can be found in numerous periodicals published in these countries throughout the period from 1880 to 1920.

IV.   History of the Brown print Processes 1885-1920

The first reference in the literature to a working brown print process, found by the writer, was published by L. P. Clerc, who attributes a

"Sepia" paper to H. Shawcross in 1889. Interestingly, this is the year in which Nicol, the inventor applied for his first patent on the kallitype I. Clerc classified the paper as a "sepia photo-copying paper," and reported it is "prepared by soaking translucent paper in a solution containing iron ammonium citrate, citric acid, and silver nitrate. Clerc wrote "the paper is widely used for making copies from tracings," but the use of the verb in the present tense suggests he is speaking about 1930, when he writes, not 1889, when the paper was first produced by Shawcross. According to Clerc, the sensitized sepia paper was exposed to light until an image was formed. The image was sufficiently visible that exposure could be judged and controlled by inspection. The fact that the paper was translucent suggests that it was placed on top of the copy material emulsion side up, and light was shone through the original and the print paper for a right reading copy. On removal from the printing frame, the paper was rinsed in water and then fixed for five minutes in a 2% solution of hypo. Finally, the print was washed "in several changes of water." It is not made clear whether Clerc's discussion of processing refers to way the Shawcross paper was processed in 1889, or the way it was processed in 1930.[5]

The first reference to an iron-silver process that produced a continuous tone sepia print by development in water alone appeared in *the Photo Annual* for 1891. *The Annual* prepared a list of the commercial printing papers that were available during the preceding year. It included a "Sepiatype" print paper in the volume for 1891 and reported that the paper was newly marketed by Sharp and Hitchmough during the year 1890. *The Annual* described the paper as follows:

> A process that requires neither developing or toning and gives all shades of tone in sepia. All that is required after printing is to soak the print in water for ten minutes and then fix [it] in a weak bath of hypo (about two drams to ten ounces of water) when it can be washed and dried. No acid bath is required at all and yet pure whites are attained.[6]

Prices were listed as 11 7s 6d for a "band" of 30 feet, 30 inches wide. A sheet 36" by 30" cost 3/—and cut sizes were available in per dozen quantities. A dozen 10" by 12" sheets, for example, cost 4-6. "Sepiatype" apparently was sold as a printing paper for regular photographic use by professionals and amateurs.

The Photo-Beacon published a report of a German patent, granted in 1895, for a "new photographic paper." The article states the paper was "intended to be used for such purposes as architect's plans and drawings; to be used instead of blue paper." In all likelihood, this is the same paper which, according to Eder, was patented by Arndt and Troost in 1895. [7] The formula for this paper, as given in *the Photo Beacon*, follows:

Arndt and Troost Brown Print Paper 1895

Ammonio-citrate of iron	80 to 100 parts
Silver nitrate	12 to 20 parts
Tartaric acid	15 to 20 parts
Gelatine	10 to 15 parts
Water	1000 parts

This early formula lists the chemicals that will be variously compounded in most brown print formulas to come. We note the ferric citrate sensitizer, the silver nitrate, the tartaric acid and the gelatin as well as the range of quantities that can be employed. We have already explained the sensitizing function of the ferric citrate and the image forming function of the silver nitrate. The tartaric acid was used to facilitate the chemical conversion of the iron image to the silver image. The gelatin facilitated the even coating of the solutions and may have had a slight warming and deepening effect on the final image. *The Photo-Beacon* reports that "after printing, the paper simply requires washing in water," The implication that no solution other than water is required to fully develop, fix, and wash the paper was probably intended. It made the simplicity of the process appear more attractive. The reader will recall that typical kallitype processes required a chemical solution for development, and after that a clearing bath and finally a fixing bath, each requiring chemicals in solution in water. *The Photo Beacon* also reported that the paper is "very sensitive to light and gives a brown image on a white ground."[8] It is apparent that the objective of Arndt and Troost was to create a simple, easy to use inexpensive silver paper to replace blue print paper, which was also processed in water alone.

The next sepia or brown print process we find in the literature was a commercial liquid preparation useful for making note paper, menus, postcards, etc. The product also provided a means of making positive photocopies, a process now accomplished by xerox and other office copying methods. The product sold was not a prepared paper, but rather a sensitizing solution in a bottle, which the user applied to any paper on which he wished to make a photo copy from a negative or stencil. Clerc has this to say about the product:

> Sensitizing solutions for use with a brush have been put on the market from time to time for the local sensitizing of drawing paper, Bristol-board, or notepaper . . . As an example, the formula of such a mixture is reproduced here, according to the analysis of a commercial preparation by E. Valenta in 1899.

Valenta was a highly respected Viennese photo-chemist who, along with research on a wide variety of other photographic subjects, did research on ferric oxalate and its double salts and their several applications.

<u>Valenta's Brown Print Sensitizing Solution</u>

A. Solution

| Ferric Ammonium Citrate (green) | 100 grms. |
| Distilled water, to make | 1000 mls. |

B. Solution

| Silver nitrate | 140 grms. |
| Distilled water, to make | 600-800 mls. |

> Add pure ammonia solution drop by drop to solution B, shaking continuously, until the brown precipitate first formed is re-dissolved. If an excess of ammonia, which can be easily recognized by its smell, has been added in this way, the mixture should be rendered odourless by adding very dilute sulphuric or citric acid drop by drop.
> The two solutions are mixed in the dark, and the sensitizer prepared in this way should be kept in the dark until required[9]

A summary of the manipulation of Valenta's process follows.

When printing from negatives lacking contrast, a few drops of a 5 per cent solution of potassium bichromate may be added to the mixture. The mixture is applied with a brush (preferably one not bound with metal of any kind) to the paper surface to be sensitized. The sensitizer can be used either plain or after it has been thickened for more even spreading. To thicken, add before use a little freshly prepared starch or arrowroot paste. Exposure is accomplished in bright daylight. When the image has attained a slightly greater depth than that finally required, it is fixed in a 3 to 5 percent solution of hypo. After remaining five minutes in this bath, the print is thoroughly washed. The sepia tone obtained in this way may be modified before or after fixation by toning with gold, this process also improving the permanence of the image. [10]

The British Journal of Photography noted the appearance of a brown print process developed by Herr Vollenbruch in its April 14, 1899 issue.[11] Four months later **The Photographic Times**, under the title of "A Cheap Substitute for Platinotype," published a complete description of Vollenbruch's process, taken from the **Deutsche Photographen Zeitung**. Vollenbruch's brown print paper required the preparation of four solutions. The first solution contained gelatine, the second, a green ammonio citrate of iron, the third, silver nitrate, and the fourth, uranium nitrate. The sensitizing solutions are compounded as follows:

Vollenbruch's Brown Print Sensitizing Solutions

No. 1
Distilled water   100 parts
Gelatine   0.5 parts
Salicylic acid   .05 parts

No. 2
Distilled water   100 parts
Green Ferric ammonium citrate   20 parts

No. 3
Distilled water   80 parts
Silver Nitrate   10 parts

No. 4
Distilled water   100 parts
Uranium nitrate   20 parts

According to the proportions of these solutions used
the results differ; more silver gives soft prints, and more
iron contrasty prints. Reduction of uranium gives browner,
and increase gives more blue-black tones. The following
proportions give the best results with a normal portrait
negative, and the tone of the picture is a grayish-black, which
cannot be distinguished from platinotype:

Vollenbruch's Working Solution For Normal Negatives

No. 1   2 parts
No. 2   2 parts
No. 3   4 parts
No. 4   4 parts

The solutions must be mixed in the above order or else
the mixture will be cloudy. The solution should be poured on
to the paper and then distributed with a pad of cotton . . . The
paper must be quickly and evenly dried, and will then keep
for four days if protected from light and damp, but it is better
to use it fresh.[12]

It should be noted that Vollenbruch's brownprint process is one of
the few which requires the use of a "developer." The recommended
developer is:

Vollenbruch's Brown Print Developer

Distilled water   700 parts
Ferrous sulphate   30
Acetic acid   10   "[13]

The Photographic Times report on Vollenbruch's process does not
directly explain the theory or action of the developer. One notes that
Vollenbruch recommends the paper be exposed only "till the half-tones
are faintly visible," so it is likely that he intended the developer to

complete the formation of the image by chemical action instead of by a longer exposure to light, as in most brown print approaches. The developer formula is included for those wishing to try a brown print approach that involves a shorter exposure and a developer.

Vollenbruch's approach to brown printing was praised as a platinum print substitute, in view of the neutral black image it produced. The Vollenbruch method of brown-printing was classified as a form of kallitype and was included in the kallitype entry in *Cassell's Cyclopedia of Photography,* published in 1911. Since this encyclopedia was one of the most authoritative and complete sources of information on print processes, Vollenbruch's approach received considerable attention at that time and since. The editor's decision to reprint it is not surprising, in view of the sophistication of the process.

Interestingly few references to Vollenbruch's process by users have appeared in the literature. Perhaps amateurs were intimidated by the short keeping time of the paper. It is worth noting that Vollenbruch's paper was offered to amateurs for individual preparation, in advance of any offering as a commercial paper. For most processes, the reverse order, commercial sale, followed by disclosure of the process, was the usual practice. It is not known whether the paper was ever sold in England or America.

Next we find a brownprint paper offered for sale by the Eastman Co. It is difficult to date, from the available sources, exactly when the sale of the product, 'Sepia Paper,' actually commenced. Advertisements for the paper have been found dated as early as February, 1900, in **Camera Magazine**. The ads claim Eastman's paper has "beautiful warm brown shadows and halftones, with mellow, creamy high lights, and are especially effective when made from broad sketchy negatives." In large, bold type, the ads proclaimed Eastman paper to be "as simple and cheap as making blue prints." The last ad for Eastman's Sepia Paper, found by the writer, appeared in the April 1904 *Photo Miniature*. The Eastman Co. also marketed blue print paper in a variety of sizes. One way it was packaged was in tightly sealed metal cans, bearing the inscription "for making post cards." Brown print paper kept no better than blue print paper, so the packaging of the commercial paper in cans was required to keep it dry.

The directions for using Eastman's Sepia paper can be found printed in numerous advertisements. They are informative.

Eastman's Sepia Paper is about three times as rapid as blue paper. It should be under rather than over exposed and is developed by washing in plain water. After two or three changes of water, fix 5 minutes in a solution of hypo, 1-1/2 grains to the ounce of water and afterwards wash thoroughly. Short fixing gives red tones; longer fixing produces a brown tone.

Incidentally, the price schedule for Eastman's Sepia paper was

| 4 by 5" | per pkg of 2 dozen | $ .15 |
| 8 by 10" | per pkg of 1 dozen | .40 [14] |

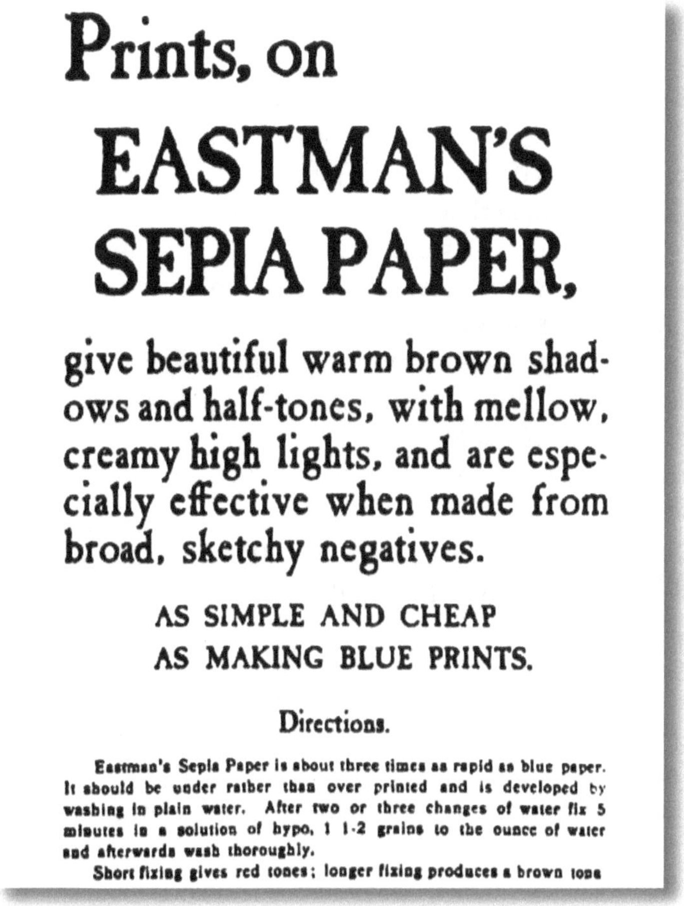

Prints, on

EASTMAN'S SEPIA PAPER,

give beautiful warm brown shadows and half-tones, with mellow, creamy high lights, and are especially effective when made from broad, sketchy negatives.

AS SIMPLE AND CHEAP
AS MAKING BLUE PRINTS.

Directions.

Eastman's Sepia Paper is about three times as rapid as blue paper. It should be under rather than over printed and is developed by washing in plain water. After two or three changes of water fix 5 minutes in a solution of hypo, 1 1-2 grains to the ounce of water and afterwards wash thoroughly.

Short fixing gives red tones; longer fixing produces a brown tone

Figure 24. Eastman Co, sepia brown print paper ad.

In February, 1901, *the Photo Beacon* published a "Report on Eastman's Sepia Paper." The name of the author was not given. The report was a written version of "a paper which had been read before the

Photographic Society of Philadelphia." The report viewed Sepia paper as a printing paper for regular use by amateurs.

> The sepia paper manufactured by the Eastman Company ... possesses many advantages which will be appreciated by the average amateur as well as the more advanced worker. The tones obtained are pleasing; it is very simple in manipulation and is cheap, the price being about the same as for blue-print paper. It is almost a "printing-out paper," as the image is quite distinct before development.[15]

The article discussed in considerable detail various ways of manipulating the color of brown prints. It reported that the color of brown print images can be varied from pink to brown by fixing the print in a weaker hypo than that recommended, that is less than 1-1/2 grams of hypo to the ounce of water, and for a lesser time than 5 minutes. The color may also be varied by "toning" in toning solutions. A weak potassium iodide solution (1 grain to the ounce of water) produces a "rather pleasing" orange brown print color. Toning the print in "a weak mercuric chloride solution" and then in a potassium iodide solution produced a somewhat cold, gray sepia. A print placed first in a mercuric chloride bath and then in the potassium iodide bath, and then into a hypo bath and last in a water bath produced a "warm, rich true sepia." The writer indicated the print color can be further changed by immersion in other baths, ammonium sulphocyanide, copper chloride, permanganate, etc. He concluded,

> These few crude attempts will show some of the possibilities of the paper, although the regular method is possibly the best, as it certainly is as simple as any one can desire. The paper is best suited to broad effects, the rough surface interfering with the reproduction of much fine detail. It gives however, more detail than does blue print paper, with which it compares in price and simplicity of manipulation.[16]

It is interesting to note how amateurs at the turn of the century were disposed to take a very simple print process and turn it into one that could be extensively manipulated. It suggests that part of the fantasy of being a photographer at the turn of the Century had to do with darkroom manipulation—what is today called "darkroom magic." And, of course such manipulation accorded with the pictorial taste for personalizing the print.

The brown print was given its most thorough description by George Brown F.I.C., in **Ferric and Heliographic Processes,** a book published in England in 1899 and in America one year later. We have already referred to Brown's description of KI and KII processes. A full report of Brown's approach to brown print also appeared in the November 1901 issue of *Amateur Photography*.[17] We hasten to note in passing that this study has found no evidence for any connection between the name of the brown print process and the name of the writer, George Brown, whose book did much to popularize the process.

George Brown, was an internationally respected writer on photography. He was well educated, held an advanced degree in chemistry, and, as noted earlier, he collaborated with Nicol, the inventor of kallitype, on a number of scientific projects. For twenty five years Brown was the editor of the prestigious **British Journal of Photography Almanac,** an annual that compiled a good deal of what was worth knowing about photographic processes. Brown also wrote numerous articles for photographic journals of the day. He possessed an international reputation as an expert on photographic processes. Interestingly, when the editor of **Cassell's Cyclopaedia** reproduced Brown's formulas in its entry on the brownprint, the editor remarked that the brown print is "a more modern process than kallitype" and characterized it as a "water developing kallitype, a good paper for home production." It is difficult to speculate about the meaning he had in mind for the word "modern": later or simpler? The commercial sale of the brown print paper antedated the sale of kallitype paper by about a year.

As Brown's book described the brown print, the sensitizer was a mixture of four stock solutions, the first containing ferric ammonium citrate, the second, tartaric acid, a third containing silver nitrate and a fourth, gelatin. Equal amounts of these stock solutions were mixed together and the resulting solution was coated on the paper and dried. After the paper was exposed, the print was "developed" in plain water. The exposed print was then fixed in a weak hypo bath for five minutes and a twenty minute wash completed the processing.

The details of Brown's process follow.

G. E. Brown's formula for the Brown Print

A. Green Ferric Ammonia Citrate	110 grains	7.2 gms
Water	1 oz.	28.4 ml
B. Tartaric Acid	18 grains	1.4 gm
Water	1 oz.	28.4 ml.
C. Silver Nitrate	46 grains	3. gm
Water	1 oz.	28.4 ml.
D. Gelatine	30 grains	2. gm
Water	1 oz.	28.4 ml.

Mix the solutions and store in separate brown bottles. Use distilled water throughout. [18]

The A and C Solutions keep well in the dark. The B solution keeps for a few days only, owing to the formation of mould. The D solution should be made just before use, the gelatine being soaked in the water and dissolved by heat.

To prepare the sensitizing solution for use, mix equal parts of A, B, C, D, as follows: Warm the gelatine solution D and add solution A and B to it, maintaining the warmth by standing the receiving vessel in hot water. Lastly add the silver, solution C, a few drops at a time, stirring it in with a glass rod. The warm solution is coated on the paper and hung to dry.

The print is exposed until the image is "vigorous" and is then washed in plain water for about two minutes during which time the image develops to greater strength and turns reddish brown in color. The purpose of this bath is to rid the print of unused chemicals which are still light sensitive and can therefore fog the paper if not removed. The print is then fixed in a weak hypo bath, 100 grains (6.4 grams) to 10 oz. water or in a 10 percent sodium sulphite solution. In either bath the print color becomes a rich brown. Finally the print is washed for about half an hour and dried.

A simple, single-solution approach to brown print, credited to George E. Brown, is provided in **Cassell's** *Cyclopedia of Photography*. The editor indicates the formula was widely used for menu, note paper, and post-card work. The entry is printed as given.

Geo. E. Brown, who has advocated it, gives the following directions: 55 grains (3.6 grams) of silver nitrate is dissolved in 4 or 5 drms.(17.5 mls) of distilled water; and liquor ammoniae (.880) diluted with an equal quantity of water, is very carefully added. As the first drop or two is added, a copious precipitate of silver oxide is thrown down in the solution. Addition of more ammonia solution will re-dissolve this precipitate; cease to add ammonia on the disappearance of the last traces of the precipitate. Weak sulphuric acid is next added drop by drop until the faint odour of ammonia disappears. 40 grains (3.6 grams) of green ferric ammonium citrate dissolved in 6 drms. (21 mls) of water is then added, and the liquid is complete. Stored in the dark or in a stone bottle, it will keep good for several months. It is applied to paper in the same way as other kallitype sensitizers, dried, and the paper treated . . . with water and "hypo." [19]

The editor of **Cassell's** concluded the account of brown printing with the observation that while brown print is "a good paper for home production, . . . the warm brown results cannot be said to equal those given by the proper kallitype formula." [20]

Cassell's included another "simple" brown print sensitizer, this one recommended by Professor Rudolfo Namias of Milan. Interestingly, Namias' process was not listed under the heading, "Brown print," or "Kallitype" but rather under the heading, "Sepia Paper." Again, the short entry follows.

Namias' Sepia Paper

A printing paper coated with a compound containing salts of silver and iron; development being effected with plain water and fixing in a weak "hypo" bath, toning being optional. The finished prints are of a good sepia tone. The Namias process is recommended; for this two solutions for sensitising are required—

A. Green ferric ammonio-citrate	44.0 grams
Citric acid	10.0 mls.
Distilled Water	100.0 mls.

    B. Silver Nitrate   11. grams
        Distilled Water   40. mls.

These are mixed together, made up to 200 ccs. [with water], and applied to the paper with a Blanchard or Buckle brush, and then dried; the whole of the operations must be carried out in the dark-room. A second coating is sometimes advisable. The paper is printed in contact with a negative in daylight in the usual way, but not very deeply, because in the washing after printing, a slight intensifying action takes place. After about five minutes washing, the print must be fixed for one to two minutes in an 8 per cent solution of "hypo." The print is finally washed for about twenty minutes and then dried. [21]

A "Blanchard brush" was a home made "brush" for applying sensitizers to paper. It was formed by wrapping some calico or other cloth around a small rectangular piece of glass, or other non-bibulous material. The "Buckle brush" served the same function, but was made with a hollow glass tube which contained a ball of cotton which did the spreading. Coating could also be done with a simple wad of cotton or a wide paint brush.

It is notable that the above brown print process involves the use of a hypo fixer to which has been added sodium sulfite. From L. P. Clerc we learn that the idea of using sulfite to deepen the brown color of the sepia print originally came from Rodolpho Namias. Clerc reports that Namias' approach, recommended in 1901, produced a "more opaque image . . . by replacing the hypo fixing solution . . . with a 15% solution of anhydrous sodium sulphite." Clerc finds this approach is "more expensive"[22]

Namias was a late nineteenth century photo-chemist who, according to Eder, "started the most important Italian photographic periodical, *Progresso Fotografico*, and headed the Instituto Chimico Fotochimico in Milan.[23]

The January 1900 issue of *Photo-Miniature,* while entitled "The Blue Print and its Variations," included several accounts of other print processes, one of which is a brown print process, called a "ferric citro-oxalate process." The formula for the sensitizer was unique in that it included both ferric oxalate and ferric citrate:

Ferro Citro-Oxalate Sensitizer

A. Solution

Ferric Citrate,	50 grains	3.25 grams
Ferric Oxalate,	25 grains	1.6 grams
Distilled Water	1 oz.	28.4 mls

B. Solution

Silver Nitrate,	25 grains	1.6 grams
Distilled water	1 oz.	28.4 mls

The A and B Solutions are mixed and bottled for later use use. The combination should be shaken and any precipitate that forms should be filtered off, so the sensitizing solution is a clear liquid. The paper is coated with sponge or tuft of absorbent cotton, dried quickly, and exposed under a negative to sunlight. The image "prints out fully," and after being washed in clear water, the print is fixed in a dilute solution of ammonia, (2 percent of ammonia to the ounce of water). One wonders if water alone was enough to fix and clear prints made with this solution, since ferric oxalate sensitizers normally require the use of chemical developers and clearing baths to avoid "yellow stain."

The source of this process is not revealed. John Tennant, the editor of *Photo Miniature* indicates that he wrote the entire monograph with the help of F. C. Lambert.[24] The above description of the ferric citro/oxalate process is reprinted in the April, 1902 issue of *the Photographic Times,* over the signature of B. C. Roloff, [25]

After 1902 the brown print as described in photographic publications tended to stabilize. The process went through a variety of minor modifications of method, most of which were concerned with changing image color. Many of the writers on brown print merely replicated the formulas of Brown and other earlier writers. We shall skip over them to James Thomson's formulas, which were, if anything, not cut of common cloth.

In 1904 James Thomson reported that he had been working for six months "with a number of formulae whereby prints could be developed by plain water instead of baths in which . . . chemicals are incorporated." He "presumes many . . . readers are familiar with Eastman's Sepia Paper" which, he states, "is a kallitype preparation." The paper he is writing about will be "similar in manipulation and color." He writes, "for the past six months I have been experimenting with a number of formulae whereby prints could be developed by plain water."

> I have succeeded in my attempt, having now three or four methods of producing prints, which while all good, differ somewhat in quality and color. . . . while the paper is primarily intended for water development, other effects may be had by using chemicals—short exposure, for instance—and development by silver nitrate made acid with citric [sic] produces good results, varying according to strength of silver solution. A print from a negative with more than enough contrast may be improved by immersion in a solution (developing) of Rochelle Salts, ten grains to the ounce of water. I simply hint at these possibilities, leaving it to others to carry along . . . [26]

Thomson's brown print process is reminiscent of the early silver chloride papers in that it involves a salting and a sizing solution, applied in two sequential applications, the paper being dried before the second solution is applied. In Thomson's process, the first application coats the "salting" solution and the second, the sensitizer. The reason the two solutions are applied separately, as with the early "salted" paper, is that if the salting and sensitizing solutions were mixed together, the silver and chloride would precipitate out of solution as silver chloride. As such the chemicals would be useless for image formation. Applied sequentially, the two coatings could be proximate to each other because the chemicals are inactive when dry. Thomson's use of the terms "salting" and "sensitising" is confusing, since the ferric sensitizers are found in his "salting" solution and the silver salts are found in his "sensitizing" solution.

Thomson's Salting Solution No. 1 (1904)

Citrate of iron and ammonia	20 grains	1.3 grams
Ferric oxalate	12 "	.775 "
Oxalate of potassium	12 "	.775 "
Chloride of Copper	6 "	.4 "
Oxalic acid	4 "	.26 "
Gum Arabic	10 "	.65 "
Distilled Water	1 "	28.4 ml.

The paper should be coated with the No. 1 solution, the salting solution, and dried. It should then be coated with the second solution called 'the sensitizer.'

Thomson's Sensitizing Solution Number 1 (1904)

Silver Nitrate	50 grains	3.24 grams
Oxalic acid	2 "	.13 "
Citric acid	20 "	1.3 "
Distilled water	1 oz.	28.4 mls

The sensitizing solution is applied evenly and the surface dried by gentle artificial heat. The print is exposed till half tones appear and is taken from the printing frame and developed in running water for 3 minutes. The prints are fixed in a solution made of 50 grains (3.24 grams) of hypo in one quart of water. Last the prints are washed in running water for 30 minutes. If more contrast is desired, more ferric citrate should be added to the sensitizing solution one grain at a time. The resulting print should be dark brown. The prints may be toned with gold or platinum.[27]

In 1910, Thomson revised his 1904 brown print formulas keeping the ingredients the same, changing only the quantities. He continued to revise his brown print formulas during the next fifteen years, making what had been a simple process into an increasingly complicated one. Thomson believed that the simpler formulas discussed above produced prints with too much contrast. He wanted to produce 1) a "paper with good gradation" and 2) a paper which printed out in the frame. In an effort to make a brown print paper with extended gradation, Thomson continued to explore his two coat approach involving a salting" solution and a second sensitizing solution coated on top of the first. In the following formulae, the 1904 and the 1910 solutions are given side by side for comparison. The 1910 formula has quantities expressed in grams as well as grains. [28] "Watertone was one of Thomson's names for brown prints.

Thomson Sepia Watertone Salting Solutions No. 1 and 2

	1904	1910	1910
Salting Solution	No. 1	No. 2	No. 2 (grams)
Ferric Ammonium Citrate	20 grn	6 grns	.39 grams

Ferric Oxalate	12	20	1.3
Potassium Oxalate	12	20	1.3
Copper Chloride	6	4	.26
Oxalic Acid	4	0	0
Gum Arabic	10	10	.65
Distilled Water	1 oz.	1 oz.	28.4 ml.

Thomson Sepia Watertone Sensitizing Solution

	1904	1910	1910
Sensitizing Solution	No. 1	No. 2	(grms.)
Silver Nitrate	50	40	2.6 gram
Oxalic Acid	2	10	.65 "
Citric Acid	20	10	.65 "
Distilled Water	1 oz.	1 oz	28.4 "[29]

The column of figures to the right is a conversion into grams of the formula No. 2 quantities.)

In 1910, Thomson revised the directions for working his brown print process. The new directions follow. The salting solution is applied by brush and dried. Then the sensitizing solution is similarly applied and dried. The paper is exposed under a negative in a printing frame for three to five minutes. After exposure, the print is given a three minute wash in running tap water to remove all soluble chemicals. It is then placed in a fixing solution (25 grains or 1.6 grams of hypo to 16 ounces of solution) for two to five minutes. The time of fixing is limited to the shortest period which will not bleach the image. The No. 2 fixing solution allows for a longer fixing time than the No 1 solution which, being stronger, was prone to bleach the image. If more contrast is desired, more ferric ammonium citrate should be added to the salting solution. If less contrast is wanted, the salting solution can be diluted with water.

Thomson wrote that the papers "I am placing before the reader are similar in manipulation and color to Eastman's "Sepia Paper." While the paper is intended for water development, "other effects may be had by using chemicals" to develop the paper. Using a chemical developer will permit a shorter exposure and permit greater control of contrast and color. The formulas given produce a "jet black kallitype that can not be distinguished from platinum and is as easy to produce as a blue print."

> As one expert kallitypist remarked, Thomson's two brown print solutions, salting and sensitizing, contained a "bewildering collection" of chemicals—"a warring array." [30]

Thomson's way of working the watertone kallitype process appears derived from Fox Talbot's early method of preparing Calotype paper with separate salting and sensitizing baths.

In Talbot's paper, the salting and sensitizing solutions were simple and they made sense. Observers of Thomson's approach to the water-developed kallitype process felt that the two coating solutions needlessly complicated the simple brown print process and that simpler formulae would be not only less expensive but more effective. Nevertheless, Thomson's approach developed a considerable following among editors of photographic periodicals on both sides of the Atlantic. They continued to enthusiastically praise the prints he sent them. His writings on brown print never failed to find publishers and readers, and his complex formulas were eagerly sought for inclusion in contemporary reference works on photography.

In 1908 A. J. Jarman published an interesting approach to the preparation of sepia brown prints which used a mercury bleaching bath. Jarman coated on paper a relatively traditional sensitizer made by combining the following solutions.

Jarman's Sensitizer

Solution 1
| Hot distilled water | 2 ounces |
| Citric acid (crystals) | 1/2 oz. |

Solution 2
| Distilled water | 4 ounces |
| Ferric ammonium citrate (green) | 1/2 oz. |

Solution 3
| Hot distilled water | 2 oz. |
| Silver nitrate | 1/2 oz. |

After the solutions are homogeneous and cool, add Solution 1 to Solution 2. Stir till the combined solution is uniform and then add solution 3 to the combination of # 1 and # 2. Filter and remove the filtrate.

Coat the paper with the mixed solution by "floating" the paper on the solution in a tray. "Floating" is a process of lowering the paper to be sensitized onto the sensitizer solution so that only the bottom surface is wetted by the liquid. There are two methods. One involves curling the paper (in the form of a" U") and raising and lowering it, wetting the paper by rolling it back and forth the surface of the sensitizer tray. Or the paper may be floated by folding up the four sides of the paper into the shape of a shallow boat. This can be done by folding up the four sides of the piece of paper, one inch or so from each end. After the paper rests on the sensitizer, and soaks up some solution, it is flattened, dried and subsequently exposed under a negative. When the image is "moderately" visible, the exposure is complete.

To complete Jarman's processing, the following solutions must be prepared.

Jarman's Mercury Bleaching Solution

Hot Water	30 oz.	1 liter
Potassium Bromide	120 grains	8 grams
Bichloride of Mercury	120	8 grams

Jarman's Fixing Solution

Water	20 oz.	500. ml.
Hypo	3 oz	9.3 grams

After the two solutions are clear and cool, place them in separate trays. The print is processed as follows.

1. Wash the exposed print till water remains clear (2 minutes).
2. Place the print in the tray of mercury solution till image completely disappears.
3. Rinse the print in clear water, approx. 1 minute.
4. Place print in hypo solution. This step produces a rich sepia print.
5. Rinse the print in water 2 minutes.

Repeating steps 2 thru 5 changes the print color from sepia to a "rich burnt umber colored print." The solutions can be reused till their strength is depleted. [31]

Jarman wrote on a variety of photographic processes during the first quarter of the twentieth century.

We shall report one final derivative of brown print to illustrate the range of development the brown print process underwent. Teresa Del Fabro invented a variation on brown print which he patented in both France and England. The method produced "a new series of blacks" "useful for copying stamps and engravings." Del Fabro, a capable chemist, worked on iron printing methods for a number of years. His patented process was abstracted in the *Journal of the Society of the Chemical Industry* which reported that a French patent, #443633, was granted to him in 1912. Del Fabro's process was also summarized in the March 1913 issue of *Camera* which reported on the English patent, #11,193, granted him on May 10, 1912. The patents were granted for a water developing iron printing paper using ferric oxalate as the sensitizer. The following solutions are prepared:

Del Fabro's Brown Print Stock Solutions

A. Solution

Ferric Oxalate	36 gram
Oxalic Acid	5 gram
Distilled Water	100 ml.

B. Solution

| Silver nitrate | 16 gram |
| Distilled water | 100 ml. |

Del Fabro's Brown Print Sensitizer for Black Tones

To make the sensitizer, combine
a. 30 mls of Solution A
b. 10 mls of solution B
c. 4 drops of a saturated solution of ferric chloride
d. 4 drops of hydrochloric acid.

Paper coated with Del Fabro's sensitizer is dried and exposed under a negative. After exposure it is developed in "simple" water or water containing sodium or potassium oxalate. The print is then fixed for 10 minutes in a 2% solution of hypo and water. Last, it is washed in running water and dried. [32]

This is a puzzling process. It appears to be a kallitype process that can be developed in water alone or with a kallitype developer. There would be definite deposits of yellow stain if the process were used as described.

V.   Conclusion.

Although writing about brown print occurred during the years from 1920 to the present, we shall conclude our report of the publications on the brown print with the year 1920. After 1920, writing on brown print became either 1) repetitive[33] or 2) preoccupied with problems of engineering an increasingly effective commercial brown print paper for the reproduction of engineering drawings. [34] Since the latter is a specialized concern having little to do with the artistic use of kallitype paper, the subject of this book, and since rewriting old ideas adds little to the storehouse of knowledge it seems reasonable to end the report on brown print with the year 1920.

To summarize, the brown print was one of four major kallitype (iron silver) print processes. During the 1880's, it was sold as a commercial printing paper and as a ready to use solution useful for making copies. The commercial paper was sold for copying drawings and plans and as a continuous tone paper for amateur use. The paper also was widely prepared by amateurs between 1890 and 1920, a time when individual preparation of printing paper was a common practice of amateur photographers.

As described by most writers, the individually prepared brown print was a simple, inexpensive, relatively easy print process to work. It produced prints that varied in color from sepia to dark brown on home coated paper that nearly printed out during exposure. Brown print paper was quite fast, about three times faster than blue print paper and about two times as fast as kallitype. It could be developed in plain water or in a suitably compounded chemical developer. The latter, if used, permitted shortened exposure times.

As a photographic process, the brown print remains, a less impressive print media than kallitype. When equally well made prints from the two processes are compared, brown prints will be seen to lack the depth and richness of tone of kallitype prints. Many have judged the image colors produced by brown print to be less pleasing than those produced by kallitype. Because of the relatively thin layer of silver metal that forms the image, brown prints are vulnerable to bleaching by strong concentrations of sodium thiosulfate (hypo) fixers. Over exposing prints that were to be fixed in hypo was the usual remedy for hypo bleaching. Brown prints when carefully made have a long scale of tones with clean whites and rich sepia to chocolate tones. They can be toned in chemical baths to a variety of colors.

The brown print process underwent a rebirth of interest after the 1950's, among artistic photographers who worked in mixed media. The brown print surface, when devoid of a gelatin layer, "takes" applied color well, whether the pigment applied is paint, pastels, water color, or whatever. Brown print has also been used recently for printing high contrast images on cloth for images used on quilts, tee shirts, large cloth prints etc. It has also been used for artistic bookmaking. Readers interested in mixed media applications are encouraged to check the bibliography under mixed media uses of brown print and kallitype.

The brown print process continues to be used as a straightforward printing process. When used in this fashion with artistry and imagination, certain delicate and charming effects can be achieved.

Those interested in working brownprint are welcome to try any of the approaches described in this chapter. If a recommendation for a first try is wanted, the writer suggests one of the approaches of George Brown.

Notes: Chapter IX. The Brown Print

1. No author given, "Some New Printing Papers," **British Journal of Photography**, November 1901, p. 693. The editor calls brown print "the Van Dyke print," and provides the explanation that the name is derived from the color of a painting pigment.

2. John Frederick William Herschel, "On Certain Improvements of Photographic Processes Described in the Former Publication . . ." **Philosophical Transactions**, June 16, 1842, p. 181.

3. Robert Hunt, **A Manual of Photography**, 3rd edition, London, John Griffin and Co., 1853, See pages 59, 134, 187; also **Researches on Light**, London, Longman, Brown, and Green, 1844, Arno Reprint, Chapter 5, pp. 137-143.

4. See John Towler, **The Silver Sunbeam**, New York, J. H. Ladd, 1864, Morgan and Morgan Reprint, pp. 271-4. Towler discusses print processes based on the light sensitivity of metals other than silver. He says of them: "they are very interesting but have not as yet been applied to any useful purpose." p. 272. He discusses Herschel's discovery that "certain persalts of iron, when exposed to light in contact with organic matter undergo decomposition and are reduced to the proto-salts . . ." p. 273.

    Also see W. H Burbank, **Photographic Printing Methods**, "A Practical Guide to the Professional and Amateur Worker," New York, Scovill Adams Co., 1891, # 22, 3rd edition., 1891. The second edition was published in 1888 and the first in 1886. Burbank has a chapter on iron printing methods. This work is scholarly and detailed with respect to historical and chemical information. It details workable processes for blue print, uranotype, and argentotype, the fore runner of kallitype and brown print. It includes methods for the preparation of the necessary chemicals for iron printing.

    Also see P. C. Duchochois' **Photographic Reproduction Processes**, New York, Scoville and Adams, #38, 1891. Duchochois lists in the appendix a number of workers who explored "pre-kallitype" iron silver processes in middle decades of the 19th Century: Borlinetto, 1853; C. J. Burnett, 1857; Godefroy, 1858; De La Blanchere, 1858; and T. L. Phipson, 1861. Each worked on a proto-kallitype process.

5. L. P. Clerc, **Photography Theory and Practice** English translation by George E. Brown, Bath, Pitman and Sons, 1930, p. 385-6.

6. No author given, **Photography Annual**, London, 1891, p. 494.

7. J. M. Eder, *History of Photography*, New York, Dover reprint, 1972, p. 543.
8. No author given, "A German Patent October," *the Photo-Beacon*, 1895, p. 409.
9. L. P. Clerc, op. cit., p. 386.
10. Ibid.
11. No author given, *British Journal of Photography*, April, 14, 1899.
12. Herr Vollenbruch, "A Cheap Substitute for Platinotype," *The Photographic Times*, Vol. xxxi, # 8, August, 1899, p, 386. This name is sometimes misspelled Vollenbach in some of the literature.
13. Ibid.
14. Advertisement by Eastman Co., *The Camera*, Vol. IV, #8, February 1900, p. 59.
15. no author, "Report on Eastman's Sepia Paper," *the Photo-Beacon*, February, 1901, p. 59.
16. Ibid., p. 60.
17. George Brown, *Ferric and Heliographic Processes*, Dawbarn and Ward, London, 1899. This work was published under the same title in America by Ward and Tennant in 1900.
18. The Brown formulas quoted here are from *Cassell's Cyclopedia of Photography*, Cassell Publisher, London, 1911, Arno Press Reprint, 1974, p. 315a.
19. Ibid. p. 31 b.
20. Ibid.
21. Ibid., p. 484.
22. L. P. Clerc, op. cit., p. 386.
23. J. M. Eder, op. cit., p. 700.
24. John Tennant, "The Blue Print and Its Variations," *The PHOTO-MINIATURE*, vol. I #10, January 1900, p. 500.
25. B. C. Roloff, "Ferric Citro Oxalate Process," *Photographic Times*, April, 1902.
26. James Thomson, "Kallitype Simplified," *The Photo-Beacon*, July, 1904, p. 203.
27. Ibid., p. 203-6.
28. James Thomson, "Kallitype and Allied Processes," *THE PHOTO-MINIATURE*, Vol. XVI, No.185, p. 224.
29. Ibid.
30. J. M. Hammond, "Variations in Iron Silver Printing," *Photographic Journal of America*," Vol. LIV, 1917, p. 293.

31. A. J. Jarman, "Preparing Paper for Sepia Printing With The Salts of Iron and Silver," *Photo Era*, March 1908, p. 133-137.

32. Teresa Del Fabro, "Improvements in the Kallitype Process," *The Camera*, March 1913, p. 184. This version contains errors. It says, "1 % solution of potassium permanganate," instead of the correct amount, 10%.

33. As an example of rewriting, consider Sigismund Blumann, "Home Sensitizers and their Application," *Camera Craft*, October, 1914, p. 495-502. Blumann reprints a sepia paper formula already reported. There are numerous such articles by Thomson, Jarman, and others, etc.

Incidentally, Blumann mentions that Eastman's Sepia paper was "taken off the market because of its poor keeping qualities." p. 498.

34. An example of the latter is Jaromir Kosar's book, *Light Sensitive Systems*, Wiley, New York, 1965. Kosar provides an extensive bibliography of modern research efforts on iron printing, most by professional chemists. It lists 20th century patents on select ferric printing processes, including brown printing.

SECTION III

The Demise and Rebirth

Chapter X

The _Denoument_ Begins 1904-1914.

I.   Media Responses and Developments 1904-1914

It's time to report the responses made to the KII processes of Hall, Brown, and Thomson detailed in earlier chapters. Publications which disclose new attitudes and methods of working KII shall also be reported. Reviewing the writing of these kallitypists will complete the picture of the decline of kallitype in the first quarter of the 20th century, a period intense concern about pictorial expression and individualized printing. The review, will show that while the level of technical creativity and enthusiasm for individualized paper remained high, there is a noticeable decline in the quality of interest. To indicate the trend, this chapter shall continue reporting, however briefly, on the kallitype publications between 1904 and 1914. The report will cover in detail those writers who made notable contributions. Whatever the level, the writing on kallitype of this period adds detail and color to the picture of the kallitype process and kallitype printers in the first decades of the twentieth century.

In 1904 "A description of a Home Made Printing Paper" was published in the **Photographic Times** by Professor Rudolpho Namias, a renowned Italian photo-chemist whose work on brown printing was discussed in the last chapter. Namias published his researches on the kallitype in **Il Progresso Fotografico**, a magazine he edited. Namias' "home made paper" was a kind of kallitype I paper that was sensitized with ferric chloride and oxalic acid and developed in a 2% bath of silver nitrate. The resulting print was cleared in a 5% solution of oxalic acid and fixed in a 5% solution of hypo. Namias' approach is exceptional in its use of a sensitizer based on ferric chloride. He ended his article with the remark that his "process is so economical and easy that it is worth trying by both amateurs and professionals of an experimental turn."[1]

In the May, 1906 issue of **Photo Era** an article on kallitype II appeared, written by Louis Fleckenstein. Fleckenstein was a well known California pictorial salonist, whose images were reproduced with some frequency in the periodicals of the time. He may be the best known American artist image-maker to write on kallitype.

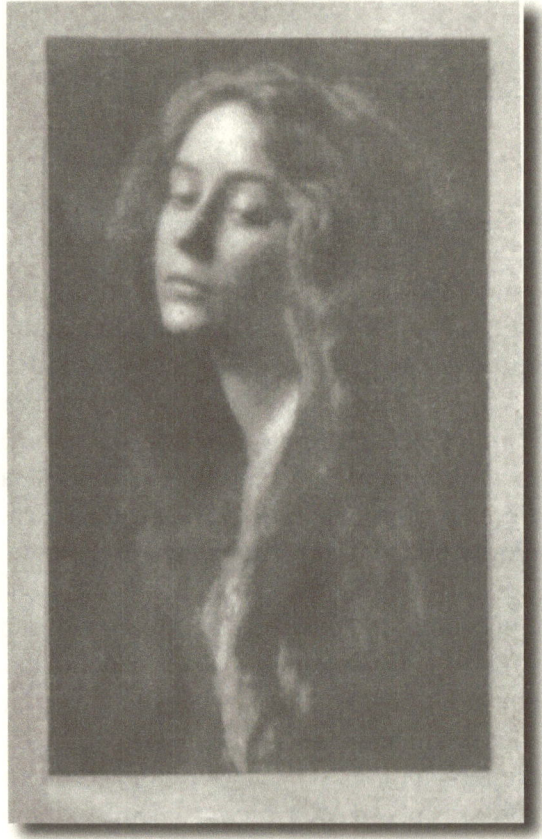

Figure 25. Louis Fleckenstein, kallitype

He wrote, "within the past few years Kallitype has sprung into prominence as one of the standard photographic processes." He was critical of Thomson's approach.

Many formulas have been published, but in most cases at such length as to confuse the mind of the uninitiated and create doubt as to the vaunted simplicity of the process. To the adept these details are very useful; but they are unessential to the person desiring only the groundwork.

Fleckenstein recommended a "simple approach," really Hall's method, which he recapitulated accurately and in detail.[2]

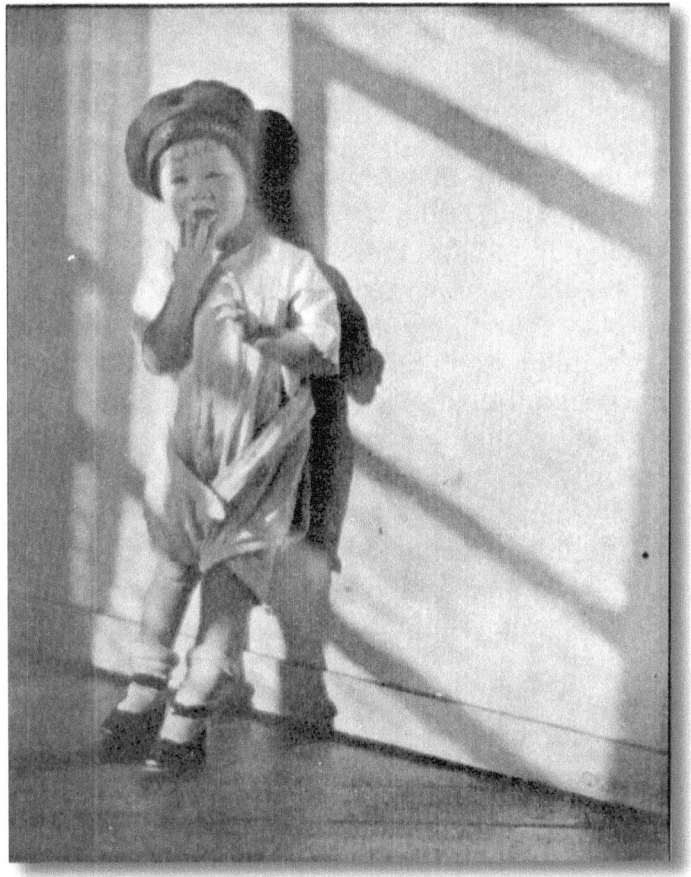

Figure. 26. Louis Fleckenstein photograph on commercial paper

On Sept 3, 1906, another demonstration of kallitype was held at the Bowes Park and District Photographic Society in England. The demonstrators were "Messrs. Cunningham and Craston," and the process shown was the Bennett-Burton approach to K II.[3]

The second edition of *The Photographic Reference Book*, a book aiming to inform amateurs about "the profusion of printing papers," appeared in 1906. The book contained a discussion of the long ignored kallitype I process, evidently copied from Nicol's original patent. No mention was made of Kallitype II. [4]

In the April 1907 issue of **Photo Era**, we find an article on Kallitype II written by James S. Escott of Louisville, Kentucky. The article, entitled "Kallitype for Winter Landscapes," spent a scant five sentences on that topic. The bulk of the article described how to work the kallitype. The article was addressed to "the progressive amateur who practices photography from the pure love of it and is usually a true artist at heart, who sooner or later chafes at the restrictions imposed upon him by the printing papers on the market." Escott "knows of no improvement over Henry Hall's formula" which he describes in detail. He concludes "there is no greater pleasure connected with the art than to watch the development of a beautiful print on home-made paper and to realize the immense possibilities and control which lie under the hand of the worker."[5]

In June, 1907 **The Photo-Beacon** published another description of the Bennett-Burton approach capably written by Charles S. Taylor, another American pictorialist. Taylor tells us that in the last fifteen years "the kallitype process has gained considerable popularity." He is impressed with how quickly the manipulation of the paper can be learned and recommends the simplest approach to it, that of Bennett and Burton. Though he has seen many excellent kallitypes made by others and has made a few good prints himself on commercial brands of kallitype paper, Taylor, believes that for amateurs, "the home-prepared variety is best . . . if one wishes to understand and appreciate the merits of the kallitype."[6]

The July 1907 edition of **Camera** abstracted an article taken from **Photo-Korrespondenz** which reported the experiments of N. Adrianow on the preparation of an iron paper. Adrianow's preferred formula is

Solution No. 1

Distilled water	200 cc.
Oxalic Acid	3 gram
Ferric Oxalate	30

Solution No. 2

| Distilled Water | 100 cc. |
| Silver Nitrate | 5 gram |

The two solutions are mixed together to make a Kallitype II sensitizer. Saxe paper is floated on the resulting solution. The prints are said to develop in plain water and are fixed in weak hypo. It is quite likely that the report of this process is incomplete, for more than plain water is necessary to develop a ferric oxalate image and clear the iron from the print. This is not the first incomplete and misleading published report on kallitype and it certainly wasn't the last.[7] The report indicates that kallitype printing was being done in Germany.

In the July issue of the *Camera and Dark Room*, an American journal, we find a formula which bears resemblances to Nicol's patented Kallitype III formula. The formula is

Anonymous K II Sensitizer

Ferric Oxalate	150 grains
Potassium Oxalate	30 grains
Silver Nitrate	30 grains
Water	2 oz.

The prints are said to "print right out." They are then put in a bath of

Anonymous K II Developer

Sodium citrate	290 grains
Citric acid	120 grains
Water	10 oz.

This bath appears to be a solvent for ferrous oxalate, the product formed when ferric oxalate is exposed to light. The solvent dissolves the ferrous image produced by exposure, permitting it to react with the silver nitrate to form an image in silver. After "development" the prints are fixed in a weak solution of ammonia .880, about 1/2 oz. to 1 gallon of water.[8] No source is credited for this publication. The ammonia fixer dates back to Nicoll's patent.

In 1907 *Camera and Dark Room*, G. Caspar Elmberger provided another account of Hall's K II process. Elmberger reports working with the process for two years before writing. His report provides extensive but familiar details of kallitype manipulation, and concludes with the comforting homily "he who combines cleanliness with carefulness (sic) will be sure of success."[9] In the next month's issue, Elmberger apologized for failing to include all the ingredients in his published formulas. Somehow he forgot to include the silver nitrate. No matter, he adds; since writing the previous article, "I have obtained splendid results by combining a suitable developing agent with the sensitizer. Alas! another KIII process is born. [10]

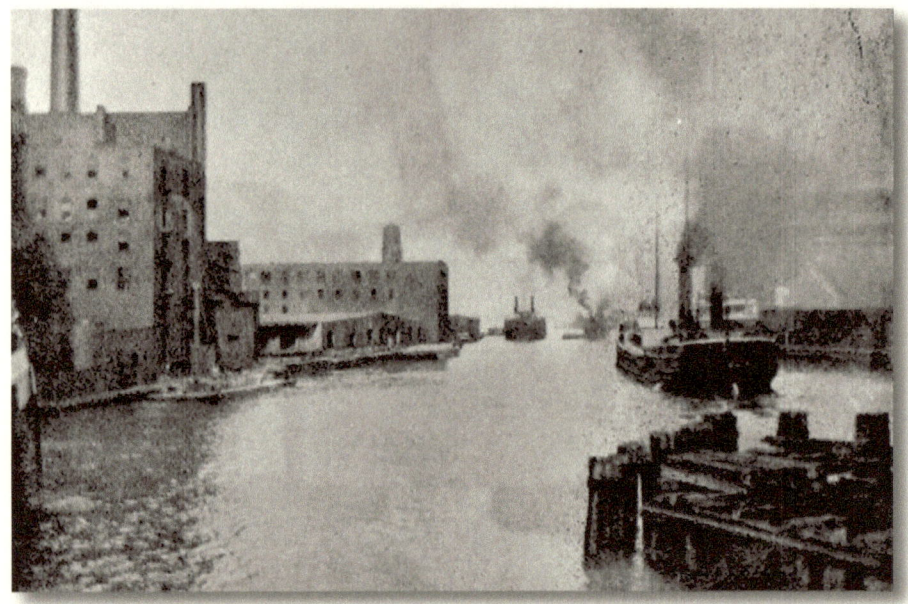

Figure 27. G. Gaspar Elmberger, manipulated kallitype

In the November 1907 issue of the **British Journal of Photography**, J. J. M. Sellors reported an "exhaustive demonstration" of kallitype given to the Croydon Camera Club. The club secretary reports Sellors "conclusively proved that the kallitype was simple, absurdly cheap, yet capable of giving most beautiful and artistic pictures." Sellors' approach to kallitypy, the Bennett-Burton approach, reflected the continuing English preference for a simple approach. [11]

In the May, 1908 issue of **Camera Craft**, published in San Francisco, J. Will Palmer wrote on "The Kallitype Process." He did not aim to write ". . . anything especially new, but rather . . . to make each step so plain and easily understood, that anyone can make Kallitype prints without the least fear of failure." He followed Hall's method, more or less. [12]

In the 1908 **American Annual**, Walter W. Lakin, another American salonist, wrote that "through the medium of the popular annual, I would

like to call the attention of the serious worker to the merits of Kallitype."
He continued

> good prints by this process cannot be told from platinum,
> and now that platinum paper has advanced so in price, it is
> well to be able to use so good a substitute, one so satisfying as
> to tones, so easy of working, and withal so cheap.

Lakin recommended the Bennett-Burton approach. [13]

In the June, 1908 **Photo Era**, Madison Phillips wrote on "Kallitype
for Making Glass Positives." Glass positives are transparencies made by
coating kallitype sensitizer on glass pre-coated with gelatin. Photographs
on glass were hung in windows for viewing. Phillips wrote:

> Kallitype printing, as it is practiced today, is a method
> the advantages of which appeal most strongly to the amateur
> worker who practices photography purely for the love of it,
> and is artist enough at heart to be willing to devote a little
> more than ordinary effort and patience to the production of
> real pictures. Enthusiastic camerists always, sooner or later,
> wish to free themselves from the limitations of the various
> commercial mediums. To such persons the control and
> opportunities for the exercise of individuality are afforded, as
> well as the beauty and variety of the results obtained, which
> are beyond the scope of the processes on the market They
> invest this printing medium with a lasting appeal.

In a jab apparently directed at Thomson's approach, Phillips
continued,

> Formulae for many sensitizers have been published,
> some of them unnecessarily complicated. I shall, therefore,
> state the solutions which, in my own experience, have been
> satisfactory; they are modifications of the formulae of Dr.
> Nicol, Henry Hall, and Dr. Frederick. [14]

Phillips' approach is very detailed and well informed, but, apart
from the idea of glass positives, it adds nothing new.

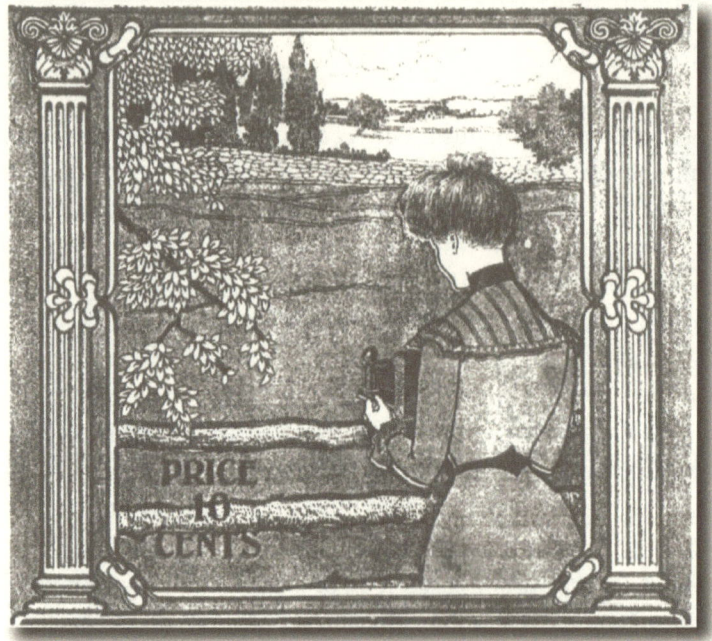

Figure 28. Portrait of a Female photographer

Eleanor W. Willard, one of several female amateur photographers who worked with kallitype, reported in the 1908 edition of the *American Annual of Photography* that she had been interested in Kallitype process for several years, having first made experiments with Mr. Hall's formulae as published in *The Photo-Miniature* #47. She continued, "very perfect and beautiful prints may be made by his method, but for simplicity the formulae recently published by Mr. James Thomson in October 1906 *Photo Era*, and . . . in the *Photo-Miniature* #81 are superior." She liked the simplicity of the Thomson's approach, especially his "two wide mouthed bottles into which the proper chemicals are literally 'dumped'." "One holds the sensitizer and the other the developer." She recommended and reproduced Thomson's "Formula 1" sensitizer and suggested that "other formulae and detailed directions may be found in the valuable articles Mr. Thomson has contributed to the magazines from time to time. She warned, "There will, of course, be troubles . . . unaccountable spots and markings, but when such occur, and careful filtering, shaking of solutions, and smooth coating does not remedy matters, it is best to throw the mixture away and begin over again." Thomson, by his own reports, did not inspire many favorable fan letters, but here, at last, he had one.[15] Incidentally, Ms. Willard provided excellent suggestions about the use of exotic papers for kallitype printing.

In 1909, a reprint of A. J. Jarman's article on brown print, described in the previous chapter, was published in *The British Journal of Photography*. In it Jarman concluded the brown or sepia print processes are "unique in appearance and novel in style, when made on cloth, while prints on paper can be secured in colour from a deep brown to a rich violet." [16]

Also in 1909, A. Leonara Kellogg, another woman, had an article "On The Kallitype Process" published in the *American Annual of Photography*. She wrote "the kallitype process has been so often and so thoroughly discussed that there remains but little if anything to be added." She was surprised that "so few amateur photographers take up this interesting and inexpensive process, for there is no other which can so fully meet the requirements of intelligent and ambitious beginners." She questioned why "complicated" instructions for the kallitype were so often recommended, when "the whole proceeding is simplicity itself." She finds "no better or more concise instructions for the Kalllitype process than those given by Henry Hall in **the** *Photo-Miniature No. 47*, and the formula by G. W. Frederick, M. D. in *The Camera*, for May 1903." She advised beginners "to obtain one or both of these and study them carefully."[17]

The Complete Self Instructing Library of Practical Photography also appeared in 1909. While this ten volume set of books aimed primarily at instructing commercial photographers, it did treat a number of topics of interest to amateurs. Vol. I, Chapter 23 contained a discussion of the kallitype process. The kallitype was recommended because it was cheap, simple, and a close imitation of the platinum process. The Library offered the Bennett-Burton approach as a "satisfactory formula for doing the process. It is interesting to note that this encyclopedia of practical photographic applications recommends the kallitype for professional use. We learn that while

> . . . this process is not generally used in photographic studios, yet it is employed by a great many commercial and landscape photographers, as the process is inexpensive.

The **Encyclopedia** provided several suggestions for the manipulation of the kallitype. An example follows.

> By noting on the backs of your test prints your observations of the effects of the different manipulations, making particular note of the various qualities of paper which produced the best results, filing all these prints in your proof-file, you will establish a record of valuable information for future reference.[18]

It is interesting to learn that as late as 1910 the kallitype was seen to have professional application.

In **1910 the *British Journal of Photography*** published an article by Thomson entitled "A Printing Paper for Black, Brown, Green, and Blue Effects." It was a reprint of an article that had already appeared in *American Photography.* The article described a printing paper "well adapted for the amateur who aspires to artistic effect." The editor of ***BJP*** observes, in the headnote, "Mr. James Thomson deals at length with a subject which he has made peculiarly his own, namely the use of the iron-silver process . . . for the production of prints of full range of tone." The original article, which this publication reprints, was published in July 1904, six years earlier. The article discussed Thomson's salting and sensitizing approach to the preparation of brown print paper. Thomson informed his readers that "from one and the same coating may be had not only brown in various shades, but greens, blues, and black, and varieties that are neither one nor the other. The worker can decide what color to make the image "after the print leaves the frame." The colors are achieved by adding various chemicals to the water the paper is developed in. For example, placing the exposed brown print in a water bath containing uranium nitrate produces a red colored print, while a blue or green print is achieved by inserting the exposed brown print in a bath containing nitric acid and potassium ferricyanide for varying periods of time (longer for blue). The basic printing process of the article is brown printing, which was referred to as "imitation kallitype" by the writer. [19]

Another article on "Glass Positives by the Kallitype Process" appeared in ***Camera***, March 1910 written by Phil M. Riley. [20]

1910 and 1911 are both lean years for kallitype. In **the *British Journal of Photography*** for January 27, 1911, James Thomson wrote an article entitled "On the preservation of Iron Salts." He estimated that "nine-tenths of the [kallitype] worker's inability to obtain satisfactory results is due to . . . deteriorated ferric oxalate." He advised careful buying from reputable sources, frequent testing, and air tight storage of the chemical. [21] Other than this, the file for 1911 contains only a formulary description of kallitype and another on "Sepia Kallitype" from *The Photographic Annual.* [22]

1912 proved to be a better year.

An important article "Variations in Iron Silver Printing," was published in the *British Journal of Photography* on June 21, 1912. Its author, J. M. Hammond was a respected authority on photographic processes. For **the Photo-Beacon** he had written an authoritative series of articles on how to work several photographic processes. In the present article Hammond summarized and critically commented on the major approaches to the kallitype that had been developed up to 1912. The article had originally been delivered as a paper before the Photographic Society of Philadelphia. In view of its quality, it merits detailed reporting.

According to Hammond, since the invention of the kallitype, three major approaches to iron-silver printing had developed. They were the "silver-in-the-developer formulae of Thomson, (which we have labelled K I) and the two silver-in-the-sensitizer approaches of Bennett-Burton and Hall (K II). Hammond's article provided a detailed description and critical commentary on each. Hammond did not discuss the brown print.

In Hammond's view, "the best silver-developing formulae" are those of James Thomson. Although we have already discussed Thomson's K I approach in the previous chapter, we shall here reprint Hammond's admirably brief and accurate summary of Thomson's four sensitizers, one developer, and notes for their use.

Thomson K I Formulae	#1	#2	#3	#4
Iron Ammonia Citrate	32	50	28	18 grains
Ferric Oxalate	16	13	23	38 "
Cupric Chloride	4	8	8	9 "
Potassium Oxalate	33	35	35	36 "
Silver Nitrate	10	16	19	18 "
Oxalic Acid	10	16	15	16 "
Gum Arabic	10	10	10	10 "
Citric Acid	4	—	—	—
Drops Potass. Bich. 5% sol.	10	5-10	5-10	5-1
in Distilled Water	1	1	1	1 oz.

The different formulas produce prints with different visual qualities.

Solution 1 gives "black and white effects"
Solution 2 gives "softer effects"
Solution 3 is suitable for thin, flat negatives
Solution 4 gives "fine gradation and delicate greys"

Thomson's K I Developer

Silver nitrate	40 grains
Citric acid	10 "
Oxalic acid	8 "
Sodium phosphate	1 1/2 "
Water	8 oz.

Hammond gives a qualified endorsement to Thomson's methods. "These formulae can be depended upon to give the results stated." However, Hammond, who knew chemistry well, has some reservations about the extensive lists of chemicals in the formulae.

Their constituents are . . . a warring lot, the cupric chloride and silver nitrate being, on the face of it, irreconcilable in solution. I have found that the copper may be left out of the sensitiszing solution and placed in the developer as cupric nitrate, and a quantity of citric acid may be substituted for the oxalic acid. The solution will then be clear after the addition of the silver nitrate.[23]

The changes Hammond recommends are motivated by the desire to prevent the silver from precipitating from the solution when it contacts either the copper chloride or the oxalic acid. The precipitated silver would be a useless waste of a costly chemical. The changes Hammond recommended would keep the silver in solution and thereby make richer prints.

Hammond recommended the following re-formulation of Thomson's Kallitype I approach and he encouraged the reader to try his revised formulas instead of Thomsons'.

Hammond's Revised Kallitype I Sensitizer Formula

Ammonium iron citrate	25 grains
Potassium oxalate	35 "

Citric acid	30 "
Ferric Oxalate	15 "
Silver nitrate	5 "
Water	1 oz.

Hammond's Revised K I Developer

Silver nitrate	40 grains
Copper nitrate	20 "
Water	8 oz.

Hammond, remarks that "No great harm will be wrought by leaving out the copper altogether." He further recommends that "addition of acid to the developing solution is not absolutely essential unless the quantity of acid in the sensitizing solution is decreased." [24] The second suggestion gets at an important idea in kallitypy. It doesn't matter whether the acid is in the sensitizer or in the developer. It is enough if there is sufficient acid present to complete the reaction of the ferrous oxalate and silver nitrate.

It will be noted that while Hammond's reworked Thomson formulas reflect Thomson's approach, Hammond's formulas differ in specifics from all of the original Thomson formulas. The quantities of almost every compound have been changed, the proportions of the chemicals to each other are varied, and the number of ingredients employed has been significantly reduced. Hammond's recommendations are sound; experience and theory support them. It is clear that he, unlike Thomson, understood chemical theory and knew how to apply it.

Regarding Kallitype II processes, in which the silver is in the sensitizer, Hammond observed "they will be preferred, no doubt, by many workers." He summarized the Bennett-Burton approach without critical comment. To solve the yellow stain problem, he suggested two clearing baths—1) a 5% solution of citric acid or 2) a 10% solution of potassium oxalate. Either "is efficient" for removing the yellow stain of iron oxide that often shows up after development in K II processing.

Hammond also gave an accurate summary of Hall's Kallitype II formulas and remarked they are an "excellent and more recent" approach than that developed by Bennett and Burton.

Finally, Hammond provided a number of incidental bits of kallitype wisdom. He offered a method for converting any platinum sensitizer formula into a kallitype sensitizer. He observed that mercuric nitrate in the K II sensitizer helps to achieve "good black tones" (contrary to its browning effect on platinum prints). He reported that paper sensitized with Ferric Sodium Oxalate solution and developed in a K I silver bath gives a beautiful steel-black; but if silver is also placed in the sensitizer, a reddish brown print color results.[25] Observations such as these attest to Hammond's extensive chemical and process knowledge.

When evaluating Hammond's contribution, it will be seen that though he presented no new conception of kallitype, he filled his article with sound information and critical insights of continuing value. Hammond's article remains one of the more comprehensive and critical pieces of writing in kallitype literature.

In 1912 George P. Swain, another salonist, wrote in the **American Annual of Photography** what he learned in six years of using Kallitype "as my almost exclusive printing medium." Swain wondered why more amateurs didn't turn to Kallitype at a time when "there is such a tremendous expenditure of time and agony in the evolution of abortions in the name of 'gum' and 'oil.' Kallitype, for Swain, is a process that gives "all the freedom possible for the expression of individuality and assures prints that are artistic to the last word." Swain advocated Hall's process as "most reliable." His article provided extensive details on manipulation gained from his long experience. [26] His suggestions will not be summarized here, since most of them have already been reported by others.

Cassell's Cyclopaedia of Photography, one of the most important reference books ever published on old photographic processes, appeared in 1912. The general editor of this publication, Bernard E. Jones, was not known for his work as a practicing photographer. However, he staffed the various departments of the encyclopedia with writers of unquestioned photographic expertise. Among them was G. E. Brown, whose book popularizing the kallitype and brown print was discussed earlier. **Cassell's Cyclopedia** remains to this day an immensely valuable source of information on the manipulation of all pictorial printing processes and on the innumerable details of chemistry that underlie them.

The kallitype entry in the *Cyclopaedia* included a brief history of the kallitype, giving due notice to Herschel and Nicol, and provided a summary account of the formulas and operations of three of the four major kallitype processes discussed in the present book: the Bennett-Burton K II sensitizer and developer; Brown's approach to the brown print; and an account of Thomson's 1904 salting and sensitizing approach to Kallitype I.[27] Elsewhere, an entry under Sepia Print details Namias' two-solution formula for the brown print.

"Cassell's kallitype entry reported, in passing, that commercial kallitype paper was withdrawn from the market by Nicol "because he was not satisfied with the permanency of the results." The credibility of this observation increases in view of the fact that Brown, one of the "chief Cassell contributors" knew Nicol well.

Cassell's Cyclopaedia remains today a useful reference work for doing the "old" processes. It provides extensive chemical, optical, and manipulative information on many photographic processes, and excellent illustrations of devices and definitions of obscure terms. This writer would not consider investigating an unknown old process without first consulting Cassell's.

The British Journal of Photography on October 11, 1912 published a brief report of a German patent granted to Teresa del Fabro of Rome for a method of using potassium permanganate in the preparation of the kallitype sensitizier ferric oxalate. Del Fabro's approach aimed at controlling the purity of the ferric oxalate solution by converting any ferrous oxalate that might be present to ferric oxalate with potassium permanganate. As we have noted, the experience of many amateur experimenters with ferric oxalate had established that "bad" ferric oxalate was a cause, if not the major cause of kallitype difficulties. "Bad" ferric oxalate was ferric oxalate that had degenerated wholly or partially into ferrous oxalate as a result of exposure to light or moisture. Del Fabro's patented invention changed ferrous oxalate back into ferric oxalate (the desirable condition) by adding potassium permanganate to it. He recommended adding a 10% solution of potassium permanganate to the ferric oxalate solution, cc. by cc., till the ferric oxalate solution effervesced. Effervescence was a sign that the ferric oxalate was no longer contaminated by ferrous.

> This method of treatment . . . makes it easy . . . to determine exactly the moment in which the oxidation takes place, that is to say, the moment in which the ferrous salt is transformed into ferric salt, especially if it is in the presence of oxalic acid. In fact, the permanganate, after having oxidised the ferrous salt, oxidises the oxalic acid, transforming it into

water and carbonic acid, which acid is set at liberty with active effervescence. This is the proof of the oxidation having taken place, without the long calculations required in other methods of oxidation, for instance, with nitric acid.[28]

The most complete description of Del Fabro's technique occured in the June 13, 1913 issue of the ***British Journal of Photography***. The technique was also reported in the **Photo Digest** section of October, 1913 issue of *Camera Craft*. In the writer's trials, the method worked. The usual test for the presence of ferrous oxalate in ferric oxalate solution is to drop some ferric oxalate solution into a solution of potassium ferricyanide. If the drops precipitated no blue deposit, then no ferrous oxalate was present and the ferric oxalate was "good"—that is devoid of ferrous. Del Fabro reported his permanganate treatment of ferrous oxalate passed the ferricyanide test.

In 1913, a very good year for Kallitype publication, an announcement of "Etral Paper, made and marketed by Gevaert Ltd, a Belgian firm appeared. Etral was "a modern form of kallitype II paper, one prepared with a sensitive iron compound in conjunction with a silver salt." According to Gevaert, the image, "acquires a rich black or warm black colour," after development in a platinum bath, due to the replacement of silver by a platinum toning solution. Simply put, Etral was a conventionally developed Kallitype II paper that was subsequently toned with platinum. It was a response to the desire to make prints that resembled platinum at a time when the scarcity of platinum metal made the price for pure platinum prints prohibitive. Etral paper took advantage of the fact that platinum toning required less platinum than platinum printing but produced an image that appeared indistinguishable from the image on a platinum print. The platinum-toned kallitype had the additional advantage of being more "permanent" than an untoned kallitype, since, after toning, the image forming metal became predominantly platinum, which is resistant to chemical changes induced by pollutants in the air. Interestingly the Gevaert Co. indicated that Etral may be used either as a kallitype paper that is platinum toned, or without any toner to produce a straightforward silver kallitype print.[29]

In the May issue of *Camera Craft* appeared one of the earlier articles of Sigismund Blumann, the last major writer on kallitype up to 1950. In this article, entitled "Laboratory Work in Photography," Blumann, a chatty, playful, writer, caricatured himself and other photographers who submit to the miseries, disappointments, and expense that accompanies the preparation of home-made paper. Blumann's lighthearted satire on working Kallitype formulae is worth sharing.

> Then those kallitype formulae. Such beautifully elaborate directions! And the list of ingredients! You fellows who buy ready mixed, concentrated developers and things like that can never guess the pleasures of laboratory work. It is the best part of photography. Any old professional can turn out good prints, but to try all the recipes that I have tried and fail to turn out a decent picture with any one of them, that is the proof of genius.

The point made here is significant. In the last couple of decades, Kallitype printing had acquired a large amount of baggage—innumerable formulas of intimidating complexity which frequently failed to produce good prints. Blumann continues:

> About Kallitype, I know something. I should. All that has been written on the subject in the English language is on my shelf. And I have tried every formula, twisted, modified, and elaborated each. I exude kallitype, when I get excited, at every pore. And the best sensitizer has proven to be one that I throw together almost haphazard with a spoon for a measure. It contains the fewest posssible ingredients and it works on any old paper that is decently free from chunks of metal and cobalt.

Poking fun at the experimentally oriented amateur photographer, Blumann suggested the whole effort may be misplaced.

> My dark room contains without exaggeration nearly five hundred dollars' worth of chemicals. I have become, yes, really, a proficient technical chemist. But I have yet to make a really fine print. I am shamed by the ignorant friends who . . . found time to make prize-winning pictures, while I, full of wisdom, was grubbing away with every new photographic idea that came along.

L'envoy. For trade: A choice and assorted lot of chemicals;
a complete collection of scales, weights, measures, graduates,
tubes, etc.; a miscellaneous and more or less complete
knowledge of chemistry . . . all, or in lots to suit, for a little
practical ability to take and make some good pictures. [30]

One wonders if this article represents, not just a change of mind in
Blumann, who continued to make kallitypes as well as write about them
with frequency, but rather a change in the attitudes of his readers, who
were beginning to tire of the complications and labor of independent
preparation on the one hand and the dearth of prize photographs produced
by laboratory methods on the other.

Contrary to his self-effacing remarks, Blumann was a capable
kallitypist, according to Thomson, who spoke favorably of the quality
of Blumann's prints.[31] It is not surprising to learn a few years later, that
the witty and enthusiastic Blumann became editor of **Camera Craft**. We
will hear from him again.

In 1913 Paul L. Anderson, a renowned pictorial printmaker, teacher,
and writer published a pair of articles on "The Choice Of A Printing
Paper, With Especial Reference to Platinum." These articles are
probably the best of many that appeared over a period of twenty years
on the advantages and disadvantages of the better known pictorial print
processes. Anderson believed that so many processes were competing
for the printmaker's attention, that the amateur needed help to decide
which to select. In his two articles gaslight bromide, P.O.P. (plain silver,
self-toning, Gelatine, Collodio-chloride, and Albumen), Kallitype, Oil and
Bromoil, Gum, Carbon, Gum-Platinum, and Platinum, in both homemade
and commercial versions. The criteria that Anderson, an engineer, applied
were: permanency; invariability and reliability; possibility of control;
color variability; textural variability; ease of manipulation; and quality.
The complete text of Anderson's evaluation of kallitype follows.

This extremely interesting medium (kallitype) fulfills all
the requirements of a high-grade printing medium except two,
the first [permanency] and the sixth [ease of manipulation]. It
has a quality fully equal to that of platinum; it is capable of
great modification in contrast, color, and texture, and results
may be duplicated with comparative ease, but it is not easy to

manipulate, and the image is so unstable that the process should be used for only the most ephemeral work. This statement will doubtless provoke violent protest from enthusiastic Kallitype workers, but it is nevertheless true. Kallitype prints may in some circumstances seem fairly durable, but there is no reason to expect this to be the case. A well-known worker writes as follows: "Kallitype prints made with care and kept in a dark drawer for a few years showed no indication which side of the paper was used to print on." It will be understood that an image of metallic silver which is fully exposed to the air must be more liable to deterioration than one which is protected by gelatine, so that on purely theoretical grounds Kallitype cannot be regarded as permanent—not even so stable as bromide—and when theory is substantiated by experience, it will be seen that this medium is not advised for work of any importance. [32]

One more time, Kallitype is hung by the rope of "impermanence." It should be noted that Anderson's argument may be based on theory and the hearsay of one unnamed worker. His quotation was, in fact, taken from *The Complete Photographer,* written by R. Childe Bayley. Bayley was a popular writer on photography, but one whose experience with kallitype was limited, quite likely to a single, poorly controlled trial he conducted, as can be seen from an examination of the context of the quote. And, if one examines the quotation from Bailey in its original location, one finds that Anderson has advantageously misquoted the source, introducing the words, "made with care" into a sentence that never had them when Bayley wrote it.[33] There is no doubt that unskilled amateurs (or professionals, for that matter) made unstable kallitypes from ineffective formulas, contaminated materials, or poor workmanship, and their images faded. But other types of silver prints, made with the inadequate craft, display similar ephemerality. Nevertheless, Anderson's charge of impermanence has continued to weigh heavily, and unfairly, against the permanency of kallitype prints.

The other part of Anderson's argument rests on the question whether a silver image open to the air, unprotected by a medium such as gelatin or albumen is more prone to deterioration than one which is not so protected. Anderson implies that a "protective" medium, such as gelatin, which prevents the contact of contaminants in the air from contacting the silver of the image, enhances permanence. The contrary argument, which was made repeatedly by writers on photography in the 1890's, and which provided the rationale why hypo was avoided in Nicol's patented processes, was that media such as gelatin and albumen radically inhibit the washing away of the contaminating residue of hypo which in time acts as a destructive influence from within the print, on the silver image. Perhaps the most judicious resolution of this issue would be to say that permanency depends more on whether a print is well cleared and washed than on whether the silver is encased in a "protective" gelatinous medium. On the matter of permanence it should be noted that, unlike Bothamley whose lab tests found the kallitype to be as stable as any silver print, Anderson reports no laboratory tests on the permanence of kallitype. When Anderson's evaluation is closely examined, it will be found to rest on questionable theory, hearsay, and doubtful quotation. The above considerations suggest Anderson's conclusions about Kallitype permanency require further substantiation before acceptance.

Anderson's approach to making kallitypes, possibly the one used in his comparison article mentioned, is reproduced below.

Anderson recommended, first, the preparation of a "General Sensitizer," one suitable for sensitizing blueprint, brown print, platinum, or kallitype papers after suitable modification. The general sensitizer is simply a 26% solution of ferric oxalate to which has been added some oxalic acid. It is made by adding to hot distilled water the ferric oxalate powder and oxalic acid. The oxalic acid is present to maintain the ferric condition of the ferric oxalate.

Paul Anderson's General Kallitype Sensitizer # 1

Water, distilled, hot	60.0 ml.
Oxalic Acid	1.1 gram
Ferric Oxalate	16.0

Care must be taken not to use water too hot or the ferric oxalate will turn partly ferrous.

To make Kallitype II sensitizer, from the above "general sensitizer," combine the following.

Paul Anderson's Kallitype II Sensitizer (using Gen'l Sensitizer above)

Water, distilled	6 ml.
Gen'l Sensitizer #1 solution	9 ml.
Silver nitrate	1 gram

Allow 4 cc. to an 8 by 10 inch sheet, and brush the solution on the paper. Anderson also provides a single solution Kallitype II Sensitizer.

Anderson's K II Sensitizer

Ferric Oxalate	1.6 grams
Oxalic acid	.17 "
Silver Nitrate	1.0 "
Distilled Water	15. ml.

After the prints are exposed, develop thoroughly in

Anderson's K II Developer

Water, Distilled	500 ml.
Rochelle Salt	50 grams
Borax	25 grams

There is no clearing bath. The prints remain in the developer until they are clear. Then they are fixed in

Anderson's K II Fixer

Water	500 ml
Hypo	25 grams

Fix for 10 minutes, and wash with extreme thoroughness.

Anderson remarked, "This is by far the simplest kallitype formula that I have ever seen, and gives excellent prints of a warm tone, especially with moderately strong negatives. It is not so well adapted to prints in a high key, or to soft negatives."

The name "W. E. Dancy," appears on Anderson's typed copy of these formulas, which resides in the collection of the International Museum of Photography in Rochester. Dancy may have been the source of the formula.[34] The quantities of the chemicals suggest the ultimate source was Bennett or Burton.

Moving on, in 1913 Thomson wrote on "Uranium as an Intensifier and Toner." He indicates that uranium can have two almost opposite effects in photography. Used one way, uranium will intensify underexposed negatives. Used in another, it can bleach overexposed prints reducing them to acceptable tone levels. Uranium-reduced prints can later be toned with another metal. [35] He found the uranium treatment is applicable to kallitype prints.

In August 1913, A. J. Jarman wrote on "Ferro-Argentum Paper for Printing Out or Development." Jarman was almost as prolific a writer as Thomson and wrote on several processes. Careful reading of this article reveals Jarman offered another version of brown print or "imitation kallitype." His sensitizer was a conventional combination of ferric ammonium citrate, tartaric acid, and silver nitrate. He offers the usual developing solution for brown printing—just plain water. Jarman indicated that print exposure would be about one half the normal time used with ferric oxalate sensitizers. Jarman's brown print process allows less control of the color of the print than conventional kallitype processing[36]

In the ***Photographic Times*** for September, 1913, Thomson published an article entitled "Platino-Argento Paper." This paper appears to be a response to Gevaert's "Etral" commercial kallitype paper put on the market early in 1913. As reported above, Etral was a kallitype paper that could be toned with platinum salts to resemble a platinum print. Thomson's formulae for making a silver platinum paper follow:

Thomson's Sensitizer for Platino-Argento Paper

Distilled water	1 oz.
Citrate of Iron and Ammonia (green)	18 grains
Potassium Oxalate	18 grains
Platinum solution (15 grains of	
Potas. Chloroplatinite in one oz. water)	18 drops

Potassium Bichromate
    (22 grains in 1 oz. water)   as required
    for contrast
Gum arabic   10 grains

Thomson's Developer for Platino-Argento Paper

Distilled Water	1 oz.
Silver Nitrate	40 grains
Citric Acid	10 grains
Oxalic Acid	8 grains
Phosphate of soda	2 grains

Except for the small amount of platinum in the sensitizer, this is a variation on a Kallitype I formula. Thomson wrote "the image obtained by this process is doubtless a silver one, the small amount of platinum originally in the sensitizer simply furnishing a vehicle whereby the final result is made. Nevertheless, the print is "in appearance identical with platinum." Thomson reports that he "had the good fortune to work out the method some half a dozen years ago." He has used it through the years since "in preference to any other." It is notable that Thomson's approach to this paper reverses the usual procedure of making a silver image first, and toning it in a platinum bath afterward. [37]

Thomson published still another article in 1913, "On Kallitype Failures," in the *American Annual of Photography.* He wrote, "many are the failures of the kallitype, and the generality of them are undoubtedly due to deteriorated ferric oxalate." He advised buying the chemical from the best suppliers, making a careful tests of each new puchase, and conducting tests of the ferric oxalate before using it in a sensitizer. He proposed that the compound may be kept fresh after purchase by coating the bottle cork with paraffin, by hermetically sealing the ferric oxalate bottle in a larger bottle with an airtight stopper.[38]

This completes our review of published reports of kallitype activity up to the beginning of the World War I.

The war influenced the direction of thinking and writing on the kallitype in several ways which we shall take up in the next chapter. For now, it seems an appropriate time to sum up the development of kallitype up to this point and make a few comments.

II: Summary and Comment

In the twenty years or so since amateurs became interested in iron silver printing, four main kallitype processes developed.

First, Bennett and Burton initiated a relatively simple Kallitype II approach by only slightly modifying Nicol's patented process. Their approach involved an invarying sensitizer which contained both ferric oxalate and silver nitrate and two developers, borax/Rochelle salt, for brown images and sodium acetate for black. Their simple formulas, compounded of dry chemicals at the time of use, continued to be popular with amateur printmakers and pictorial photographers up to the war.

The second approach to Kallitype II evolved from studies prepared by Frederick and Hall. This approach promised flexibility and variability, made possible by the employment of a number of stock solutions. The stock solutions could be variously compounded to produce a variety of different colors and contrasts. The Frederick-Hall approach was as popular with amateur printmakers as that of Bennett and Burton.

The third kallitype approach, the "brown print," was popularized by G. E. Brown, John Thomson, and A. J. Jarman. The brown print used a different ferric compound in the sensitizer, ferric ammonium citrate, than that used by the kallitype (ferric oxalate) and the print was developed in plain water. While it was never regarded as the equal of the kallitype I and II processes, during the period leading to world war I, the brown print drew an active following, especially among beginners on the one hand and modern experimental artists on the other. The brown print was variously modified. It was recommended to advanced amateurs for its ease of production, its delicate images made with applied color, and later, its use for montage and multiple image effects.

The fourth kallitype process evolved from Nicol's original Kallitype I process. In K I, ferric oxalate is coated on paper as a sensitizer and after exposure, the image is developed by floating the print on an acidified silver nitrate solution. The K I process, after being ignored for its early failures, received renewed attention by amateurs, during the period leading up to the World War I, most extensively by James Thomson, who developed a variety of K I formulae. In Thomson's hands, the K I process became far more complex than Nicol's relatively simple patented K I process.

There was one more kallitype process, K III—a paper that was pre-coated with sensitizing, developing, fixer and clearing chemical compounds. After exposure, the paper needed only water for processing.

While theoretically conceivable, a successful KIII paper never came into being. Nicol and others tried to make it happen, but no KIII paper was reported to work outside the experimenter's laboratory.

In the literature on kallitype from 1904 to 1915, we have watched the four kallitype processes develop and receive publication by a number of writers of varying credibility. As time passed, the amount of writing on kallitype varied; in some years a surprising amount was produced, while in others there was little. Once introduced, all of the processes continued to attract readers to try them. Interestingly all of the processes had advocates and all had critics. None of the four approaches dominated the periodical literature or won out over the others during the interval we have been observing. If any process received more criticism than the others, it was the K I process. The problem there was Thomson's lengthy and undependable formulas, not the process itself.

It should be noted that none of the major writers presented the kallitype as if all of its problems were solved. They continued to mention the same problems that plagued early kallitype workers. And the writers recommended the same solutions—the need for intelligence, care, and creativity in use of any formula. That and careful mixing of fresh chemicals

Looking over the period reviewed in this chapter, it is evident a considerable number of articles appeared on the kallitype. Some writers were knowledgeable and some were not. Some understood chemistry and clarified issues, others did not and served only to muddy the waters. The kallitype appears to have been more popular in America than in England, and more alive in England than on the continent. There is evidence of kallitype activity throughout the photographic world; articles were reported from Germany, France, Belgium, Italy, and Japan. From many sources a literature was produced that continued to provide worthwhile information, helpful theory, and practical suggestions. It must be admitted, however, that as the literature on kallitype neared its third decade, most publications contained little that was really new. Much of the writing rehearsed what was already well known, simply reprinting earlier methods. Later writing was produced for readers who presumably had not read about the kallitype during the first decade, when truly innovative writing on kallitype prevailed

Considered broadly, the literature on kallitype indicates some movement in the status of the process and the "process movement" in general. Up to World War I, it is evident that the kallitype was a perennial "draw" for photography magazines on both sides of the Atlantic. Every photographic journal showed interest in publications about the four processes at one time or another between the dates of 1885 and 1914. Some journals, *The Photo-Miniature,* t*he Photo-Beacon*, and *The British Journal of Photography* actively supported the kallitype throughout the period. Others began their support later, e.g. *Camera* and *Photo Era*. It should be pointed out that photographic periodicals exploited many print processes other than kallitype in their effort to reach and maintain an audience committed to pictorial values. Throughout this period readers of photographic publications continued their fascination with a wide gamut of photographic processes. Were this a history of print processes in general, we would have discussed a great many articles on platinotype, gum-bichromate, carbon, bromoil, and silver bromide and chloride printing. Some of these processes may have had as many or more articles published about them than were published on kallitype.

Two strong influences on kallitype developments during this period were the lack of availability of platinum and the continuing hold of pictorial values. Under the first heading, much of the period's interest in kallitype had to do with its possibilities as a platinum substitute. Under the second, the pictorial values of personal artistic expression and artistic printmaking by hand control, continued to support kallitype use.

We conclude that publications on Kallitype can be viewed as a weathervane indicating the direction and intensity of amateur artistic concerns. If so, we conclude the period reviewed in this chapter was a time when hand controlled photographic print processes were a major concern of of a large body of amateur photographers who sought a means of artistic expression in kallitype. Amateurs of the time were also still very much interested in technical knowledge, control of process, and the ability to produce and display individually manipulated hand influenced prints. Many were fascinated by the chemical "magic" of do it yourself printmaking. There is evident a strong feeling against manufactured papers. among many pictorialist printers. Manufactured papers are repeatedly blamed for being anathema to the production of expressive artistic work.

The period we have been observing was a time when photographers were wooed to accept manufactured materials for their convenience, reliability, and low cost. The continued interest in kallitype suggests amateurs were reluctant to part with the heritage of personal investigation and independent exploration of the means of making photographic art.

The men and women who wrote the articles on kallitype and those who read them did not yet want to give up the personal satisfactions associated with individually prepared materials for the highly controllable but unvariable commercial papers and processes offered by manufacturers. Yet, hearing the compaints one wonders if the rejection by pictorial artists of manufactured paper amounted to little more than prejudice against technologically sound commercial papers.

But the winds of change were blowing as the 20th century progressed. Fanned by the efforts of eager marketers, photography was moving in a direction in which photographic materials would increasingly be mass produced. The amateur photographer would be pressured to become a consumer of materials engineered to perform in predetermined ways. Engineered printing papers were by now well made in many respects. They were convenient, reliable, easy to use, and the processing was simple and invariably effective. In spite of these advantages, Pictorial writers continued to regard manufactured papers as limiting, common, and boring. They persisted in wanting to preserve the older way of preparing one's own materials, regardless of challenges met in making unique, expressive prints. This may be the underlying meaning of the amateur process movement, as a whole, and the significance of the kallitype at this time. Something like the creative person rising against the limit imposed by standardized products made available by the impersonal manufacturer.

While the kallitype was not the most successful paper in salons (we have already indicated that major salonists predominantly preferred platinum), nevertheless, the kallitype and the discussions about it clearly supported fundamental esthetic values of the artistic amateur.

The period just finished suggests the dominant values of artistic photography continued to be pictorial. They emphasized personal response, expressive intent, and prints that clearly showed personal control of process for expressive purpose. The straight photograph, the unmanipulated print was routinely snubbed.

While these values were maintained during the period we have been observing, we must confess a discernible lowering of the quality of investigation and writing on the kallitype. It appears that the longer the process was written about, the more the essential ideas were diluted by rewriters who added little to the lore as they passed the original ideas along. This decline in quality may be an index of a similar change that was occurring in the quality of the amateurs themselves. Early on, the amateur had been well educated and motivated to investigate and contribute. By 1910 one senses there are more opportunists writing about the kallitype than investigators and real experimenters. There is also discernible tendency to reduce the kallitype to a simple procedure, an approach to be followed rather than explored. Last, there was Blumann's article ridiculing the whole enterprise of making individual artistic photographs by chemical adventures which use up time better spent taking great pictures.

Clearly, there were signs that the denoument of kallitype had begun. In the next chapter, we shall see if the decline continued.

The period we will examine in the next chapter, from 1914 to 1925, is a transition period that will initiate further changes in photographic and printmaking attitudes. Taste during the coming period will vacillate between earlier and later values, but it begins a definite drift in the direction of the future, that is away from the subjective emphasis of individually controlled print values and toward a new esthetic of straight print objectivity. As we approach 1925, the delicate balance between the old ways and new ways that the journals had helped to maintain before the war, begins to totter. Mortenson and Weston will soon square off; Pictorialism will be challenged by a succession of new esthetics—the New Objectivity, Formalism, and Experimentalism—the straight print, the formal print, and the god knows what print will bury the pictorial print esthetic. Professional, museum certified, artist/photographers will replace the magazine supported print manipulator.

Photographers whose work emphasized process will continue to exist. But they will no longer flourish. All along they had been accused of being "process mongers," persons more interested in the dark mysteries of chemistry than in the production of art. Now the process mongers are satirized with increasing frequency, often by former pictorialists. We have already seen Blumann make fun of himself being lost in complex chemical printing problems. The very magazines that formerly elevated the process oriented photographer to the status of hero now sold copies by presenting him as a confused and ineffectual laborer who pursues quixotic and nonsensical goals. The concept of the artist as pictorialist—a person concerned with sensitive personal response to experience which is translated into unique, hand made prints—this idea continues, but it seems, tired, and less glamorous.

After a deluge of fuzzy brown toned prints, an esthetic of indistinct rural images, artistic photographers appear ready for a new direction.

Photographers are ready to support a print esthetic that takes advantage of the efficiency, reliability, and machined look available in well made, brilliant, manufactured papers.

It can be said, from the vantage point of the 21st century, that the destiny of the new century was the mechanization of photography. Clearly, the time when the change will be fully accomplished is decades away. After 1914, the kallitype (and other individually controlled paper processes) will continue to excite writers and readers of articles and makers of kallitype and other hand influenced prints, and they will suggest that the idea of pictorial image-making is still attractive and alive. But signs will become increasingly clear that the revolutionary vitality has gone out of the process movement. We will see fewer genuine investigations of the kallitype, or any other process for that matter. Articles will continue, but in great part they will be rehashed explorations. Or they will be nostalgic looks at beauties of a past commitment no longer able to inspire vital, fresh, art.

In 1890, when Nicol announced the kallitype and the first photojournalists wrote about the kallitype, they felt assured a period of great expectations had begun. Their oft repeated judgment was the kallitype "had a future before it." In 1925 the kallitype began to be viewed an "old process," one no longer having the potential to shape the present and future of photography. For at least a generation after 1925 the process would wither in progressive disregard. The period from 1925 through the fifties displayed little more for the kallitype than a quiet old age.

Notes: Chapter X. The Denoument Begins

1. Rudolpho Namias, "A Home Made Printing Paper," *Photographic Times*, June 2, 1904, p. 274.
2. Louis Fleckenstein, "Kallitype," *Photo Era*, May 1906, p. 321-325.
3. H. C. Bird, "The Kallitype Process," *Photography*, Sept.18, 1906, pp. 235-6.
4. E. J. McIntosh, *The Photographic Reference Book*, Iliffe and Sons, London, 1906, pp. 147-8.
5. James S. Escott, "Kallitype for Winter Landscapes," *Photo Era*, April, 1907, pp. 190-
6. Charles S. Taylor, "The Kallitype Process," *Photo-Beacon*, June, 1907, p. 182-3.
7. No author, "Questions and Answers," *The Camera*, July, 1907, p. 267.
8. No author, "Kallitype Process," *The Camera and Dark Room*, July, 1907, p. 207.
9. G. Caspar Elmberger, "The Kallitype Process," *The Camera and Darkroom,* Vol. 7, August, 1907, pp. 249-253.
10. G. Caspar Elmberger, "The Kallitype Process Appendix," *The Camera and Darkroom,* September 1907, p. 326-7.
11. J. J. M. Sellors, "Croydon Camera Club," *The British Journal of Photography*, Nov.15, 1907, p. 873.
12. Will Palmer, "The Kallitype Process," *Camera Craft*, May, 1908, pp. 165-169.
13. Walter W. Lakin, "The Kallitype Process," *American Annual of Photography*, 1908, pp. 61-62.
14. Madison Phillips, "Kallitype for Making Glass Positives," *Photo Era,* June 1908, pp. 297-301.
15. Eleanor W. Willard, "The Possibilities of Kallitype," *American Annual of Photography*, 1908, pp. 21-24.
16. A. J. Jarman, "Preparing Paper for Sepia Printing with the Salts of Iron and Silver," *British Journal of Photography*, June 4, 1909, p. 442-3.
17. A. Leonara Kellogg, "Kallitype," *American Annual of Photography*, 1909, pp. 230-1.

18. J. B. Schriever, ed., *Complete Self Instructing Library of Practical Photography*, Vol. I, Elementary School of Art and Photograph, Scranton, 1909, pp. 271-77.
19. James Thomson, "A Printing Paper for Black, Brown, Green, and Blue Effects," *British Journal of Photography*, March 1910, pp. 157-160.
20. Phil M. Riley, "Glass Positives by the Kallitype Method," *Camera*, March 1910, pp.103-4.
21. James Thomson, On the Preservation of Iron Salts, *British Journal of Photography*, Jan. 27, 1911, pp. 61-3.
22. *Photographic Annual Incorporating Figures, Facts, And Formulae Of Photography*, London, George Routledge and Sons, 1911, p. 220-1.
23. J. M. Hammond, "Variations in Iron Silver Printing," *British Journal of Photography*, June 21, 1912, pp. 481.
24. Ibid. p. 482.
25. Ibid.
26. George P. Swain, "Why Not Kallitype?" *American Annual of Photography*, 1912, pp. 182-190.
27. Bernard E. Jones, ed., *Cassell's Encyclopaedia of Photography*, Cassell and Co., London, 1912. Kallitype entry, pp. 314-317.
28. Teresa del Fabro, "Report of Patent on Preparation of Ferric Oxalate," *British Journal of Photography*, June 13, 1913, p. 461. The *BJP*, October 11, 1912 announced a german patent was granted to Teresa del Fabro for using potassium permanganate in the preparation of the kallitype sensitizing solution. "The Photographic Digest" in *Camera Craft*, Oct. 1913, pp. 487-8 also reported an abstract of the patent.
29. Announcement, no author, "Etral Paper," *British Journal of Photography*, March 13, 1913, p. 199.
30. Sigismund Blumann, *Camera Craft*, May 1913, p. 224-6.
31. James Thomson, "Kallitype," *American Annual of Photography*, 1922, p. 62.
32. Paul L. Anderson, "The Choice of a Printing Paper With Especial Reference to Platinum*,"* *American Photography*, June 1913. Kallitype is discussed on p. 342. The July issue completed the article.
33. R. Childe Bayley, *The Complete Photographer*, Frederick A. Stokes, New York, 1926, p. 381.

34. Paul Anderson, Kallitype Formulas. I am indebted to the International Museum of Photography, Rochester New York, for Andersons' kallitype formulas.
35. J. Thomson, "Uranium As An Intensifier and Toner," *Photo Era*, August 1913, pp. 67-8.
36. A. J. Jarman, "Ferro Argentum Paper for Printing Out or Development," *The Camera*, August 1913, pp. 457-466.
37. J. Thomson, "Platino Argento," *Photographic Times*, Sept., 1913, pp. 18-23.
38. J. Thomson, "On Kallitype Failures," *American Annual of Photography*, 1913, p. 106-7.

Chapter XI

The Demise and Rebirth

I.   Concluding this History of the Kallitype

During the years between 1915 and 1925, new publications on
kallitype continued to appear and their content shall be reported in this
chapter. By 1925, the last two important writers on kallitype ended their
contributions. After them the frequency as well as the quality of published
material on kallitype declined precipitously. While occasional articles
on kallitype appeared after 1925, there are too few of them to justify a
year to year account. The articles that were published contributed little
new information on process or commentary. Therefore, we shall end
detailed treatment of kallitype with the year 1925. We shall make only
brief remarks on kallitype developments after that time as we conclude
this history.

II.   The Demise? 1915-1925.

In 1914 the World War I began, and, perhaps, because of it, there was
less writing on the kallitype. In Britain there are indications that kallitype
continued to sustain interest, in spite of the war. F. W. Horn, reported
on a kallitype demonstration given to the South Suburban Photographic
Society on Oct. 1914. The report revealed attitudes about kallitype were
changing from those found in in previous years. Horn's remarks suggest
that kallitype activity may be going nowhere.

> The kallitype process is one that the average photographer
> reads a lot about, but rarely if ever, uses, although, judging
> from the American journals, every amateur in the States
> appears to have tried his hand at it, with perhaps no greater

success than altering the formulae given and concocting some
of his own, which he does not hesitate to have published.

He continued "the process is little known and very much less used
on this side." Horn is described by the secretary of the society as "an
enthusiastic student of chemistry who goes to the bottom of things. The
secretary records that "many members seemed restless and confused
while Horn filled the blackboard with chemical symbols, none failed to
understand the working details." Horn had "commenced by condemning
some of the American workers, who in his opinion, did their best to
complicate formulae and make the process more difficult than it need be."
The sensitizer Horn demonstrated was Burton's and the developers were
those "recommended in Mr. [G.E.] Brown's book." Horn, apparently "led
several to try their hands at the process, which after all, is not so difficult
as the bulk of literature, particularly American, would have us believe."[1]

Horn's report raises some questions about what was happening to the
prevailing concept of the kallitype as a result of the articles by Thomson,
Jarman and others. Horn believed that recent writing on kallitype had so
emphasized the need for complex formulas and meticulous workmanship
that amateurs were intimidated and repelled. Horn's remarks are worth
our attention. After Burton, kallitype formulas and procedures had grown
in complexity. Early writers on kallitype, had promised that excellent
prints could be made by any amateur who could follow simple directions.
Such claims had, over time, created a sizeable kallitype following. The
recent emphasis by writers on the complexity and difficulties attendant
on working the process could only slow the tide of kallitype popularity.
The perception that the kallitype process was unnecessarily complex,
tedious, and difficult could only get worse as major manufacturers of
gelatin silver paper made their products simpler and easier to use.

In the June, 1914 issue of *Camera*, Jarman wrote about the use of
American art papers for sensitizing of individually prepared prints, since
French and German papers were no longer available. He recommended
Angora white, made by the Whiting Paper Mfg. Co., Oronoco Bond,
and some Strathmore papers. (Strathmore still markets 100% pure rag
papers which have excellent characteristics for making kallitypes today.)
The process that Jarman recommended was a conventional approach to
brown print.[2]

In the October, 1914 *Camera Craft*, Blumann wrote on "Home
Sensitizers and their Application." Blumann specified his reader as "the

amateur pictorialist" who "has never tired of making his own paper." He described first, an approach to making blue prints and brown prints, the easier processes, and finished with formulas for kallitype II. As usual his writing was chatty, effervescent, and full of enthusiasm for kallitype.

> All this time I have been teasing myself, keeping back from my favorite process: Kallitype. It is a beautiful and exciting way of getting just exactly what you want . . . It is a wonderful medium, this periodically rediscovered Kallitype . . . My esteemed friend Clute [the editor of *Camera Craft*] says my middle name is Kallitype—and the formula I have boiled the whole of my knowledge down to is one that Mr. Thomson modestly gives with the fewest directions and least number of ingredients.

Unfortunately, the formula Blumann presented was Henry Hall's. Unaware of his mistake, Blumann ended his cheerful discussion with, "Verily, this Kallitype is "some process." Of course, Blumann received letters.[3]

Publications in 1915, as in the previous war year, were concerned with making substitutes for papers made unaffordable or unavailable by the war. Thomson in *American Photography* had published a process for making "Silver Platinum Printing Paper." The article indicated that the availability of platinum was very uncertain.

> With platinum, the metal, at a price just double that of gold, the outlook for a boom in pictures based upon [it] seems a thing uncertain and remote. [4]

The article mentioned that, as a result of the scarcity of platinum and rising prices, Willis and Clement placed upon the market in 1913 a combined silver and platinum paper. Thomson also responded to the shortage. He offered a formula for the individual preparation of a platinum silver paper he had worked out a dozen years earlier. Thomson claimed he invented his platinum silver paper before before Willis and Clement produced theirs. Examination of Thomson's claim suggests it may be in error. Willis' original platinotype paper had involved the use of both silver and platinum in the sensitizer and that paper came on the market in 1873, long before Thomson's first iron-printing experience. In any case, Thomson's approach to silver-platinum printing was not very different from his approach to Kallitype I. The only difference was the addition of a small amount of platinum to the sensitizer. [4]

In 1915 two more articles on silver-platinum appeared in **Camera Craft**. Both described substitutes based on kallitype methodology. Both articles reprinted "Dr. Vollenbach's"[sic] brown print process originally published in 1899 as a platinum look-alike printing process. Dr. Vollenbruch's process has already been reported in the chapter on brown print. The re-publication of the process is a testimony to the continuing hold platinum esthetics had on amateur tastes in printmaking.[5] It suggests the lesser role the kallitype played at this time: substitute.

In 1916 we find a report written by Mr. J. M. Sellors of a demonstration of the Bennett-Burton K II process, given by W. H. Smith to the Croydon Camera Club in England. Sellors, the secretary of the club, reported that Smith, a respected English expert on printing processes, demonstrated the complete manipulation of the KII process. Smith showed how modifications of the Borax-Rochelle salt developer achieved different colors in the print. During the demonstration, Smith claimed the permanence of kallitype prints was greatly improved by the use of hypo as the fixer instead of ammonia. He also made the interesting observation that in kallitype printing it is almost impossible to obtain uniform print color when producing a batch of prints because "the developer alters with each print passed through it." Smith concluded that color variability was more a problem for the professional than it was for the amateur who rarely produced more than a few prints from the same negative.[6] This observation provides, another explanation for the failure of the kallitype to penetrate the commercial market.

Also in 1916, Blumann published an interview with Nelson C. Hawks, an American who manufactured and marketed a kallitype II paper called "Polychrome" from 1900 to 1904.

Hawks' venture into kallitype manufacture antedated Blumann's articles by fifteen years, and the Polychrome paper had been off the market for more than ten years when Blumann conducted the interview. We have already recounted Hawks' story and formulas in Chapter III which dealt with the history of the commercial sale of kallitype II paper.[7]

Unfortunately, Blumann's article gave little historical information about Hawks' kallitype manufacturing and marketing business or how much material he sold, etc. Nor did Blumann report Hawks' interpretation of why kallitype failed to remain on the market. An advertisement for commercial Polychrome paper is reproduced above, on page 118, and can be found in various issues of *Photo-Miniature* between January 1900 and July 1903. [8]

In 1917, Colin N. Bennett, F.C.C., F.R.P.S., wrote in the *British Journal of Photography*, about his "Trials of Kallitype." His reason for turning to kallitype was "I wanted the nearest thing I could get to platinotype effects, and I wanted them on the cheap." His "nearly invariable experience with other home-made papers . . . [was] that none of them . . . turned out to be quite good enough." "Kallitype seems to be the exception." Bennett followed G. E. Brown's account of the process in *Ferric and Heliographic Processes*. Bennett recommended an improved means of exposure and a way of assuring the dryness of kallitype paper. He dried the sensitized paper "with an electric fan and exposed it with arc lighting and mercury vapor lamps. The latter light was preferable because it was a cool light source." In spite of Brown's advice to the contrary, Colin Bennett used ammonia as his fixer. As of the date of publication, Bennett wrote there was no reason to doubt the permanence of his prints.[9]

In 1917 *The Photographic Journal of America*, a relatively new journal, reprinted Hammond's excellent article on "The Variations in Iron-Silver Printing," originally published in 1912.[10] Hammond's detailed and critical article was reported at length in the last chapter.

In 1918, still responding to the "prohibitive cost of platinum papers," James Thomson wrote on "Plain Salted Paper" in *American Photography* and on "Kallitype and Modifications Thereof" in *Camera and Darkroom*. In the former article Thomson wrote that salted paper, "worked at its best, is the peer of Kallitype; it has the added advantage of being a print-out paper." [11]

In the latter publication Thomson introduced his "least complex and cheapest" kallitype I formula. It was the same formula he introduced in October 1904. Also in this article, Thomson reprinted his formula for a salting and sensitizing approach to the brown print. He wrote, "Kallitype is . . . yearly becoming better known for the qualities usually associated with the more expensive platinotype." "Kallitype at its best cannot be distinguished from platinum." [12] Apparently, Thomson was not bothered by the fact that kallitype was losing ground as a process worth doing for its own sake.

Thomson again presented kallitype as a "Possible Substitute for the Platinum Print" in *American Photography*, November 1918. He reported that "the government at present . . . is commandeering all unmanufactured platinum in the country at a price of one hundred five dollars an ounce." In the article he considered replacement metals—palladium, iridium, and gold, and combinations of them with platinum and each other—and concluded "from an artistic point of view, kallitype should prove the most available candidate for the place of the platinotype." He ended the article with a warning: "the kallitype process is . . . at best one of uncertainties, so much depending upon the quality of the chemicals employed . . . particularly the ferric oxalate: problem."[13]

In the 1918 article, Thomson reprinted his "first formula for black and white effects," which was originally published in 1904. He reported that prints made "12 years ago with it are in as good condition as when made." He also reprinted Formula No. 3, originally published in 1912, which "gives a more visible provisional image." In this article Thomson revealed, what many had wondered about, the role of the copper chloride in his sensitizers. He wrote, "a trifle more of the copper tends to finer grain, but flatness." Apparently he believed that the way copper chloride reduced contrast was a useful property for printing contrasty negatives. He ended the article with a short explanation of the failure of kallitype to achieve commercial acceptance.

"Kallitype . . . as a process . . . has never been taken up by any of the big interests, which of course today is necessary to make any mode of printing a success.[14]

He was in error. The Eastman Co. had marketed a brown print paper for amateur use for a number of years after 1900. But the paper had failed to catch on. One report said that the packaged paper was susceptible to dampness and had a short shelf life.

After publishing Thomson on K I in its June, 1918 issue, *the Camera and Darkroom* published in July an account of the Bennett-Burton K II approach taken directly from G. E. Brown's book. It is a simple reprint with no added material or commentary.[15] The two articles, published back to back suggest that articles on kallitype were becoming for publishers a commodity. Any kallitype formula would do to sell magazines.

Another article describing the Bennett-Burton approach to K II appeared in the *Photographic Journal of America*. The article included a description of how to do brown printing, a process it considered suitable for a worker's first experience with individually prepared printing media. The author commented "the kallitype process has lately been interesting photographers."[16] One suspects the reason for the upsurge in interest is war-induced platinum shortages.

The year 1919 saw few publications on Kallitype. One article, written by the renowned American studio photographer, David Bachrach, appeared in the *British Journal of Photography*. Bachrach discussed, among other things, the permanence of the various kinds of individually made prints, including the kallitype print. Without indicating the basis for his opinion, Bachrach asserted "the weakest [least permanent] of all photographic prints are the P.O.P. gelatine prints, and next to them are the old salted and silvered prints not toned with gold. Bachrach found even salt prints achieve greater permanence: when they are "toned with gold, fixed in fresh hypo, and well washed." Bachrach believed "a print, made on modern commercially sold developing papers, is the most permanent of any silver image." "The kallitype print is of about the same permanence as the old plain paper prints not toned." He concluded:

> Now if the Kallitype prints can be made by some modification that will allow them to be toned by the sulphide process, they can be made about as permanent as the developed prints.[17]

Apparently Bachrach's position on kallitype permanence was that toning (conversion of the silver in a print to a more stable compound) conferred additional permanence to silver prints, above and beyond the permanence achieved by thorough processing with fresh chemicals. Bachrach added that kallitypes, when toned, are as permanent as any modern gelatin-silver print.[17]

It is interesting to find a commercial photographer of Bachrach's stature who was interested enough to comment on the keeping qualities of prints made by the older silver and iron printing processes.

In 1920 another embarrassing publication on kallitype appeared. Blumann, in the July issue of *Camera Craft*, wrote an article entitled "More About Kallitype," in which he paid tribute to Thomson, saying that "after years of experimenting, spoiling good chemicals and paper, and wasting much time, I have graduated into the class of adepts in Kallitype, and now use one sensitizer, and one developer." His choice was the method Thomson described in the *"Post Script to Photo-Miniature #69."* Blumann praised Thomson's method, calling it "the final lesson." But when Blumann published the formula he failed to include two chemicals, one of them the image forming chemical, silver nitrate. Blumann then compounded his mistake by insisting the reader follow his directions exactly.

> Mix these in the following way, and no other. Your way may be as good, and this way may seem foolish, but this works out just right and will give you no cause to write in that you did not get results.[18]

Predictably, in the following issue, Blumann had to eat crow. He began his apology by thanking his readers for "the compliment implied by the great number of protests." He continued with ". . . it transpires that I did not give enough about Kallitype by two important ingredients." He apologized "to the reader and to Mr. Thomson" and he concluded by encouraging the reader "to follow Mr. Thomson's directions." Perhaps as a diversionary measure, he confided, "I have been able to get a full pound of ferric oxalate in a lump crystal form that apparently is not generally available to others." "This form of the chemical," Blumann wrote, "does not deteriorate like the flake or spangle and probably is richer in acid." No other writer has mentioned this fabulous form of ferric oxalate.[19]

In 1920 Ruthven Flint's book, *Chemistry For Photographers* appeared. It contained a chapter on "Printing with the Salts of Iron," which described in detail how to make kallitype prints. Flint's approach to kallitype was identical to Hall's stock solution approach to making K II. But while Hall advised changing the chemicals in the sensitizer to vary the print, Flint advised "it is generally better to use a uniform sensitizer and exercise the necessary control by modifications in printing and development." Flint recommended use of Hall's formulas for clearing and fixing baths exactly as Hall wrote them, but modified the developing bath slightly. Overall, Flint provided a clear account of Hall's process, one quite serviceable for workers wanting to try.

Flint, who held a Ph. D. degree in chemistry, responded to Anderson's and Bachrach's negative assessments of the permanence of kallitypes relative to other silver prints.

> There is no possibility of doubt but that a silver print by the kallitype process may be made equally permanent with silver prints by any other method. When properly worked, the

process leaves an image in pure silver upon a substratum of pure paper, which is the condition essential to durability.[20]

Apparently Flint was not concerned about the effect of aerial pollutants on the exposed silver. After remarking that his prints have lasted "without appreciable alteration for seven years" Flint concluded,

> Any worker who does not get both permanent results and exceptionally pleasing pictures with kallitype ought to test his chemicals thoroughly and, especially, examine carefully his methods of working, for somewhere in these he is sure to discover the cause. [21]

Incidentally, Flint calculated the cost of making 5 by 7" kallitype prints in 1920. The cost was 10 cents a dozen.

In 1921, three years after the war ended, the file holds only one item on kallitype, an article about a demonstration given to the South Suburban Photographic Society in England "by Master Nixon, a sixteen year old photo-chemical student, who last year surprised the members of the Society by giving a learned talk on the chemistry of printing processes in general."

> Master Nixon . . . devoted an evening to Kallitype, giving his experiences with all the published formulae, demonstrating where possible and showing results—possibly the finest Kallitypes ever seen—obtained by the many systems of working. Master Nixon appears to have quite mastered the process, and although he was not able to publish anything really new concerning the preparation of the paper and development, his remarks were eagerly followed by many of the older hands who had failed to get the results expected.[22]

Nixon employed the Bennett-Burton formula for sensitizer and used what appears to be Hall's sodium acetate developer for black tones. He made his own ferric oxalate from ammonia, ferric ammonium sulfate, oxalic acid and iron wire.

While it is wonderful to learn of the 16 year old prodigy's performance, there is an air of the side show in this report.

1922 is a significant year because of the publication of Thomson's monograph, "Kallitype and Allied Processes," in *Photo-Miniature* No.

185.[23] Both the *Photo Era* and the *British Journal of Photography*, competing publications, announced this publication to their readers. The *BJP* article briefly commented upon Thomson's career—"he has made the technique of these processes peculiarly his own, and has published many papers in the periodical press giving the results of his original experiments." The *BJP* also praised the *Photo-Miniature* for compiling the many iron printing methods of recent years and the many formulas of Thomson into a single work of "exceedingly compact form."[24]

In his historically oriented introduction to Thomson's monograph, Tennant characterized the kallitype as "a simple process for making beautiful prints which is now thirty years old and has flowered in innumerable varieties to the keen delight of countless amateurs here and abroad." He reminded his readers that Thomson was originally made aware of kallitype in 1900 when he read in Vol. I, Number 10, *the Photo-Miniature's* first publication on the process, "The Blue Print and its Variations."

Tennant emphasized that *Photo-Miniature* had been connected with the history of the kallitype ever since. It published Hall's monograph on kallitype, which appeared in 1903. That issue sold over five thousand copies and led to the formation of a society of amateurs who circulated a portfolio of kallitype prints. *Photo-Miniature* also made contributions to kallitype in Numbers 67 and 81 which published Thomson's early approaches to kallitype. Tennant concluded his history of involvement with the kallitype by remarking that "now Number 185 provides an up to date compendium of all previous important approaches to kallitype I." Tennant added remarks on Thomson's general contribution to kallitype. Thomson, he wrote, "has devoted much of his leisure during the . . . [last] twenty years" to the development of the kallitype. He then praised the quality of Thomson's prints. The many kallitypes Thomson has sent

> show the adaptability of the kallitype for different classes
> of subjects, the remarkable variation of colors and effects
> possible by using different paper bases and formulas and . . .
> justify the claim that no photographic printing process yields
> results more beautiful than Kallitype.

Tennant also characterized Thomson's writing on kallitype: he "writes as a practical worker, directly and to the point." He "is as eager to tell about the methods and formulas of other workers as about his own."[25] Both remarks accurately describe the writing in the monograph, which consists of little more than a list of formulas and working methods. Thomson was not a theoretical or analytical writer. His remarks on the theory of the kallitype, a process he investigated for twenty years, required but seven lines. Thomson summarized the formulas and working methods of all whom he felt had figured prominently in the history of the kallitype: Bennett, Burton, Frederick, Hall, Hawks, Jarman, Blumann and himself. Sad to say, Thomson provided little reflection on or criticism of any of the approaches to the process. He merely compiled them. Like others, he reported his own formulas in a compendious fashion, without the illumination of theory or comment.

In his account of his own work in **P-M 185**, Thomson indicated he began by publishing in the 1903 **Photo-Beacon** a "modification of Nicol's K II formula. This was the only time in a long career that Thomson worked the kallitype II process.

In 1904 Thomson developed a kallitype I approach, in which the silver is in the developer. He concluded the K I approach was easier and less expensive to work than Kallitype II. Interestingly, he called his original approach to the K I process 'the Thomson method.' He revised it over the years into four formulas he numbered 1,2,3, and 4. The first three formulas allowed the worker to adjust and control the visibility of the print in the frame, the color of the print, and the contrast of the print. In **P-M** #185, Thomson provided formulas and the working details for his first three Kallitype I approaches. The details are identical to those summarized by Hammond in the previous chapter.

In **Photo-Miniature #185** Thomson also compiled several methods of working the brown print, which he called "the watertone process." All of his brown print methods employed a ferric ammonium citrate sensitizer to which an organic acid (tartaric acid) and silver nitrate are added. All are developed in water. Thomson also summarized Brown's brown print formulas. These formulas have already been reported in earlier chapters.

Finally, Thomson compiled in the monograph his own and others' formulas for making silver-platinum prints, either by substitution methods or by toning. He also included an assortment of printing methods which produced images in platinum, uranium, copper, and ferricyanide. It would take too much space to summarize all these formulas here. Fortunately, for those who wish to study them, copies of the monograph are available in libraries or from used photographic book dealers.[26]

It may be worthwhile to provide a few critical remarks on Thomson's major contribution to the kallitype in the *Photo-Miniature*, No. 185 monograph. The present writer believes that Thomson's monograph, "Kallitype and Allied Processes" has the value of a compendium of formulas and methods that is only rarely enlightened by commentary of any kind, historical, chemical, or photographic. While reading it, the reader hungers for discussion that would illuminate the pages of information but finds only formulas, quantities, and directions. Thomson's compendium teaches only how, not why. The reader reaches for insight through explanation and receives only another variation. That is the work's weakness. Its strength lies in the way it details many processes which continue to have value for photographers and scholars.

There is another question that must be asked: did the monograph have any impact on its time? Sad to say, in the many years since its publication, Thomson's monograph has been rarely mentioned by later workers. The book's impersonal and compendious manner may explain part of the indifference. The rest may be the result of a growing shift of photographic taste, a general disaffection with individualized printing processes which began to occur about the time the book was published. We shall discuss this shift in taste and value shortly.

Thomson's last publication was a four page article which appeared in **the American Annual for 1922**. Reading it, one watches an aging Thomson respond one last time to a request for an article. He began the article by confessing that he had nothing to add to his already voluminous writings. "In as much as I have done no experimenting along kallitype lines for the last few years, I have nothing new to offer."[27] Nevertheless, he recommended "a new and less complicated formula which in the hands of the novice is less likely to give trouble." Because this is Thomson's last approach to working kallitype, we shall present the details.

Thomson's final Kallitype I Formula	
Distilled Water	1 oz.
Ferric Ammonium Citrate	20 grains
Ferric Oxalate	20 grains
Potassium Oxalate	18 "
Gum Arabic	10 "
Platinum Solution	10 drops

Bichromate solution 5% sol   4-10 drops

Stir and wait 24 hours before using. Filter off the gritty particles. Retain the brown sediment. Apply a thin coat to a well sized water-color paper."

Thomson does not indicate how to filter off gritty particles while keeping the brown sediment or what to do with them.

The formula for the platinum solution is:

Platinum Solution
Distilled water   2 oz.
Potassium Chloroplatinite   15 grains
50% phosphoric acid   2 drams

Add the platinum salt to the water, then add the acid.
After the prints are coated, dried, and exposed, they are developed in

Thomson's final KI Developer

Distilled Water   1 oz.
Silver Nitrate   40 grains
Citric acid   10   "
Oxalic acid   10   "

This makes a stock solution which should be diluted 1 part with 7 parts of water for use. Use a porcelain tray and develop "a trifle longer than it takes to fully form the image." Then fix in

Thomson's Final KI Fixer

Water   32 oz.
Sodium Thiosulfate   50 grains

Fix for five minutes. Wash for 1/2 hour in running water. [28]

This article is accompanied by reproductions of two of Thomson's kallitype prints. The two images, both close-ups of flowers, give a sense of Thomson's craft.[29] The flowers show gradation of light and sensitive composition. (Thomson had worked as a designer, before taking up kallitype journalism). Both images are printed on an attractive laid paper. This brief, final article may prove to be Thomson's best single communication on how to work kallitype; from it the reader gains as much insight into Thomson's way of working kallitype as he gets from the thirty-five page monograph.

Reader response to Thomson's last two publications was strangely quiet. Careful research in the periodical literature of the next ten years has turned up no response to either piece of writing. The few articles on the kallitype that appeared in the succeeding years are not responsive to Thomson. It is interesting, that when he revised Woodbury's *Photographic Amusements* in 1922, editor Frank Fraprie included a thorough presentation of Hall's kallitype II formulas, "an old and well tried method of considerable interest"—and totally ignored those of his fellow Bostonian, Thomson.[30] In 1923 Fraprie published a version of the Bennett-Burton approach to kallitype, not Thomson's, in *Practical Printing Processes*, a ten volume series.[31] Fraprie does not tell us what is practical about the K II process he chose to reprint, or what was impractical about the K I processes he avoided. In the following year, 1924, when Blumann, now the editor of *Camera Craft*, wrote an article on making Christmas cards with Kallitype process, he recommended the Bennett-Burton approach. Blumann mentioned Thomson in the article, referring to him as "the final authority on the "Iron Processes," but avoided any mention of his formulas.[32] These are the only two publications on Kallitype that have been found after a diligent search of the literature in the four years after Thomson's compilation. The lack of response suggests either that the monograph failed to light many fires or that reader interest in kallitype was at a thirty-five year low. Probably both.

We have arrived at 1925, the year set as the terminus for this historical report on the literature of the kallitype. In the year 1925 the frequency of published articles and books on kallitype rapidly tapered off. One suspects that this date marks a turning point in amateur photographic taste and activity. Thereafter from 1925 to 1975, kallitype publications are few and far between, limited almost completely to formularies in reference books and encyclopedias. Not until the last quarter of the twentieth century does interest in kallitype revive again. Then such works as Kent E. Wade's *Alternative Photographic Processes*,[33] William Crawford's *Keepers of the Light*,[34] Nancy Howell Koehler's *Photographic Art Processes*,[35] and Jan Arnow *Handbook of Alternative Processes*[36] suddenly appear and reignite the old interest in process. These publications indicate that by 1975 a new interest in alternative photographic processes was rising, one fueled by a source of energy quite distinct from historical amateur pictorialism. We shall comment on this renewed interest shortly.

Before making our closing remarks, we shall briefly recapitulate the history of the kallitype since its inception. In an earlier chapter, we observed that the kallitype was one innovation of a wide search undertaken during the nineteenth century for improved photographic products of all kinds—cameras, films, papers, and processes. The search for a successful iron-silver paper was conducted by a number of ingenious experimenters in England, many of whom were "amateurs." After a generation of early explorers had tried without success, W. W. J. Nicol, a young university chemistry professor, patented a working process. Nicol's intention was the creation of a more variable and beautiful paper than the current workhorse, albumen-silver, and a cheaper alternative to the expensive platinum print. The Birmingham Photographic Co., under the leadership of John Lewis, attempted to commercially exploit Nicol's photographic paper, KI, following the enthusiastic reports the paper received at the hands of knowledgeable reviewers of the patent. In spite of their unanimous prediction of 'a great future,' we saw how commercial paper, offered successively in two forms, K I and K II, quickly failed in the market place.

The repeated high praise of the Kallitype's beautiful tones, ease of working and the unanimous predictions of its success, did not produce photographic acceptance and the company stopped selling the product by 1905. We looked for an explanation of the failure and found evidence of

1) insufficient product development before placing the paper on the market;
2) difficulties with quality control of the product after sales had commenced;
3) questionable marketing techniques when introducing successive products, and
4) intense competition by attractive rival papers and processes.

We concluded that it was unlikely that the commercial difficulties of the kallitype were attributable to the iron silver process itself, since the platinotype, a very similar process, enjoyed market leadership for over thirty years, during the period of time when both Kallitype papers made efforts to succeed. We concluded that commercial kallitype failed in England because Nicol and Lewis were not as competent in engineering and marketing kallitype products as Willis and Co. were in making and marketing the platinotype. We also suggested the product may have developed greater acceptance if it had been invented and marketed twenty years earlier when competitive processes were less advanced.

Subsequent to the failure of commercial Kallitype I and II papers in England, a number entrepreneurs made and sold kallitype paper in America,—in New York, Chicago, and San Francisco—between 1895 and 1905 under the names like Vici, Mirrotype, Celerite, and Polychrome. These squatter businesses had no tie to Nicol or the Birmingham Photographic Co. The American entrepreneurs proved to be no more capable than Nicol and Lewis, with the result their businesses were short lived and not particularly remunerative.

The manufacture and sale of kallitype paper ceased about 1904, and the process was abandoned. The abandonment of the kallitype by commercial interests in England and America was followed by an unprecedented "take-over" of the process by amateurs. Even before manufacture of kallitype paper ceased, interested amateurs saw an opportunity to improve what they viewed as an incomplete or imperfect process. and began publishing accounts of how to make kallitypes in newly popular photography magazines. They encouraged others to do likewise. Their enthusiastic accounts were followed by strong reader responses and a chain reaction of amateur publications occurred. Written by amateurs widely separated in background, occupation, and education, the articles detailed how to mix solutions and work the paper. Writers argued that kallitype was an easy paper to make and gave reasons why amateurs artists should make their own. Between 1895 and 1925, a dialogue lasting 30 years ensued among intelligent, knowledgeable, and resourceful amateurs who shared their methods and their thoughts on kallitypy in the photographic journals of the time.

In the last six chapters we have reported the dialogue of amateurs who in effect, coopted the process. We have provided detailed reports of the major processes that were described during the dialogue so readers could understand and, if they choose, try to work them. Beside sketching the details of the evolution of individualized kallitype processes and indicating the response which they drew from readers, we have tried to discover any cumulative development. Out of the many investigations made by amateurs, we found four distinct approaches to kallitype process evolved. A brief recap of each follows.

Bennett and Burton, modifying Nicol's original K II formula provided in his patent, developed the first amateur formulas for what is called the kallitype II approach. The kallitype II process combined ferric oxalate and silver nitrate in the sensitizer and used either a Borax Rochelle salt or a Sodium Acetate developer. The result of Bennett's modification of Nicol's original K II process, and Burton's modification of Bennett's approach was a simple, invariable sensitizer and two developers, all of which required few chemicals. The working of this first amateur approach to kallitype was easy to understand and simple to do. We have shown how the Bennett-Burton approach to Kallitype II was subsequently rewritten over a period of 30 years by many amateurs, who rarely credited their source. The formula continues to be republished in reference works to this day.

Next, Frederick and Hall, in successive investigations developed an approach based on the use of stock solutions. Their method provided great flexibility for individual research and encouraged the making of "print variations." The Frederick and Hall approach received wide acceptance, especially in America, and was often recommended by workers for the ease of modifying the sensitizer and the developer to suit the needs of the negative or the taste of the printmaker.

These two innovative approaches, which we labeled, the Bennett-Burton and the Frederick-Hall approaches, were followed by a series of published responses which advocated now one and now the other approach, often with minor modifications.

Next, we detailed the contribution of George E. Brown and James Thomson who popularized versions of the brown print. The brown print had been marketed between 1895 and 1910 as a paper for copying engineering and architectural plans by a number of firms in England, Germany, and the United States. Brown print paper had also been sold by several manufacturers as an amateur printing paper. The brown print process utilized a ferric ammonium citrate and silver nitrate sensitizer and required only water to develop the exposed paper. Working in England and America, Brown and Thomson successively refined the technique for making brown prints. They popularized a "modified kallitype" process that was cheap and simple enough for beginning amateur printers to work. Brown's sound and simple formulas for the individual preparation of brown print paper were described along with Thomson's more complicated formulas. Thomson's approach was frequently criticized in the photographic press for needless complexity and, several times, for questionable chemistry.

Next we documented how James Thomson directed attention away from amateur preparation of kallitype II paper and focused it on the preparation of Kallitype I paper, which had been ignored by amateur printmakers since 1891. Nicol's Kallitype I process sensitized the paper with ferric oxalate alone and developed the paper in a bath of silver nitrate. Thomson, a fertile and persistent investigator, published a seemingly endless series of complex variations of the K I process over a period of twenty years.

Finally, we described thirty years of amateur efforts to modify earlier kallitype formulas and produce papers that combined silver with other metals, notably platinum, palladium, uranium, and iridium, in various combinations. By these efforts experimenting artist printers hoped to find substitutes for platinum paper, the favorite of the most successful photographic artists of the time.

This completes our summary of thirty years of publications on commercial and amateur development of iron-silver papers.

III.   The Progress of the Kallitype Process 1890-1925

We shall now draw what conclusions we consider reasonable about the development of the several kallitype processes.

During the thirty-five year dialogue of processes, that is from 1890 to 1925, no single kallitype process won a decisive victory over the others. Throughout the years we have investigated, a given approach occasionally developed a momentary advantage, but such advances soon cancelled each other out. The process that seems to be most in evidence today in reference works was the Bennett-Burton approach to the KII process, which utilizes a simple ferric oxalate and silver nitrate sensitizer, a Rochelle salt developer, and a weak sodium thiosulfate fixer (2-5%). Most current books recommend kallitype II formulas, however disguised, and most of the modern recommendations appear to be variants of the Bennett-Burton process. The reason for the success of this process, in all likelihood, is its simplicity and effectiveness. It requires fewer chemicals and simpler solutions than the others and the invarying process produces good results for the uninitiated. The brown print, as described by G. E. Brown, is also commonly recommended today for the same reasons. The K I approach had few recommendations after 1925, possibly because of the complicated state in which Thomson left it, and it continues to be ignored today. The Frederick and Hall approach, which introduced stock solutions, was the process most often recommended in America between 1915 and 1925, probably because of the flexibility, control, and convenience that follows from the use of stock solutions. This approach was also valued for its production of variations in print color. But the stock solution approach has not received much attention since. It's limitation in modern times is that it appears intimidating to modern printers who often possess little appetite for or experience in compounding solutions.

In addition to describing the processes, we discussed ideas and events—social, technological, and artistic—which we believe influenced and gave meaning to the development of the kallitype. We noted the interests in science and habits of research possessed by well off and well educated early Victorians, and their need to make needed supplies and equipment crucial at the beginning of photography. We suggested that these "habits" of mind and practice influenced a later generation of upper and middle class amateurs to take pleasure in exploring photographic processes. We noted that the availability of the portable camera, dry plates, and roll film stimulated the growth of amateur photography in the 1880's and 90's. We suggested the increased numbers of amateurs led to a proliferation of photographic periodicals, photographic societies, and local, national, and international competitive exhibitions. We noted that as photography became easy, it became popular resulting in mass production of boring images. Critics begrudged common snapshooters for their habit of making banal images. Sensitive photographers stated their preference for images that interpreted and expressed as opposed to chemical and process mongering. Artistic Photographers eagerly joined the rebellion of impressionist painters against traditional academic restriction and methods. We suggested that all of these influences played a part in the development of new attitudes about the art of photography. Photographers began to support new approaches to personally controlled printmaking that expressed artistic taste and feeling We suggested that these ideas led to the development of a new photographic esthetic,

Pictorialism. Pictorial values supported a variety of individually prepared, hand-manipulated printing effects, among which the kallitype was recommended for its ease, cost, variability and charm. Manufactured papers had serious limitations of variability, texture, and color. Pictorial esthetics gave life and a mission to the kallitype. In return the kallitype supported the pictorial values of print variabililty, freedom of process, and the full involvement of the artist.

We also made some observations about the the users of the kallitype process during the period 1895 to 1925. We noted that after 1895 the kallitype process was adopted by a class of amateurs who were intelligent, well educated, curious, experimental, in comfortable circumstances, and committed to competitive local exhibitions and national and international salons. We also noted that none of the major artists of the time (eg. Stieglitz, Strand, White, Coburn, etc.) was a committed kallitypist. We did find evidence that a few of the successful salonists, notably Anderson and Fleckenstein, had tried the process and had written about it. A number of lesser salonists also wrote to recommend Kallitype printing.

We noted that the most competent kallitypists were not recognized in their time, or ours, as major photographic artists. The explanation given for this anomaly was that most of the major artists of the period worked with platinotype, if they worked with an iron-sensitized paper at all. The explanation given for their indifference to kallitype was that only the platinotype contained all that was valued in a print at that time. Platinotype was universally admired for its permanence, its fine black image, its matt surface, and its variability by personal control. The kallitype resembled platinotype in these particulars, with the exception of comparable permanence and ease of use. Its unique print values were greater print variability at a price one fifth the cost of platinum. Price, apparently, was not a compelling concern for financially comfortable pictorial artists and salon competitors while platinum paper was available for purchase during of the first quarter of the twentieth century.

IV.   Interpreting the Demise of the Kallitype and the Pictorial Esthetic

We have reserved till now the question of why, after 30 years of intense and sustained amateur interest, the kallitype lost acceptance so abruptly after 1925.

To answer the question we shall give a brief resume of the rise and fall of the pictorial esthetic which supported the kallitype and other individually prepared print processes.

Earlier, we observed that the development of the kallitype was a part of a larger movement in photographic taste, commonly called pictorialism, which began sometime before the last quarter of the nineteenth century and captivated popular photographic taste until approximately 1925. The pictorial esthetic—a complex of esthetic ideas, attitudes, values, and practices involving Art, Artists, Photography, Expression, Subjectivity, and Photographic Printing, etc.—endorsed a number of photographic printing concerns. Two of its main concerns were an interest in personal expression and an interest in individualized photographic printing. The essential concerns of the pictorial esthetic were summed up by one of the most successful pictorial artist-photographers, Robert Demachy, who worked creatively with brush and pigment on gum bichromate prints

> Now, speaking of graphic methods only, what are the distinctive qualities of a work of art? A work of art must be a transcription, not a copy of nature. The beauty of . . . nature has nothing to do with the quality that makes a work of art. This special quality is given by the artist's way of expressing himself. In other words, there is not a particle of art in the most beautiful scene of nature. The art is man's alone, it is subjective, not objective . . . .
>
> Pictorial photography owes its birth to the universal dissatisfaction of artist photographers . . . [with] the photographic errors of the straight print. Its false values, its lack of accents, its equal delineation of things important and useless, were universally recognized and deplored by a host of malcontents. There was a general cry for liberty of treatment and liberty of correction. Glycerin-developed platinotype and gum bichromate were soon after hailed with enthusiasm as liberators; today the oil process opens outer and inner doors to personal treatment.[37]

Earlier, in 1892 H. P. Robinson wrote "a method that will not admit of the modifications of the artist cannot be an art"[38] Robinson was successful at photo-collage.

Pictorial photographers believed individually prepared papers permitted a fuller expression of an artist's ideas and feelings than did manufactured papers. In past chapters we suggested that a variety of individually prepared printing processes were supported by pictorial photographers in the belief that they helped distinguish the artistic photographer from the mere copyist. William Mortenson was a major commercial and fine artist who continued to represent pictorial values and methods—in opposition to Edward Weston and others. He defended pictorial methods and ideas well into tlhe 1940's.

The highest goal of the pictorialists was the establishment of photography as an art. They believed that goal could best be reached by emphasizing the plasticity of the photographic medium. They encouraged photographers to express ideas, feelings, and taste by evident manipulation and supported use of highly variable media. That is the simple explanation why kallitype, a highly variable printing process and other manipulatable print processes appealed to artistic photographers in the quarter century before and after the year 1900. The broad acceptance of pictorial values explains the rise of all individually prepared and controlled printing processes, in general, and the sustained interest in kallitype in particular—a major concern in this book.

But the fantasy that photography was a highly expressive and creative art employing a variable and highly manipulable set of processes, as with all esthetic ideals, did not continue without change. After 1925, the pictorial ideal began to tarnish as a result of a number of excesses and a new ideal for photography began to take shape. As so often happens in historical progression, the dialectic of taste swung from thesis to anti-thesis. The new fantasy of what photography should be was less concerned with subjectivity and manipulation and more concerned with objectivity, formal beauty, and refined control of less variable (straight) printing. It asserted that photography was and should remain a "pure" medium, one which must deny evident hand manipulation and ostentatious craft. Edward Weston was one of the most articulate spokesmen of the new "purist" point of view. In an essay entitled "What is Photographic Beauty?" Weston wrote,

> The photographer's power lies in his ability to recreate his
> subject in terms of its basic reality, and present this re-creation
> in such form that the spectator feels that he is seeing not just

a symbol for the object, but the thing itself revealed for the first time. [39]

Weston had written earlier about the form current photographs should take: they . . . should be sharply focused, clearly defined from edge to edge, from nearest object to most distant . . . They should have smooth or glossy surface to better reveal the amazing textures and details to be found only in the photograph.

Finally, print values "should be clear-cut, subtle or brilliant, never veiled."[40]

This brief statement of the esthetic views before and after the pictorial period obviously oversimplifies an esthetic reorientation that had many dimensions. To further describe and document the rise and fall of pictorialism and the new esthetic that later supplanted it would require more space than is available. However the esthetic that succeeded pictorialism might be defined, it quickly gained adherents throughout the photographic world. After 1925, photographic prints with characteristics quite different from the pictorial model became the goal for many photographers. The esthetic described above by Weston supported the making of "straight" prints on manufactured glossy gelatin-silver paper, preferably by contact printing a large negative. It advised the use of large cameras and sharp lenses to show the infinite detail and the full range of tones photography was capable of displaying. It decreed that in "pure" or "straight" photography, handcraft should be "invisible," hidden.

Such statements clearly suggest that "pictorial" printing that resulted from obvious manipulation was on the way out by 1935.

Art in photography, as the post pictorial esthetic defined it, referred to what the artists "saw" in the real world and how well they composed and disposed the photograph to reveal the "essence" of what was "seen." Craft involved subtle variations in lighting, developing and printing that manifested the range of tones possible with the new highly refined commercial silver bromide papers. The new age of realistic formalism, sometimes called the "new objectivity" had begun. It gained momentum as technology offered the small camera capable of capturing real instants. Smaller fine grained films and controlled development emphasized basic, direct photographic values that opposed ostentatious manipulations hand wrought on the negative or print. The interest in pictorial feeling, scenic subject matter, and hand treatment did not suddenly stop. Adherents, such as William Mortenson continued to wage a visual battle against adherents of the new, direct esthetic.

Mortenson's prints at their extreme, displayed heavily shaded representations of poetic subjects, hand engendered material, printing screens, with various hand manipulations wrought on manufactured paper. A full report of the battle between the two esthetics would be interesting, but would take us too far afield.

As the new vision took hold, the means for achieving the old dream, the various hand intensive processes, were left without a driving purpose. Rather quickly obvious hand-work on photographic prints was questioned and criticized as an unphotographic intrusion into the native realism of photography. Adherents of the *process* esthetic responded, but their efforts, as far as the kallitype was concerned, were limited to tired reiterations of old values and labor intensive process methodology. It was inevitable that readers would tire of attempts to pump life into a dream that had outlived its time.

We conclude that the decline of artist's interest in the kallitype and similar processes after 1925 can be attributed, at least in part, to the decline of vitality, purpose and credibility of the pictorial esthetic which had supported it.

As a movement comes to an end, it slows and sputters. Like an engine running out of gas, it continues "to rev" for a while, producing noise, but little energy or motion. So with the kallitype between 1920 and 1930. Articles in this period continued to be written, but their number declines. Those that were published had less to offer and the writing addressed issues of less importance and urgency. The writers no longer had great expectations for the kallitype; they did not see pictorial ideas as influencing the course of photography in important ways, as writers had in past years. Nor did the later writers display much investigative insight into kallitype's persistent technical problems. Close study of the articles produced after 1920 indicates that most writers repeated without innovation what past writers had already written on esthetic issues or craft. After 1925, kallitype writing became tired, repetitive, trivial and banal. Not surprisingly, the ideal tarnished and the interest in kallitype waned.

Our argument is that the pictorial values that had supported kallitype progressively lost their energy after 1920. The trouble laden process movement, of which kallitype was but one example, was left without a justification for its complexities, costs in time and energy, and chemical difficulties. There are several kinds of evidence that support this conclusion. One indication is the decline of publications about the kallitype. (There was a similar decline in writing about all types of plastic print media, gum, bromoil, etc., but a full discussion of that topic would again further lengthen this book.) We do not mean to suggest that after 1925, writing about individually prepared prints ceased or that individually prepared prints were no longer made. There were conservative individuals who tried to prolong indefinitely the pictorial esthetic—witness the Pictorial Society of America, the sponsor of salons and camera clubs, which has an active following to this day. The photographic press, which for so long had supported the craft-intensive prints, was not so loyal. Photo magazines stopped restating the old values and ceased republishing the process descriptions and imagery that supported the old ideal. Editors, learning that the winds of taste were changing, began to publish articles emphasizing new artistic values and new process concerns allied to them: the miniature cameras, new fast roll films, new simple methods of tank development of miniature films, and finally printing with enlargers. Magazines printed advertisements promoting these new processes, not the old. And of course, the photographs they honored by publication were of the new objective style. For example, as early as 1900 the *Photo-Miniature* published an entire monograph on "Street Photography." Without the support of the press that had sustained the pictorial craft approach, the kallitype along with other intensive hand processes ceased to be what they had been—vital photographic process that inspired and energized artistic photographic activity.

In subtle ways the articles we summarized in the beginning of this chapter suggest this decline of interest, quality, and substance. We cite the article on the use of kallitype for making Christmas cards, on the making of kallitype window transparencies, and the article about the sixteen year old printing genius as notable examples of writing uninspired by high artistic concerns. The lack of inspiration and substance justify 1925 as a turning point. After that date almost nothing of importance was published on kallitype for 50 years.

V.   Changing Times and Changing Taste: 1925-1950.

The reports we have been studying suggest that after 1925 photographic esthetics progressively changed. A number of changes occurred in the photographic scene that specify the decline of pictorial taste. We shall now sketch a few of these broad photographic changes and inquire about the possibility of their progressive positive influence on "purism" or their repressive influence on pictorialism.

After 1925 photo-taking and photo-making increasingly developed in the direction of being, objective, fast, mechanized, miniaturized, convenient, unproblematic, and quick. Picture taking became increasingly attracted to the spontaneous recording, and to what began to be called "straight" photography. There was Weston's f/64 movement in the direction of the deliberately "previsualized" compositions taken with the large view camera, Also there was the emphasis on printing the unadulterated negative and controlling printing by the light cast by a single bare bulb. Both searches opened up an area of objectivity that had been more or less displaced by the subjective hand modified print emphases that accompanied the rise of pictorial manipulation. In response to the new attitudes, different kinds of photographic film, paper, and chemical solutions were developed and marketed on the basis of their effectiveness, fine grain, reliability, speed, ease, and convenience. As time passed, photographic processing became "pure," routinized and standard. Eventually darkroom work was reduced to "processing,"—an automatic service that could be performed at the drugstore by a machine on the one hand, or by Adams' zone system of control of negative and graded paper print processing. As 1950 approached, photographic processing was viewed less as a magical occasion for personally controlling the mysteries of photo-chemistry and more as a process to be engineered, made straightforward, efficient, and—in a word, artless. Photographic printing paper became increasingly a product manufactured and sold by large corporations which performed well only in the uses for which it was engineered. Its lack of variability, according to pictorially inspired photographers denied the artist much opportunity for personal expression through manipulation.

As equipment, materials, and practices changed, the expectations and values of amateurs evolved away from the heroic pictorial model originally cast in the early 1900's. The revolutionary amateurs who had aspired to understand and to control the total photographic process and who were motivated to establish photography as a high art by making prints that were incontestably individual, expressive, and beautifully crafted—such amateurs witnessed the development of a new kind of photographic hero-model. However the identity of the new photographer might be described, and regardless of which photographers might be chosen as models—Stieglitz, Strand, Coburn, or Weston and Adams, the concept of the artistic photographer for the generation after 1925 became different from what it had been. And the change was progressive in the generation of photographers who followed the early "purists", as the names, Walker Evans, Robert Frank, Gary Winogrand, and Lee Friedlander, suggest. The new heroes had a different identity than either the amateur-experimenters photography inherited from the mid nineteenth century or the pictorial amateurs who revolutionized artistic photography at the turn of the century. This is not to say that there was no continuity of ideas and values from 1900 to 1925 and from 1925 to 1950. In a number of respects many qualities of current art photographs possess qualities of the Edwardian realists and pictorialists. But there are also differences in esthetics, craft, and values that separate the Linked Ring pictorialists, the f64 photographers, and the small camera realists. It is likely that the differences have more than a little to do with the developments in cameras, lenses, film, paper and enlargers.

After 1950, almost all photographic paper was manufactured by large corporations which employed highly trained scientific, technical, and marketing personnel. As a result of the decisions of entrepreneurs and the efforts of experts, printing paper after 1950 evolved into a convenient, highly engineered utilitarian product with incredible material qualities, many of which had to do with resistance to tear and other kinds of stress, quickness of processing, and reliability.[41] By midcentury modern mass-produced photographic paper reached a high level of achievement when evaluated by the goals set by the manufacturers. But during this period photographers watched the removal from the market of papers which had exquisite texture, color, and tone—papers with names like Cykora, Indiatone, and Opal Suede. After the removal of these "pictorial" papers, in which the "vehicle" had a grace of its own, manufacturers applied themselves increasingly to the development of utilitarian photographic papers—papers which processed quickly, dried flat, and possessed carefully measured and variable contrast. When put on the market, the new papers had no particular applicability to artistic expression beyond clear depiction, contrast variation, speed, and ease of working. With the introduction of resin coated paper, artistic photographers could only shrink their creative expectations to the level of the paper that manufacturers gave them to work with. Reduced to the role of consumer, and lacking access to anything better, many artists adapted their values and practice to the only papers that were available. Photography enjoyed a period of objectivity in which the old pictorial considerations were unsupported. For a generation papers lost in uniqueness and charm what they gained in utility and faithful, brilliant, depiction. And, of course the arrival of color printing, whose difficulties of control were even more resistant to personal manipulation, exerted its influence on purist taste and practice.

Other changes after 1925 probably influenced the decline of pictorial taste. Since the first quarter of this century, photographic magazines have changed their emphasis as the most casual examination of editorial content of such magazines as *Mini Cam, U. S. Camera, Camera 35; Popular Photography,* and *Modern Photography* will reveal. It is a common observation today that most current photographic magazines exist to announce, introduce, test, and sell photographic products. It is also evident that most display little interest in stimulating amateurs to high achievement of the scientific, technological, or artistic kind that 19th century journals once did. A mass consumer audience and an emphasis on sales has dominated photographic periodicals since 1935, not the advancement of the art or science of photography. If anything, most modern photographic publications have democratized photographic taste, reducing aspiration and narrowing the focus of what a photographer can be. The turn of the century pictorial magazines had names that signaled the idealistic mission of the publications of that time: **American Photography, Photo Beacon. Camera Art, Camera Craft and Photo Era.** The longest selling photo magazine in recent years is called "Popular Photography."

The level of aspiration of amateur photographic societies has also fallen. As the century progressed, photographic societies became camera clubs. The quality of the concerns of photographic organizations has declined progressively from the level established by photographic societies popular in the mid 19th century. Based on the writer's experience with them, today's camera clubs are preoccupied with competition for ribbons, discussion of hardware and software, social interaction, and the sharing of information about pleasant travel and picture making at established pictorial locations. These are lower goals indeed than those sought by former photographic societies: the extension of photographic understanding, the development of technical mastery, and the development of the highest level of artistic expression.

The changes in the concept of photographic art, in the content and purpose of photographic periodicals, in the aims and activities of photographic societies, and in the seductive convenience of manufactured products in the years after 1925, all have had their effect on the declining popularity of individualized printing processes and on the low value attached to prints produced by them. With influences such as these, it is not surprising that interest in kallitype, once so vigorous, has progressively withered. Now, photographic publications can go a decade without an article on how to prepare one's own photographic paper. To find a demonstration of an alternative printing process at a camera club one might have to search even longer. The sense of what a photographer is and does as well as the sense of what a photograph should be and look like have simply changed a great deal.

VI. The Rebirth of Kallitype and Allied Processes.

Now a look in the opposite direction. About 1980 there were signs of a rebirth of interest in expressive photographic printing. Several manufacturers placed on the market a class of photographic papers called "exhibition paper." This higher quality silver-gelatin paper was designed to appeal to the eye of photographic artists. The paper had no more charm (texture, tonality, etc.) than standard commercial papers, but it did make measurably richer toned silver prints.

A pervasive positive influence on photographic taste and practice dates from 1970 to the present: the academization of photography. Since midcentury photography has been increasingly "academicized," that is, has been widely taught in colleges and universities by "art" teachers who themselves were university taught. Higher education in photography, as offered in art departments, (it had previously been found in departments of physics and journalism) has instilled new life and rekindled some of the old values of pictorial printmaking. A growing number of photography teachers have undertaken extensive investigations into long-ignored photographic processes—salt print, platinotype, collotype, gum bichromate, and kallitype. And a number of teacher-artists use individually prepared papers in their art. One teacher who wrote a book on alternative photographic process cited reasons for the new interest. Among them were "artists' dissatisfaction with the rather homogeneous results realized from conventional photographic materials" and the wish of artist-photographers "to return to the roots of photography."[42]

Art museums and galleries have shared in the reaffirmation of older more pictorial approaches to printing and the celebration of older concepts of the image, even while they were committed to supporting the new experimentalism so evident in "art" photography after 1960. While photographic exhibitions of "old process" photography have not been frequent, there have been a sufficient number of exhibitions to bring before the public examples of both historical and contemporary alternate print media. There has been some slight movement in curatorial institutions to collect and preserve examples of old printing methods.[43] We have already reported the increasing frequency of books published on "alternate" printing methods. It may bear mentioning that summer workshops in photography commonly include instruction in one or more old processes. Finally, kits and chemicals for working a variety of old processes are today readily available from several commercial suppliers. All of these indications point to a rebirth of interest in process in general. A growing interest in kallitype is particularly notable from 1970 to the present.

It must be admitted that the renewed interest in alternative printing processes, even in the universities, is short of a bonanza. In colleges a minority of artist-instructors teach courses on alternate photographic processes. In recent years "old processes" have been used to produce books, mixed media prints, and such art objects as clothing, quilts, images on ceramics, installations, and the like. It is probably true that some photographers, in the enlightened environment of the university, approach alternate processes only close enough to lend to their work an unusual or trendy media look. The present attitude toward "old process" photography in many art departments has been suggested by Nancy Howell-Koehler in her book entitled ***Photographic Art Processes***. She writes

> Today artists and photographers share the same arena. Society's preconceived concept of what constitutes a photograph has been shaken. Photographic images are no longer an end in themselves."[44]

Howell-Koehler's statement suggests that the battle fought three quarters of a century ago by the Pictorialists, the Linked Ring, and the Photo Secessionists and Artistic "expermenters" for the acceptance of photography as an expressive, plastic, art medium—may not have been lost! But her statement that the photographic image is no longer an end in itself raises some questions about whether the victory has been an undivided triumph.

Today, all may not be well with photography in its newly acquired status as an accepted art media. Presently, in art departments of universities across the land, the achievement of understanding and control of hand made photographic printmaking is rarely pursued as a valuable discipline—with the same respect as, say, etching or lithography. In current art departments, it may be far more valuable for the teacher to develop art that attracts notice for its "creativity and innovation" than for its mastery of process. Superficial control of a process employed merely to make a personal statement is often enough to attract notice and lengthen the teacher's exhibition record. An age that almost universally accepts and rejects entries for photographic exhibitions on the basis of computer reproductions or slides demonstrates a profound indifference to the subtleties of print tone, texture, and surface quality. Current artistic value appears to reside, almost exclusively, in the personal vision or concept of the image maker, rather than in the skill or technique required to realize that vision in a well made photograph. Fortunately, a few teacher-artists reach for the heights of perfection possible in the art of hand individualized photographic printing.

VII.   Modern Scientific Investigation of the Kallitype

In recent years, the most sophisticated investigation of the Kallitype process is done not by amateurs and not by artists, but rather by scientists. It has been done by professional chemists who have managed to provide an ultimate chemical description of iron-silver photochemical processes. Such investigators do not write about the kallitype in the way photographic artists did, nor do they spend much time in the darkroom making art. If they make prints at all, they do not exhibit them. They study the process of iron printing from the lofty heights of chemistry laboratories and photo-chemical theory. They write the equations which describe ideal photo-chemical reactions. I have in mind here the monumental work done by Glafkides[45] and Kozar.[46] The following short passage will suggest the quality of their concern and writing, and the level of their expertise.

> With organic polybasic acids Iron forms complexes of the formula $(Fe^y An^a)^x$ in which y is the valency of iron, a the valency of the organic acid and x that of the resulting complex

ion: the relation x= an-y must apply, n being the number of molecules of acid.[47]

These scientists view the kallitype as one example of iron photochemistry which itself is but one branch of the general chemistry of salts of metals.

Their descriptions of kallitype and other photographic processes based on the light sensitivity of iron provide valuable information even though it may not be readily usable by the typical amateur or artistic photographer. Seeing scientific descriptions of the iron process, contemporary artistic photographers may wonder if their untrained efforts can ever contribute anything worthwhile to the understanding and control of any photographic process. Nevertheless, Thomson's remark, quoted earlier, that the worker sometimes provides the observation that suggests the equation that the scientist will write. Photographers who can take advantage of scientific investigations will save time, energy, and money when learning process. But this history bears out the truth that enlightened empirical photographers, patiently, if blindly working their art, can achieve some control of a complex process. And, when all the investigations of phenomena are over, it is the photographer who must make the art print.

VIII. Kallitype in the 21st Century

As I write this conclusion in October, 2012, one last major process development requires attention—that of kallitype in the age of computers. I will make only a few brief remarks on this interesting topic leaving for others the analysis and the details. The question arises: 'Is the kallitype alive or dead now, when the chemical approach to photography has been largely replaced by digital electronic processes. And if alive, what life does the kallitype have?

There can be no doubt that the computer has brought incredible change to the way photographs are conceived, taken, processed, and displayed. We now have digital cameras that create images with unbelievable ease, accuracy and control. We have computer software that provides almost unlimited means for manipulation and computers that provide almost infinite control of photographic printing. We ask whether hand controlled photographic processes and particularly kallitype printing has been able to survive in such an environment? We are inclined to suppose that the kallitype must now, surely, be dead. We ask, can the firebird process still thrive in spite of such miraculous electronic development? If so, where and how does the kallitype survive in the computer age?

The answer is "Why, on the internet, of course." As any computer literate person knows, the way to find out about anything in the 21st Century is to make a Google Search. So let's make a google search of "kallitype" and see what we can learn about the kallitype in 2012.

What is revealed almost defies expectation. The first surprise is the sheer quantity of kallitype entries: hundreds. Among these entries are found listings for a surprising number of recent writings on the kallitype and allied iron processes. The postings appear to be well researched by competent process investigators and artistic printmakers. The internet search also reveals a expanding list of contemporary iron process descriptions which go into considerable detail about how to work the process as a result of the writer's newly made investigations. The search reveals dozens of internet "galleries" presenting the work of individual artists who make iron silver prints of various kinds for sale. They are offered by artists from around the world—for example, China, Japan, France, Italy, Germany, Mexico. Unfortunately some of these are written in the language of the country, so what they say is not always readable. The internet also has websites where artists can buy fresh, reliable supplies for making kallitype and other alternate processes.

At a time when great manufacturers of silver photo paper have shut down their silver paper production, a surprising number of photographers are again making kallitypes and writing about them. In blogs they write with an enthusiasm similar to that of the early amateurs who coopted the kallitype at its beginning. They write with obvious photo-chemical awareness and responsibility and their process descriptions appear trustworthy like those of the earliest kallitype writers. Their attitudes about the kallitype are various. Some writers proudly call themselve pictorialists indicating a connection with old values, which current art or consumer photo-magazines no longer cater to. Photographers with such values concentrate their efforts on traditional subjects such as landscapes, nudes, portraits, etc. photographed in beautiful traditional compositions. Other kallitypists follow modern art values and methods, and make creative "experimental" photographs that involve a variety of contemporary techniques such as collage, multiples, writing on images, applied color etc. A quick google search among the pages of entries quickly reveals the variety and vitality of the artistic uses of kallitype process today.

There are other print process developments to be observed on the internet. One finds examples of colleges that have classroom outlines for the study of kallitype photography along side ads for summer workshops that teach the process. One finds kallitype chat room websites where groups of people share their interest, their kallitypes, and their information.

The internet provides ample evidence of the vitality and quality of 21st Century amateur artistic activity with hand controlled photographic processes and with the kallitype in particular. From reading just a sample of the the entries it is evident that what is happening on the internet relates to the old amateur motive of making unique and beautiful photographs that display the individual ideas and creative taste. The websites make it clear that contemporary kallitype devotees have applied considerable knowledge, talent, and energy to no profitable end beyond their love of making charming photographs through carefully worked out process investigations.

Together the plethora of internet entries indicate that amateur involvement with the kallitype process in 2012 is vigorous and healthy. Kallitype investigations are currently produced by a quantity of individuals with various professional backgrounds, educational experiences, and artistic temperaments, who publish high quality studies of kallitype technique reminiscent of the best writing in the past. The number of artists presenting their kallitypes prints for all to see gives convincing evidence of artistic accomplishment. One cannot help concluding that what underlies all this activity is the fascination of many kinds of people with a magical process. They are deeply energized and intrigued by controlling a set of chemical events to produce a protean image that has rich artistic attraction. They are fascinated with a challenging hand process that when brought under control provides not only a fine print but the joy and personal satisfaction of having expressed oneself.

We cannot help but admit that the kallitype is indeed, like the firebird, alive again, experiencing another rebirth. It must be said that, nourished by fascinated amateurs, the kallitype thrives in the computer age.

Notes: *CHAPTER XI. The Demise and Rebirth*

1. F. W. Horn, "The Kallitype Process," *British Journal of Photography*, October 23, 1914, p. 796-7.
2. A. J. Jarman, "Photographic Prints on American Made Paper*,*" *The Camera*, June 1914, pp. 326-334.
3. S. Blumann, "Home Sensitizers and Their Application," *Camera Craft*, October, 1914, pp. 495-502.
4. J. Thomson, "A Silver Platinum Printing Paper," *American Photography*, November 1915, p. 630-5.
5. Unsigned, "A Substitute for Platinum," *The Camera*, February 1915, p. 128. Also, "A Way of Making Prints Resemble Platinum," *The Camera*, October 1915. p. 622.
6. W. H. Sellors, "The Kallitype Process," *British Journal of Photography*, May 26, 1916, p. 305.
7. S. Blumann, "A Patriarch in Kallitype," *Camera Craft*, June 1916, pp. 262-269.
8. John Tennant, editor, *Photo-Miniature*, Issues for January 1900 to 1902, unpaged advertisements.
9. Colin N. Bennett, "Some Trials of Kallitype," *British Journal of Photography*, July 20, 1917, pp. 378-9.
10. J. M. Hammond, "Variations on Iron Silver Printing," *Photographic Journal of America*, 1917,
11. J. Thomson, "Plain Salted Paper," *American Photography*, February, 12, 1918, p. 64.
12. J. Thomson, "Kallitype and Modifications Thereof," *Camera and Darkroom*, June, 1918.
13. J. Thomson, "Possible Substitutes For the Platinum Print," *American Photography*, November, 1918, pp. 642-649.
14. Ibid., p. 648.
15. no author given, "Kallitype," reprint from G. E. Brown in The Amateur Photographer, in *Camera and Dark Room*, July, 1918, pp. 263-66.
16. no author given; "Kallitype Printing," *Photographic Journal of America*, Thos. Coke, editor, 1918, pp. 402-03.
17. David Bachrach, "Permanence of Kallitype Prints," *British Journal of Photography*, November, 22, 1919, p. 685. Reprinted from Abel's Weekly.

18. S. Blumann, "More About Kallitype," *Camera Craft*, July 1920, pp. 215-218.
19. S. Blumann, "More About Callitypes," [sic], *Camera Craft*, August, 1920, p. 265-6.
20. Ruthven Flint, *Chemistry for Photographers*, American Photographic Publishing Co., Boston, 1920, pp. 147-152.
21. Ibid., p. 147.
22. no author given, "Report of Meeting of the South Suburban Photographic Society," *British Journal of Photography*, March 18, 1921, p. 160.
23. John Thomson, "Kallitype and Allied Processes," *Photo-Miniature*, Vol. xvi, No.185, January 1922, pp. 210-245.
24. no author given, **Photo Era**, June 1922, p. 346; see also *British Journal of Photography*, Feb. 24, 1922, p. 116.
25. John Tennant, Introduction, *Photo-Miniature,* vol. xvi, No.185, Jan. 1922, p. 209-210.
26. J. Thomson, Ibid, p. 210-245.
27. J. Thomson, "Kallitype," *American Annual of Photography*, p. 62.
28. J. Thomson, Ibid., p. 62-66.
29. Ibid.
30. Frank Fraprie and W. E. Woodbury, *Photographic Amusements*, 10th edition, American Photographic Publishing Co., Boston, 1931, pp. 129-131. Editions of this work were published since 1896.
31. Frank Fraprie, *Practical Printing Processes,* American Photographic Publishing Co., Boston, 1923, pp. 50-51.
32. S. Blumann, "Letterhead Holiday Greetings," *Camera Craft,* #12, December 1924.
33. Kent E. Wade, *Alternative Photographic Processes*, Morgan and Morgan, Inc., New York, 1978.
34. William Crawford, *Keepers of the Light,* Morgan and Morgan, Dobbs Ferry, N. Y. 1979.
35. Nancy Howell Koehler, *Photographic Art Processes*, Davis Publications, Worcester, Mass., 1980.
36. Jan Arnow, *Handbook of Alternative Processes*, Van Nostrand Reinhold Co., New York, 1982.
37. Robert Demachy, *Camera Work*, July 19, a1907, pp. 21-24.
38. H. P. Robinson, "Paradoxes of Art, Science, and Photography," *Wilson's Photographic Magazine,* vol. xxix, 1892, p. 254.

39. Edward Weston, "What is Photographic Beauty," *Camera Craft*, Vol. xlvi, No. 5, June 1939, p. 254.

40. Edward Weston, "Photography," *Enjoy Your Museum*, Esto Publications, Pasadena Calif., 1934, p. 12.

41. I learned of this point of view at a talk given by a representative of the Eastman Kodak Co. to the Midwest Society for Photographic Education held at Ohio State, University, in November, 1980. The point of the talk was a criticism by dissatisfied teacher-artists of the design and qualities of photographic papers made by the Eastman Kodak Co. Unfortunately I no longer have the speaker's name.

42. Jan Arnow, ibid., Preface, unpaged.

43. The University of Louisville Archives, located in the William F. Ekstrom Library, Louisville, Kentucky has a collection of prints made by of various photographic print processes.

44. Nancy Howell Koehler, Ibid., p. 7.

45. Pierre Glafkides, *Photographic Chemistry*, Vol. I, Fountain Press, London, 1958, translated from the French by Keith M. Hormsby, FRPS, pp. 423-432.

46. Jaromir Kosar, *Light Sensitive Systems*, Wiley, New York, 1965.

47. Glafkides, Ibid., p. 422.

Illustrations

1.   Birmingham Co. Kallitype advertisement p.41

2.   Vici Paper ad...p.113

3.   Polychrome ad. p.118

4.   Aristo Gold Post card ad. p.127

5.   Mimosa, Cyko, Velox, Eastman paper ads. p.128-9

6.   Henry Hall Kallitype, "Pussy's Portrait" p.194

7.   Henry Hall Kallitype, "Swinging" p.204

8.   Kodak's Famous Ad. "You press the button" p.223

9.   J.M. Mc Corckle, pinhole photo "Flatiron Bldg." p.226

10. Pictorial Photo Instruction Book p.228

11. *Photo-Miniature* Monograph, "Chemical Notions for
Photographers" p.230

12. Tennant & Ward Process Monographs p.231

13. Example of Photomanipulation by Brushing Oil Pigment p.235

14. Robert Demachy, Gum Bichromate Print...........................p.236

15. Oil Pigment & brush for painting on Photos
and Negative p.237

16. R. Demachy. Brushwork on a photographic print p.238

17. J. F. Strauss, The Bridge. Popular heavy dark tones............p.239

18. Typical impressionistic "fuzzy" rural image p.240

19. Whistler, impressionistic painting that influenced
photographers p.241

20. William Mortenson, hi key and low key Images...............p.243

21.   Paul Anderson, manipulated pictorial print p.247

22.   Platinum paper ad emphasizing Quality p.250

23.   James Thomson kallitype, self portrait p.258

24.   Eastman Co, sepia brown print paper ad. p.283

25.   Louis Fleckenstein, kallitype p.304

26.   Louis Fleckenstein photograph on commercial paper p.305

27.   C. Gaspar Elmberger, manipulated kallitype p.308

28.   Portrait of a Female photographer p.310

Bibliography of Books on Kallitype and Allied Topics

ABNEY, WILLIAM DE WIVELESLIE, and Clark, Lionel, **Platinum, Its Preparation and Manipulation**, New York, Anthony and Scovill, 1898.

ADAMS, W. I. LINCOLN, **Amateur Photography**, fifth edition, revised and enlarged, New York, The Baker & Taylor Co., 1899

American Annual of Photography and Photographic Times Almanac, New York, 1895

American Annual of Photography and Photographic Times Almanac, New York, 1896

American Annual of Photography, 1908

American Annual of Photography, 1909

American Annual of Photography, 1912

American Annual of Photography, 1913

American Annual of Photography, ed Percy Y. Howe, vol. xxxvi, New York, The **American Annual of Photography**, Inc., 1921.

ANDERSON, PAUL, **The Technique of Pictorial Photography**, Philadelphia, J. B. Lippincott, 1939. first ed. publ. in 1917.

Anthony's Photography Annual, 1891

Anthony's Photography Annual, 1892

ARNOW, JAN, **Handbook of Alternative Processes**, New York, Van Nostrand Co., 1982.

ARMSTRONG, T. N., **Guide to Practical Photography**, London, Dawbarn and Ward, Ltd., 1898.

BARROW, THOMAS F., ARMITAGE, SHELLEY, and TYDEMAN, WILLIAM E., **Reading Into Photography**, Selected Essays, 1959-1980, University of New Mexico Press, Albuquerque, 1982.

BAYLEY, R. CHILD, **The Complete Photographer**, New York, Frederick Stokes & Co., 1926.

BLUMANN, SIGISMUND, F.R.P.S., **Photographic Handbook**, San Francisco, Photo Art Publisher, 1935.

BOTHAMLEY, C. H., **The Ilford Manual of Photography**, London, Ilford Ltd, 1912.

British Association for The Advancement of Science, 1894 Annual Report

BROTHERS, ALFRED, **Photography: Its History, Processes, Aproaches, and Materials**, 2nd revised edition, London, Chas. Griffin and Co., 1899.

BROWN, GEORGE, E., **Ferric and Heliographic Processes**, London, Dawbarn and Ward, Ltd., 1899.

BURBANK, W. H., **Photographic Printing Methods,** A Practical Guide to the Professional and Amateur Worker, 3rd edition, New York, Scovill Adams Co., 1891.

BURGESS, N. G., **The Photograph Manual**, New York: D. Appleton & Co., 1863; Reprint Edition, New York, Arno Press,1973.

BURTON, W. K., and PRINGLE, ANDREW, **The Processes of Pure Photography**, New York: The Scovill and Adams Co., 1889.

CAFFIN, CHARLES, H., **Photography as a Fine Art**, New York, Doubleday, Page and Co., 1901.

Cassel's Cyclopaedia of Photography, Bernard D. Jones, editor, London, Cassell Co., 1911. Edition used was a facsimile reprint, published by the Arno Press, New York Times Co., New York, 1974.

CLERC, L. P. **Photography, Theory and Practice**, Bath, Pittman and Sons, 1930.

COE, BRIAN AND BOOTH, MARK HAWORTH, **A Guide to Early Photographic Processes**, London, Hartwood Press in association with the Victoria and Albert Museum, 1983.

CRAWFORD, WILLIAM, **Keepers of the Light**, Morgan and Morgan, Dobbs Ferry, N. Y. 1979.

DOTY, ROBERT, **Photo Secession**, Rochester, N.Y., George Eastman House, 1960.

DUCHOCHOIS, P. C., **Photographic Reproduction Processes**, "A Practical Treatise of the Photo-Impressions Without Silver Salts," New York, The Scovill and Adams Co., 1891

EASTMAN KODAK CO., **Elementary Photographic Chemistry**, Rochester, New York, 1919.

—, **The Modern Way in Picture Making**, Published as an aid to the amateur photographer, Rochester, N. Y., 1905.

—, **Encyclopedia of PracticalPhotography**, edited by and published for the Eastman Kodak Co. by the American Photographic Book Publishing Co., Garden City, New York,1977.

EDER, JOSEF MARIA, **History of Photography**, translated by Edward Epstean, New York, Columbia University Press, 1945; Dover Publications reprint, N. Y. 1972.

EMERSON, P. H., **Naturalistic Photography for students of the Art**, second edition revised, New York: E. & F. Spon, 1890; An Amphoto Facsimile Book, New York, N. Y., American Photographic Book Publishing Co., Inc. 1972.

The Encyclopedia of Photography, ed. by Willard D. Morgan, New York, Greystone Press, 1971.

English Patents, Printed for Her Majesty's Stationary Office, London: Darling & Son, Ltd., 1890. See later editions for 1891 and 1892.

ESTABROOKE, E. M., **Photography in The Studio and in the Field**, New York: E. & H. T. Anthony & Co., 1887.

FLINT, RUTHVEN, **Chemistry for Photographers**, Boston, American Photographic Publishing Co., 1920

FREUND, GISELE, **Photography And Society**, Boston, David R. Godine, 1980

FRAPRIE, FRANK**, Practical Printing Processes**, American Photographic Publishing Co., 1914, 1923.

FRAPRIE, FRANK, AND WOODBURY, W. E., **Photographic Amusements,** 10th edition, Boston, American Photographic Publishing Co., 1931.

FULLER, PAT, **Control Processes**, unpublished notes for a catalog of an exhibition held at the International Museum of Photography.

GASSAN, ARNOLD, **Handbook for Contemporary Photography**, 4th edition, Rochester, N. Y., Light Impressions, 1977.

GATCHEL, W. D., **Catalogue Illustrated**, Louisville, Kentucky, 1888, Facsimile edition, no current publisher given.

GILLIES, JOHN, **Principles of Pictorial Photography**, New York, Falk Publishing Co., Reprint edition, Arno Press Inc., 1973.

GLAFKIDES, PIERRE, Photographic Chemistry, Vol. I, London, Fountain Press, 1958; translated from the French by Keith M. Hormsby,

GOODSALL, ROBERT H., **Pictorial Photography**, London: The Fountain Press, no date of publication given.

HAMMOND, ARTHUR, **Pictorial Composition in Photography**, Boston, Mass., American Photographic Publishing Co., 1920, fourth edition.

HARRISON, W. JEROME, **The Chemistry of Photography**, New York, Scovill and Adams Co., 1892.

—, **A History of Photography**, New York, Scovill and Adams Co.

HAFFEY, JOHN AND SHILLEA, TOM, **The Platinum Print**, Rochester, N. Y., Graphic Arts Research Center, Rochester Institute of Technology, 1979.

HEPWORTH, T. C., **Photography for Amateurs**; Llondon, Cassell & Co., Ltd: 1884.

HUNT, ROBERT, **A Popular Treatise on the Art of Photography**, including daguerreotype and all the new methods of producing pictures by the chemical agency of light, Facsimile edition, with introduction and notes by James Yinghpeh Tong; Athens, Ohio; Ohio University Press, 1973.

—**A Manual of Photography**, 3rd edition, London, John Griffin and Co., 1853.

—, **Researches on Light, London**; Longman, Brown, Green, and Longmans; 1844, Arno Reprint, New York, 1973.

JAMES, THOMAS HOWARD, with HIGGINS, G.C., **Fundamentals of Photographic Theory**, New York, Wiley & Co., 1948.

JOHN, DAVID, HUGH, **A Textbook of Photographic Chemistry**, London, Chapman and Hall, publisher, 1963.

JONES, CHAPMAN, **The Science and Practice of Photography**, fourth edition, London: Iliffe & Sons, Ltd., 1904.

JORDAN, FRANKLIN I., **Photographic Control Processes**, New York, Galleon Publishers, Inc., 1937.

KOEHLER, NANCY HOWELL, **Photographic Art Processes**, Worcester, Mass., Davis Publications, 1980.

KOSAR, JAROMIR, **Light Sensitive Systems**, New York, Wiley and Co., 1965.

LAMBERT, F. C. AND CUMMINGS, THOS. HARRISON, editors, **Pictorial Printing, Part II, The Pictorial Work of Percy Lewis**, The Practical Photographer Series No. 19, Chicago, Burke and James Co., 1905.

LANDWEBER, ELLEN, **Some Various Processes** As Demonstrated at the National Conference of the Society for Photographic Education, Printed at Humboldt State University, Arcata, California, 1978.

LIETZE, ERNST, **Modern Heliographic Processes**, Rochester, N. Y., Visual Studies Workshop Press, 1974. reprint.

MCINTOSH, E. J., **The Photographic Reference Book**, London, Iliffe and Sons, 1906.

MELDOLA, RAPHAEL, **The Chemistry of Photography**, London, Macmillan Co., 1889.

MORGAN, WILLARD D., general editor, **The Complete Photographer,** Chicago, Ill, National Educational Alliance, Inc., 1942.

MORTIMER, F. J., Editor, **Wall's Dictionary of Photography**, sixteenth edition, revised and largely re-written by A. L. M. Sowerby, Boston Massachusetts, American Photographic Publishing Co., 1944.

NAEF, WESTON, J. **The Collection of Alred Stieglitz**, New York, A Studio Book, The Metropolitan Museum of Art, The Viking Press, 1976.

NEWHALL, BEAUMONT**, Latent Image, "The Discovery of Photography**," Garden City, New York, Anchor Books, Doubleday and Co.,1967

No author given. **Photographic Annual** Incorporating Figures, Facts, and Formulae Of Photography, London, George Routledge and Sons, 1911.

PALTRIDGE, GEO. H., **Photographic Instruction Text**, Chicago, Burke and James, 1900.

Photo Times Co., **Encyclopaedic Dictionary of Photography**,1897.

PIZZIGHELLI, CAPTAIN AND HUBL, BARON A., **Platinotype**, London: Harrison and Sons, 1886, reprinted in Nonsilver Printing Processes, New York, Arno Press, 1973.

POORE, H. R., **Pictorial Composition**, New York, The Knickerbocker Press, fourteenth edition, revised, 1903.

REILLY, JAMES M. **The Albumen and Salted Paper Book**, "The History and Practice of Photographic Printing," Rochester, New York, 1980.

ROBINSON, HENRY PEACH, **Pictorial Effect in Photography**, London, Piper and Carter, 1869.

ROEBUCK, J. R., **Science and Practice of Photography**, New York, Appleton, 1918.

SCHRIEVER, J. B., ed., **Complete Self Instructing Library of Practical Photography**, Vol. I, Elementary School of Art and Photography, Scranton, 1909.

SCOPICK, DAVID, **The Gum Bichromate Book**, Rochester, N. Y., Light Impressions, 1978.

SEIBERLING, GRACE, **Amateur Photography and the Mid-Victorian Imagination**, Chicago, The University of Chicago Press, Chicago, 1896.

SENNETT, ROBERT S., **The Nineteenth Century Photographic Press, A Study Guide,** New York, and London, Garland Publishing, Inc., 1987.

STIEFEL, HENRY, C., **Plates and Papers**, London, POercy Lund,& Co., 1896.

STORY, ALFRED T., **The Story of Photography**, New York, S. S. McClure, Co., 1908.

TAFT, ROBERT, **Photography and the American Scene**, "A Social History 1839-1889," New York, Macmillan Co., 1938.

TAYLOR, JOHN, **Pictorial Photography in Britain**, 1900-1920, Arts Council of Great Britain in Association with the Royal Photographic Society, 1978.

TODD, F. DUNDAS, **Reference Book of Practical Photography**, Chicago, Photo-Beacon Co., 1900.

TOWLER, JOHN, M. D., **The Silver Sunbeam**, New York, J. H. Ladd, 1864, Facsimile Reprint by Morgan and Morgan Co., Hastings-on-Hudson, N. Y., 1969.

TOWNSEND, CHARLES F., **Chemistry for Photographers**, London, no date, but about 1890.

VOGEL, HERMANN, DR., **The Chemistry of Light and Photography**, New York: D. Appleton and Co., 1895.

WADE, KENT E., **Alternative Photographic Processes**, Morgan and Morgan, Inc., 1978.

WELLING, WILLIAM, **Collectors' Guide to Nineteenth-Century Photographs**, New York, Collier Books, Macmillan Publishing Co., Inc., 1976.

WELLING, WILLIAM, **Photography in America, The Formative Years**, 1839-1900, New York, N. Y., Thos. Y. Crowell Co., 1978

WESTON, EDWARD, **Enjoy Your Museum, "Photography,"** Pasadena Calif., Esto Publications, 1934.

WHEELER, OWEN, **Photographic Printing Processes**, Boston, American Photographic Publishing Co., 1936.

WILSON, EDWARD L., **Photographic Mosaics**, An Annual Record of Photographic Progress, New York, Edward L. Wilson Co., 1901.

WILSON, EDWARD L., **Wilson's Cyclopaedic Photography**, New York, Edward L. Wilson Co., 1894. **Practical Guide to Photographic Printing**, Scovill and Adams Co., 1892

NUTTING, WALLACE, **Photographic Secrets**, work can be found in the Boston Public Library, bk no. 8147.05 / 101. The publisher, location, and date are unknown.

This bibliography lists the book length works from which this book emerged. It attempts to be inclusive rather than brief for those scholars who for whatever reason might wish to comb the sources for themselves. When I started writing this book there was no bibliography on kallitype that extended beyond seven or eight items. This list will provide a good start for future scholars. The aim in compiling it has been to include all works on the topics relevant the development of the kallitype

Bibliography of Periodicals

The following periodicals have proved to be useful sources of articles on the development and the significance of the kallitype. The list is unavoidably confusing because of the many short lived periodicals with similar names that were published during the time when the kallitype was a popular print process. The bibliography is presented as a relatively complete list of periodicals in which may be found discussions of the kallitype or of concerns that relate to it.

The list provides, where possible, the name of the editor of the publication during the period when the kallitype was most active. When more than one editor played a part in the development of the kallitype, more than one editor is listed. The bibliography makes an attempt to indicate the lineage of photographic publications since a number of the magazines were incorporated in later periodicals. The information in the following list is as complete as it was possible to make it, under rather trying conditions. The difficulties encountered in making it are the usual ones: bibliographic sources provide conflicting data and many periodicals are not readily available for checking. Incomplete entries have been included, rather than removed, when it was not possible to complete them.

The list is intended as an aid to researchers who are unfamiliar with the periodical publications of the last one hundred years, which bear on the kallitype process. Unfortunately, no library contains even a majority of the publications, much less complete runs of them. William Welling, in **Photography in America**, The Formative Years, (see book bibliography) provides a table of the "Principle Holdings of Nineteenth-Century American Photographic Journals" in fifty American libraries. The table (on pages 414-5) facilitates location of copies of individual publications. Recently microfilm copies of some periodicals have become available.

Amateur Photographer, London, Oct 1884-June 1918, A. Horsley Hinton ed. Incorporated **Photography**. The publication was then called, **The Amateur Photographer and Photography**. Two well known editors of the combined publication were R. Child Bayley and F. J. Mortimer.

The American Annual of Photography and Photographic Times Almanac, New York, 1889-1907, Frank Fraprie, ed., organ of Photographic Pictorialists of Buffalo and Boston Photography Clan. Other editors were Dr. John Nicol and F. C. Beach.

The American Annual of Photography and Photographic Times Almanac, Boston/New York, 1887-1953, The Scovill and Adams Publishing Co. C. W. Canfield, ed.; later editors: Juan C. Abel, W. I. Lincoln Adams, and John Tennant.

Photography Annual of 1891, London

Photography Annual of 1892, London

American Photography, superseded **Photo Beacon** Boston, Mass.,1907 ff.

Anthony's Photographic Bulletin, New York, 187-1902. merged with Photographic Times.

British Journal of Photography, London, (formerly Liverpool), Jan 14, 1854 to present.

The Camera, Philadelphia, 1897-1910 Frank V. Chambers and John Bartlett, editors.

Camera Craft, San Francisco, 1893-1934, Fayette Clute, ed; later Sigismund Blumann.

The Camera and Dark-room, Brooklyn, N. Y., 1899-1906, J. P. Chalmers ed.

Camera Notes, New York, 1897-1903.

Camera Work, New York, N.Y., 1903-1917. Alfred Stieglitz, ed.

International Annual of Anthony's Photographic Bulletin, New York, 1888-1902

Journal of the Camera Club o London, 1886-1903.

Journal of the Society of the Chemical Industry

Mason College Magazine, Birmingham, England.

Modern Photography, July, 1981, Julia Scully, editor, "Processes: Kallitype, Van Dyke and Brown Print," Dick Stevens, p 86 ff.

Philadelphia Photographer, Philadelphia, Pa., 1864-1888, E. W. Wilson ed.

Philosophical Transactions, Vol. 130, 1-60, London, 1840.

The Photo-American Review, Stamford, Conn. and N. Y., 1889-1907.

Photography Annual of 1891, London

Photography Annual of 1892, London

Photo Times Co., **Encyclopaedic Dictionary of Photography**,1897.

Photo-Beacon, Chicago, Ill, Jan., 1889-June 1907; edited by F. Dundas Todd., superseded by American Photography.

Photo Era, Boston, Mass, May 1898-March 1932, vol. 1-68, Wilfred A. French ed. and publ.

Photographic Archiv, Elberfeld, Germany, Jan. 1, 1860-1897.

Photographic Annual, London, 1891-1912, A supplement to Photography.

Photography Annual of 1892, London

Photographic Journal of America, Philadelphia, Pa., 1864. Began as **The Philadelphia Photographer**, Philadelphia 1864-1888. Became **Wilson's Photographic Magazine**, New York, from 1889-1914 and finally, **Photographic Journal Of America**, New York, from 1915 to 1923. This publication was sometimes referred to as the **"American Journal of Photography"**. E. W. Wilson, editor; later L. W. Wilson.

Photography Journal and Transactions of the Royal Photographic Society of Great Britain, 1897-1919, London, W. W. Abney editor.

Photograms of the Year, London, Dawbarn & Ward, Ltd. 1895-1909, H. Snowden Ward, editor, F.R.P.S.

The Photographer, New York, organ of various camera clubs, 1904-1907.

Photographic Mosaics, Philadelplhia,1866-1879; New York 1880-1901.

Photographic News, London; Sept 10,1858-May 12,1908. Incorporated in **Amateur Photographer**, 1908.

Photographic Review, Chicago, Aug. 1909-1921.

Photographic Times, Scovill Mfg. Co., New York, 1871-1915. Merged with **Popular Photographer** in 1915. W. I. Lincoln Adams ed.

Photography, London, incorporated in **Amateur Photographer** in 1908.

PHOTO-MINIATURE, New York, April, 1899-1932, vol. 1-18, Nos. 1-205, John Tennant ed. The following numbers were especially useful:

"The Blue Print and its Variations," Vol. I, No. 10, January 1900.

   "Chemical Notions for Photographers," Vol. II, No. 18. September 1900
   "Photographic Manipulations," Vol. II, No. 23, February, 1901.

"Platinotype Modifications," Vol. !V, No. 40, July, 1903.
"Photographic Chemicals," Vol. IV No. 43, Oct. 1903
"Kallitype Process," Vol. IV, No. 47, February, 1904.
"Postscript to No. 47," Vol. VI No. 69 December, 1904
"Kallitype to Date," Vol. VII, no. 81 September 1907 "Kallitype and Allied Processes," Vol. XVI, No. 185, January 1922.

Photo Times Co., Encyclopaedic Dictionary of Photography,1897.

Proceedings of the Royal Society of Edinburgh, No. 49, 1928-9.

Saint Louis and Canadian Practical Photographer, St. Louis, 1877-1910.

Scientific American Supplement, #815; New York, 1891.

Western Camera Notes, Minneapolis, Organ of the Minnesota Camera Club, 1899-1907.

Wilson's Photographic Magazine, Philadelphia 889-1913, E. W. Wilson editor. L. W. Wilson later editor.

Index

A

a better knowledge of the kallitype is long overdue. 140
Abney and Clark 205
a brief history of the kallitype, 316
academization of photography 363
a chronological presentation of the published articles 160
a creative less objective image 227
Adams, Charlotte 232
advantages of kallitype II paper 94
advertising logos of papers 127
After 1925, the pictorial ideal began to tarnish 356
Aftermath of the Bennett-Burton Revisions 179
A. J. Jarmans 311
all of the processes had advocates and all had critics 327
Amateur became process 153
amateur making ferric oxalate 57
amateur photographers of the 19th century 52
amateur photographic societies 54
Amateur Photography and the Mid-Victorian Imagination. 145
amateur preparation of Kallitype paper 56
amateurs coopted the kallitype 141
amateurs experiment with KII 137
ammonia and hypo fixers 187
ammonia as a fixer 185

ammonia fixer 339
Anderson, Paul L.,a renowned pictorial printmaker 320
Andersons charge of impermanence 321
a new realistic esthetic 149
a new taste in "pure" photography 138
Announcement of Commercial Kallitype II & III 79
a process in the domain of amateurs. 166
Armand Rubin Fr. Kallitypist 118
Arndt and Troost 278
Arnow Jan Handbook of Alternative Processes 349
artistic photographers 225
artist should mind his image, not his paper 254

B

Bachrach,David 341
Bachrach on KII permanence 341
bad directions and print failure 264
Bartletts demonstration 116
Bayley, R.Childe on 249
Bayley, R.Childeon 249
B. C. Roloff on KII 201
Beddings Charge of Unjustified Claims 35
Bennett,Colin N. demo 339
Bennett response to Burton 177
Bennetts contribution 171

Bennetts greatest service 169
Bennett's purpose 167
Bennetts' suggestions on working
    the K II process. 170
Birmingham Photographic Co 51
Black, Brown, Green, and Blue
    Effects 312
Blanchard brush 288
Bothamley 90
Bothamley KII permanence 92
Bothamley KII review 90
Bothamley on permanence of KI 64
Bothamleys report on KI 33
Bothamleys Report on K I 32
bronzing of the shadows 213
Brookes Hypo Fixer 185
Brown 188
brown fingers 125
Brown Print , a commercial
    preparation by E. Valenta in
    1899. 279
Brown Print, A. J. Jarman 293
Brown Print commercial liquid
    preparation 278
Brown Print formulas by James
    Thomson 289
Brown Print manipulation of
    Valentas 279
Brown Print, Namias Sepia Paper 287
brown print paper offered by
    Eastman Co. 282
Brown Print, process by George
    Brown 285
Brown Print Process by Namias,
    Rodolpho Namias 288
Brown Print Process by
    NamiasRodolpho Namias 288
BrownPrint,rebirthofinterestaftterthel
    950s 297
Brown Print Sharp and
    Hitchmough1891 277

Brown Print, simplest and easiest
    individually prepared printing
    paper 271
Brown Print Solutions by G. Brown
    285
Brown Print , the easiest to learn and
    the cheapest to work. 272
Brown Print various names 271
brown print, version of kallitype 272
Brown print, Vollenbruch's process
    282
Brown Prlint ProcessF. C. Lambert
    289
Brown Prlint process, Teresa Del
    Fabro 295
Browns Description of Kallitype I
    Paper and Process 68
Brown's formulas 188
Browns Resolution of the Claims
    Controversy 66
Burtons experiments 176
Burton tests Bennetts method 174
by J. S. Hodson, 42

C

Calotype, Name of process 100, 200
Cassells Cyclopaediaa useful
    reference 317
Cassells Cyclopaediaa useful
    referenmce 317
Cassells Cyclopaedia of Photography
    316
Cassells Cyclopaedia reproduced
    Browns formula 285
Cassells Cyclopedia of Photography,
    published in 1911. 282
Catalog for An Exhibition of Control
    Processes 148
C..D. West, 177
change ferrous oxalate back into
    ferric oxalate 317

Charles S. Taylor 306
C. H. Bothamley 32
chemical reactions of the kallitype
        process 60
Claffin 242
combined silver and platinum paper
        337
Commercial Kallitype II Paper in
        America 112
Commercial Sale of Kallitype Ends
        In England 111
Comparison of Nicols and Bennetts
        K II Process 168
Comparison of the Process of
        Bennett and Burton 173
cost of making KII Prints 343
Crawford, Wm.Keepers of the Light
        349
Cunningham and Craston, demo 305
curators of collections of
        photographs in Art Museums
        139

D

Dancy,W.E. 324
decline of kallitype in the first
        quarter of the 20th century
        303
del FabroTeresa 317
DeMachy 244
derivation of the name, Kallitype. 99
descriptions of KII chemical change
        100
Designations of Kallitype Processes
        83
details of the Kallitype II process
        176
details useful for making KI paper 58
dialectic between convention and
        individual expression 234
dichromate 105

dichromate, contrasst control 91
dichromate, Instructions 105
Dr. Roland Smith KI demonstration
        47

E

Early Explanations of The Wane of
        Favor. 122
Eastmans Sepia Paper 276
Eder, History of Photography 138
Eder, Joseph Maria, 50
Eleanor W. Willard 310
Etral commercial kallitype paper
        324
Etral Paper, Belgian 318
exotic papers for kallitype printing
        311
experiencing difficulty. 109
experimentally minded amateur
        photographers 53
Experimentally minded amateurs
        229
Explanation of failure of KI 68
explanation of the failure 349
explanation of the failure of kallitype
        340
explanation of the kallitypes failure
        123

F

fading interest in Kallitype paper
        110
fatal facility of the commercial print
        processes 157
female amateur photographers 310
ferric ammonium citrate 274
Ferric and Heliographic Processes.
        203
Ferro-Argentum Paper 324
first international salon for art
        photography 222

first ref. to working Br. Prnt processs
277
first response to the patent on
Kallitype I 26
fixer,2 percent solution of ammonia
(.880) 121
flawed communication. 157
Flintonpermanence 342
Flint, Ruthven 342
flowering of amateurism 144
Formulae unnecessarily complicated
309
four main kallitype processes 326
Franklin Jordan 223
Fraprie, Frank 348
full page advertisement for Kallitype
102

G

G. Caspar Elmberger 307
gentleman amateur 145
George Bernard Shaw 221
George Davison 253
George P. Swain 316
Glass positives 309
Glass Positives by the Kallitype
Process 312
Grace Seiberling 145
G. W. Frederick 154, 193

H

Haddons scientific study of hypo
200
Halls Stock Solutions for Making K
II 206
Hammond 242
Hammond, J.M. 313
Harrison on KII developer 102
Harrison's 100
Hawks K II Polychrome Solutions
119

Hawks story and formulas in Chapter
III 338
H. C. L. Bloxam KII demo 182
Henry Hall 193
Henry Halls full-scale, monograph
on kallitype II 200
Henry J. Newton, 144
he role amateurs played 143
Herschel 22
Herschel John Frederick William
275
Herschel performed experiments on
iron sensitized paper 275
Historical Context of the Kallitype
1890-1925 220
historical kallitype workers 142
Historically Important Writers on
Kallitype 155
Home Sensitizers and their
Application. 336
Hopkins ads 114
Horn,F.W. demo 335
Howell Koehler, Nancy,Photographic
Art Processes 349
hypo as a fixing agent 97

I

I am indebted to former
experimenters. 80
imperfect condition, 45
impermanence. 321
impermanence. Anderson 321
impressionistic or indistinct
treatment 238
improvements recently introduced
79
impure ferric oxalate 180
indifference to the study of Kallitypy
139
indifference to the study of
photographic printing

processes 138
Indistinguishable from platinum 28
individually manipulated variable
    print processes 238
Individual Preparation of Kallitype I
    Paper 51
Individual Preparation of K I Paper
    57
ink drawings from kallitypes. 181
Instructions for Kallitype No. 2
    Birmingham Photo Co. 103
internet entries indicate that
    amateur involvement with the
    kallitype 368
Is photography an art 224

J

James H. Stebbins Jr. 59
James H Stebbins Jr new K2 process
    81
James S. Escott 306
James. Thomson, explanation of the
    kallitypes failure 123
James Thomsons formulas 289
J. C. Strauss 123
J. J. M. Sellors. demo. 308
John Lewis and Co 39
John Martin Hammond 123
John Martin Hammond Amer. KII
    expert 123
J. Will Palmer 308

K

Kallitype 3, a print-out paper, 80
Kallitype Advertisement 41
Kallitype,an amateur process 143
Kallitype and Photographic
    Publications 152
Kallitype and the Amateur 137
Kallitype avoided by big interests
    340

Kallitype can be viewed as a
    weathervane 328
kallitype developed a second life
    129
kallitype for professional use 311
Kallitype II and The War of Papers
    106
Kallitype II manipulation 91
Kallitype II process detailed 83
Kallitype II responses 90
Kallitype in the 21st Century 366
Kallitype means 200
Kallitype meant 200
Kallitype patent #5374 19
kallitype prints were not found 142
kallitype process is...at best one of
    uncertainties 340
kallitype was not the most successful
    paper in salons 329
kallitypists follow modern art values
    and methods 367
Kellogg, A. Leonara 193
Kellogg,, A. Leonara 311
K I advantages 71
KI and brown fingers 75
KI and commercial studio operators,
    73
KII amateur responses to 92
KII Amer paper 120
KII Amer. paperCelerite Paper. 114
KII Amer. paper, Mirrotype 115
KII Amer Paper Polychrome 117
KII Amer paper, Verotype 120
KII A Neglected Printing Process.
    117
KII Belgian paper Simili Platine
    120
KII Burton details of the K II process
    176
K II by James Thomson 213
KII Frederick-Hall method 208

KII Frederick's main obstacles to
    successful employment of the
    kallitype 195
KII Frederick's proper manipulation
    194
KII, Fredericks solutions 196
KII Frederick'sspecially prepared
    stock solutions 197
KII Hawks Polychrome formulas
    119
KII Henry Hall's Stock Solution
    Approach 202
K III a print-out paper, but 87
KIII Experimentors 87
K III formula never succeeded 87
KIII my ideal paper 80
KII late arrival in the market 126
KII paper, amount of use 121
KII paper, early responses 90
KII paper, Satista Papers, 122
KII process by Henry Hall. 203
KII processing Times, Instructions
    105
KII process is sure to win friends 93
KII resemblance to platinum prints 94
KIIthe Bennett-Burton approach 174
KII, W. K. Burtons Refinement of
    BennettsApproach 171
KI no impact on the market 73
KI problems with the early
    manufacture of 72
kits and chemicals for working a old
    processes are readily available
    364
knoll [sic] of albumen 96
knoll [sic] of albumen paper 96

L

late-Victorian amateurs 146
letters, queries, and answers 65
liberal practices of printmaking 235

Linked Ring Brotherhood 222
look-like platinum papers, made with
    silver 251
Louis Fleckenstein 304
L. P. Clerc 98

M

Madison Phillips 309
major photographic artists 246
Making Kallitypes, A Definitive
    Guide 57
manufactured papers ,a strong
    feeling against 328
Margaret Walpole 165
mass produced paper 245
Master Nixon, a sixteen year old
    photo-chemical student,demo
    343
methods in a historical perspective 160
M. H. Wilde 120
Modern Scientific Investigation of
    the Kallitype 365
movement in curatorial institutions
    to collect and preserve 364
Mr. Farrow demonstration 45
Mr. Herbert Thompson KII
    demonstration 95
Mr. H. Wild. 186
Mr. Van Loo Belgian paper maker
    120

N

N. Adrianow 306
Need for Historical Study 137
need to discover unknown KII
    processes 140
Nelson C. Hawks 117
new esthetic conception of
    photography 225
N. Gray Bartlett 116
N. Gray Bartletts demonstration 116

Nicol' Kallitype I Process 29
Nicol left Mason College 111
Nicol & Lewis abandon Kallitype 137
Nicoll ,KIII process 82
Nicol's claims 22
Nicols claims for invention 22
Nicols claims For K II 82
Nicols claims of invention 22
Nicols specified solutions for K III 88
no known vintage kallitypes prints 142

O

On Kallitype Failures 325
On the preservation of Iron Salts 312
O. Prescott Bennett 167

P

Patent No. 7312Details 81
Pat Fuller 148
permanence of kallitype prints 95, 183, 338
Peter Duchochois 61, 113
photographers of Victorian 59
photographers of Victorian times 59
photographic journals 1890 to 1920 55
photographic periodicals expanded 152
photographic press published a good deal on Kallitype 154
photography was defined as an expressive art. 244
Photo-Miniature contributions to kallitype 344
Pictorialism 237
Pictorialism and Photographic Technique 232

Pictorialism will be challenged by new esthetics 330
pictorial photographic printing 241
Pictorial prints 229
Pizzighelli and Hubl 22
Platinotype 354
platinum paper substitutes 250
Platinum Printing and the Kallitype 248
Platinum Print, Substitute for 339
Poking fun at the experimentally oriented amateur photographer 319
Polychrome 117
potassium dichromate 95
printing paper manufacturers 53
Print processes that permitted variation 244
problem with directions and bad chemicals 267
Professor Boivin, French KII experimentor 97
proliferation of photographic societies 222

R

R. Childe Bayley, 182
Reader response to Thomsons last two publications 348
reaffirmation of older more pictorial approaches to printing 364
recap the history of the kallitype 349
reply to Bedding 38
Reports of Kallitype I Use 42
Responses and Developments 1904-1914. 303
Responses to Bennett-Burton KII process 200
Responses to Kallitype II in the Market 89

ridicule.& Thomson 263
Riley, Phil M. 312
rise in amateur photographic
        publications 152
Robinson 242
Royal Society of Photography 221
RoyalSociety of Photography 54
Rudolpho Namias 303
RuthvenFlintsbook, Chemistry For
        Photographers 342

S

Sale of Kallitype I Paper 39
sale of kallitype paper ceased about
        1904 350
satire on working Kallitype
        formulae 319
Schnauss,Dr. Julius 179
Seiberlings smallest group of late
        Victorian amateurs 147
Sellors, J.M. demo. 338
Sigismund Bluman editor Camera
        Craft 118
Sigismund Blumann 318
significance of individual preparation
        52
SmithW.H. demo 338
Snyd, Ernest, Permanence 49
Social History and the Amateur 150
solarization, 174
spurious ferric oxalate, 57
Stiefel, H.C. 180
sufficient acid present to complete
        the reaction 315

T

Taft 144
tailored approach 202
Teresa Del Fabro 295
The Beginnings of the Brown Print
        275

the best silver-developing formulae
        313
The Brown Print 271
The Chemistry and Processing of the
        Brown print 273
The Chemistry of Photography on
        KII 99
The Choice Of A Printing Paper 320
The Claims Controversy and Its
        Resolution 36
The Complete Photographer 183
the denoument of kallitype had
        begun 330
the development of the various
        control processes, 149
The disastrous commencement. 50
the distinction of the early amateurs
        146
The early amateurs 145
the fatal mistake 52
the first false step 47
The First False Step 42
the incessant nocturnes 240
The Joys and Sorrows of Photo-
        Experimentation 258
the kallitype is alive again 368
the kallitype offered 150
the Kodak Research Library 141
the leadership of amateurs 129
the London Salon 244
The Lost War of Process:
        Commentary 124
the new esthetic of the
        unmanipulated photograph
        150
the origins of amateurism 147
The PHOTO-MINIATURE 189
The Pictorial Amateur 148
the Pictorialist movement 148
The pictorial movement 147
the post pictorial esthetic 357

The problem of Vintage Kallitypes
    142
The "Progress" of the Kallitype
    Process 156
The Progress of the Kallitype
    Process: 1890-1925 352
The Rebirth of Kallitype and Allied
    Processes 363
the standard of admission will be one
    of art 201
the "tailored" approach 216
the variety of non-silver print media
    148
Thmson, J., reasons for including
    chemicals 214
Thomson and the Kallitype I. 260
Thomson James,on the Kl process
    212
Thomson, J., reasons for including
    chemicals 214
Thomsons final Kallitype I Formula
    346
Thomsons monograph, Kallitype and
    Allied Processes, 343
Thomson's the watertone process 345
Thomson 's writing 344
three major approaches to iron-silver
    printing 313
T. N. Armstrong 149
typographical errors 159
typographical errors or omitted
    critical components 157

V

Van Dyke 242
variability of the Kallitype 214
Variations in Iron Silver Printing 339
variety of photo Processes 138
VICI, ad. 113
Vollenbruchs brown print paper 280
Vollenbruch's process 282

W

Wade Kent E .Alternative
    Photographic Processes, 348
Walter W. Lakin 308
W. G. Oppenheim Amer KII ad. 112
Whistler, James M . 240
W. H. Smith 122
Wild.,H., demo 186
Will Cadby, 240
William Mortenson, Wm. 242, 243,
    356, 358
Willis Platinum POP paper 96
Will J. Brooke on KIII failure 87
W. J. Brooke 183
W. Jerome Harrisons 99
W. K. Burtons Refinement of
    BennettsApproach 171
workshops in photography 364
works on the history and processes
    of printing 138
works on the history and processes
    of printing 138
works on the the history and
    processes of printing 138

Y

yellow stain 25, 45, 60, 61, 74, 91,
    93, 101, 102, 105, 125, 177,
    196, 199, 210, 289, 296, 315
yellow stain after development 60
yellow stain cause of 104
yellow stain. Harrison on 102
You push the button, we do the rest,
    145

www.ingramcontent.com/pod-product-compliance
Lightning Source LLC
Chambersburg PA
CBHW031815170526
45157CB00001B/71